Configurations of Points and Lines

Configurations of Points and Lines

Branko Grünbaum

Graduate Studies
in Mathematics

Volume 103

American Mathematical Society
Providence, Rhode Island

2000 *Mathematics Subject Classification.* Primary 01A55, 01A60, 05–03, 05B30, 05C62, 51–03, 51A20, 51A45, 51E30, 52C30.

For additional information and updates on this book, visit
www.ams.org/bookpages/gsm-103

Library of Congress Cataloging-in-Publication Data

Grünbaum, Branko.
 Configurations of points and lines / Branko Grünbaum.
  p. cm. — (Graduate studies in mathematics ; v. 103)
 Includes bibliographical references and index.
 ISBN 978-0-8218-4308-6 (alk. paper)
 1. Configurations. I. Title.

QA607.G875 2009
516′.15—dc22

2009000303

Dedicated to the memory of my teachers:

Milenko Vučkić (1911–1981)
Stanko Bilinski (1909–1998)
Abraham Halevi (Adolf) Fraenkel (1891–1965)
Aryeh Dvoretzky (1916–2008)

Contents

Preface

It is easy to explain the concept of a configuration of points and lines to any ten-years-old youngster. Why then a book on this topic in a graduate series? There are several good reasons:

- First and foremost, configurations are mathematically challenging even though easily accessible.

- The study of configurations leans on many fields: classical geometry, combinatorics, topology, algebraic geometry, computing, and even analysis and number theory.

- There is a visual appeal to many types of configurations.

- There are opportunities for serious innovation that do not rely on long years of preliminary study.

The truly remarkable aspect of configurations is the scarcity of results in a field that was explicitly started well over a century ago, and informally much earlier. One of the foremost aims of the present text is to make available, essentially for the first time ever, a coherent account of the material.

Historical aspects are presented in order to enable the reader to follow the advances (as well as the occasional retreats) of the understanding of configurations. As explained more fully in the text, an initial burst of enthusiasm in the late nineteenth century produced several basic results. For almost a century, these were not matched by any comparably important new achievements. But near the end of the last century it turned out that the early results were incorrect, and this became part of the impetus for a reinvigorated study of configurations.

The recent realization that symmetries may play an important role in the investigations of configurations provided additional points of view on configurations. Together with the increased ability to actually draw configurations—made possible by advances in computer graphics—the stage was set for renewed efforts in correcting the ancient mistakes and to studying configurations that were never contemplated in the past.

This text relies very heavily on the graphical presentation of configurations. This is practically inevitable considering the topic and greatly simplifies the description of the many types of configurations dealt with. Most of the diagrams have been crafted using Mathematica®, Geometers Sketchpad®, and ClarisDraw®, often in combination.

In many respects this is a "natural history" of configurations—the properties and methods of generation depend to a large extent on the kind of configuration, and we present them in separate chapters and sections.

We have avoided insisting on proofs of properties that are visually obvious to such an extent that formal proofs would needlessly lengthen the exposition and make it quite boring. We firmly believe that an appropriate diagram is as much of a valid argument as a pedantic verbal explanation, besides being more readily understandable. It is hoped that the reader will agree!

The text is narrowly restricted to the topic of its title. There are many other kinds of configurations that might have been included. However, the nature of such configurations, for example, of points and planes, or of various higher-dimensional flats, is totally different from our topic. It is well possible that the early attempts to cover all possibilities led to very general definitions followed by very meager results.

Two exceptions to the restricted character of the presentation concern combinatorial configurations and topological configurations. The former are essential to the theory of geometric configurations, and we present the topic with this aim in mind. We do not enlarge on the various more general aspects of combinatorial designs and finite geometries—there are many excellent texts on these matters. Completely different is the situation regarding topological configurations. Very little is known about them, and the present text collects most of what is available.

One other aspect not covered here is the detailed investigations of the hierarchies of some special configurations. It seems that at one time it was fashionable to start with a simple result, such as the theorem of Pascal, and generate a whole family of objects by permuting the starting elements, then considering all the intersections of the resulting lines and the lines generated by the obtained points, etc. This way one could secure a family of points or lines or whatever to be attached to one's name. The interested reader may gain access to this literature through other means.

For almost all the material covered, we provided as ample and detailed references as we were able to find. However, we did not give details concerning the programs that produced the various computer-generated enumerations. The reason—besides lack of competence—is that the programs and the computers on which they run change too rapidly for any printed information to be of lasting value. The interested reader should contact the authors of these results to obtain the most up-to-date status.

Results for which no reference is given are the author's and appear here for the first time. Also, as noted in appropriate places, several colleagues have been kind enough to allow the inclusion of their unpublished results—I am greatly indebted to them for this courtesy.

My gratitude goes to several other people and institutions. The American Mathematical Society was extremely helpful at all stages of the preparation of this text; in particular, allowing the illustrations to be in color has greatly increased the appeal of the book, as well as its instructional value. I greatly appreciate the attention of the editorial staff to detail and consistency, and the generous help they gave me through all stages of publication. The Department of Mathematics of the University of Washington supported my efforts in a variety of ways, both during the several times I gave graduate courses about configurations and in the preparation of the manuscript later on. The staff of the Mathematics Research Library at the university was very helpful in obtaining for me many of the old papers and books (and some new ones) and in guiding me through the labyrinths of the world of digital books and journals. Several stays at the Helen Riaboff Whiteley Center at the Friday Harbor Laboratories of the University of Washington provided the pleasant atmosphere and conditions conducive to work on this book.

Special thanks go to my friends and coauthors of recent papers on configurations—L. W. Berman, M. Boben, J. Bokowski, T. Pisanski, and L. Schewe. Their insights and comments, as well as results, encouraged me greatly while adding to the joint enterprises. The students of the courses I gave on configurations have earned my gratitude for their interest in the topic, which inspired me to investigate many questions and write up material in lecture notes.

Last—but certainly not least— my thanks go to my wife Zdenka, not only for her patience and forbearance over the long haul of my study of configurations and the preparation of the manuscript of this book, but even more for her love and support during well over half a century.

Branko Grünbaum

Seattle, October 23, 2008

Chapter 1

Beginnings

1.1. Introduction

The word "configuration" has many meanings in both the colloquial and technical settings. In the present work, however, it will be used with one meaning only, although with several nuances which will be explained soon. By a *k-configuration*, specifically an (n_k) configuration, we shall always mean a set of *n points* and *n lines* such that every point lies on precisely k of these lines and every line contains precisely k of the points. The variants of the meaning will concern the interpretation of "point" and "line", with additional distinctions regarding the space in which the points and lines are taken. However, in this Introduction it is best to interpret the words at their most basic meaning—points and lines in the Euclidean plane. It is probably surprising that even with this simple interpretation there is sufficient material to consider writing a book and that there are many problems that are easily stated but still unsolved. For a quick orientation, in Figure 1.1.1 we give three examples of well-known configurations (n_3), about which much has been written and which are known by the names of specific mathematicians. Each will appear several times in our discussions. Much less known are 4-configurations; three examples are shown in Figure 1.1.2.

A configuration (50_5) is illustrated in Figure 1.1.3.

It is both clear and natural that, with increasing k, the images of k-configurations become more complicated. In fact, the smallest n for which a configuration (n_6) is known to exist has a value of $n = 110$. (This topic will be discussed in detail in Chapter 4.) One concern that can be answered easily is whether for an arbitrary integer k there exists a configuration (n_k). Indeed, taking in the k-dimensional Euclidean space a "box" consisting of

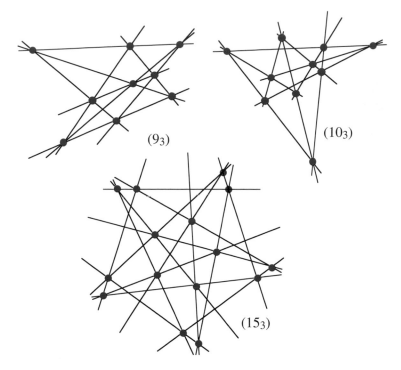

Figure 1.1.1. The 3-configurations of Pappus (9_3), Desargues (10_3), and Cremona-Richmond (15_3).

$n = k^k$ points of the integer lattice, together with the $n = k^k$ lines through them that are parallel to the coordinate axes, we see that there exists a configuration (n_k) with points and lines in the k-dimensional space. But then an appropriate projection onto a suitable plane yields the required configuration in the plane. We shall make repeated use of this configuration; hence we give it the special symbol $LC(k)$. In [**182**], T. Pisanski calls these the "generalized Gray configurations". The drawback of this construction is, obviously, that already for $k = 7$ this yields $n = 7^7 = 823,543$—a rather unwieldy number. One may expect that with some ingenuity this number can be reduced, just as the corresponding $6^6 = 46,656$ has been reduced to 130. A different construction of some k-configurations with arbitrary k was proposed by Kantor [**130**].

Some very specific questions that can be asked for any k, but which we shall illustrate here for $k = 3$ only, are as follows:

(A) For which n do configurations (n_3) exist?

(B) For each n such that configurations (n_3) exist, determine all distinct ones.

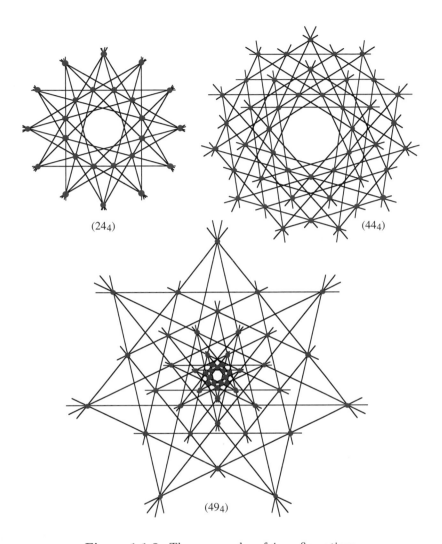

Figure 1.1.2. Three examples of 4-configurations.

(C) How can given geometric configurations be represented symbolically? Given a type of symbolical representation, how can one decide whether it corresponds to a geometric configuration, and if it does, how can one draw it?

As we shall see in Section 2.1, question (A) has a simple answer: Configurations (n_3) exist if and only if $n \geq 9$. However, the question becomes much harder for configurations (n_k) with $k > 3$; in that form it was first posed by Reye [**188**] in 1882 and is listed as Problem 12 in Section 7.2 of [**32**]. We shall consider these cases in Chapters 3 and 4.

In contrast, to answer question (B), we first have to decide under what circumstances two configurations are considered to be the same, that is, not

Figure 1.1.3. A configuration (50_5) in the projective (extended Euclidean) plane. There are ten points at infinity, in the direction of the sets of five parallel lines in the diagram. The smallest (n_5) configuration known is the (48_5) configuration shown in Figure 4.1.4.

to be distinguished from each other for the purposes of the intended classification. As it turns out, in analogy to many other geometric topics, there are several sensible ways of classification, each leading to its own answer to question (B).

As an illustration of these differences we consider the case of configurations (9_3), which include the Pappus configuration from Figure 1.1.1. The three configurations in Figure 1.1.4 are obtained from each other by a simple affine transformation. They are considered the same under the so-called *projective* equivalence, which assigns to the same class configurations obtained as affine (or, more generally, projective) images of each other.

The three configurations in Figure 1.1.5 are not projectively equivalent but have the same *incidences*.

Here and until further notice, incidences are defined between points and lines, and an incidence means that the point lies on the line or, equivalently,

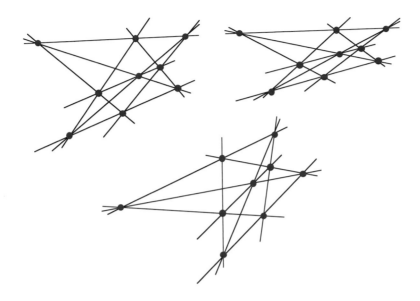

Figure 1.1.4. Three affinely equivalent configurations (9_3).

that the line passes through the point. Two configurations have the same incidences provided their points and lines can be given such labels that a point and a line are incident in one of them if and only if they are incident in the other. Since affine (and projective) transformations preserve lines and incidences, it is obvious that classification by incidences is coarser than the projective classification. The labels attached to the configurations show that they have the same incidences.

Concerning (C) we shall see that the available resources are rather modest. Some of the approaches will be discussed in the appropriate sections of the book. However, there is practically nothing relevant to configuration (n_k) with $k \geq 5$.

Three configurations (9_3) that do not have the same incidences are shown in Figure 1.1.6. While it may appear that proving their difference may be a staggering task, we shall see in Section 2.2 that—using appropriate tools—it can be accomplished in a few seconds. In fact, with just slightly greater effort, it can be shown that in this classification there are precisely three distinct configurations; one of each equivalence class is shown in Figure 1.1.6.

To simplify the expressions used in the sequel, we shall say that two configurations are *isomorphic* (or *combinatorially equivalent*, or of the *same combinatorial type*) if and only if they have the same incidences. In this terminology, all configurations in Figures 1.1.4, 1.1.5, and 1.1.6(a) have the same combinatorial type, different from the types in Figure 1.1.6(b) and (c).

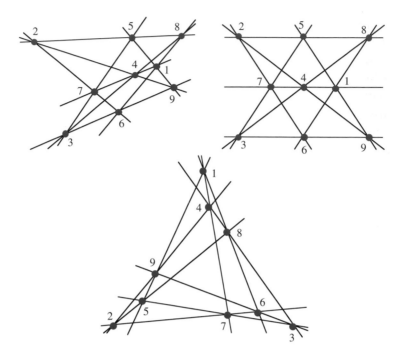

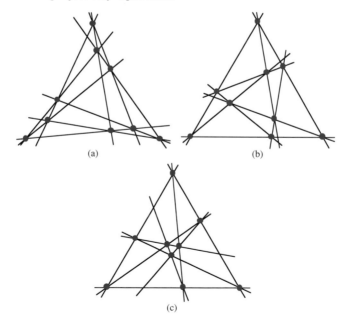

Figure 1.1.5. Three configurations (9_3) that have the same incidences but are not projectively equivalent.

Figure 1.1.6. Three configurations (9_3) that have different incidences.

In the next section we shall give an informal historical survey of the theory of configurations. In Section 1.3 we shall give formal definitions of the various concepts that are used in the book. Later sections of this chapter will present a selection of tools that have been found useful in the study of configurations. Chapter 2 will be devoted to a detailed study of 3-configurations, and Chapter 3 will deal with 4-configurations. Chapter 4 will present information about k-configurations for $k \geq 5$, as well as some other kinds of configurations. Chapter 5 will discuss known results on various properties of configurations (among them connectivity, Hamiltonicity, movability). In many sections we shall also pay attention to *combinatorial configurations* and to *configurations of pseudolines*. (The italicized concepts will be explained in the following sections.)

* * * * * *

A number of other directions of investigation start with families of lines and/or points in the plane but have distinct aims from the study of configurations. The closest one to configuration has only recently been named, although some of its results go back close to two centuries. In an attempt to find a common framework for the various results that are more or less well known, the term "aggregate of lines (or points)" has been proposed. The topics covered in this discipline deal—as in configurations—with incidences of lines and points but without the assumption of equal numbers for all lines and all points. The most famous among them are known as "orchard problems" and "Sylvester's problem". The former typically asks to locate a certain given number of points so that a maximal possible number of lines are incident with precisely 3 (or some other chosen number) of these points. For references see [**32**, Chapter 7], [**31**], [**53**, Section F12], and [**128**].

Sylvester's problem is to show that if a family of n lines is such that they are not in a pencil (that is, all incident with a single point), then there is an *ordinary point*—a point incident with precisely two of the lines. In fact, the number of points is always greater than one, and a longstanding conjecture is that there are at least $[n/2]$ such points. For more details about this problem and related ones see [**31**], [**32**, Chapter 7], or [**53**, Section F12].

* * * * * *

A guiding principle of this book is the conviction that mathematics should be both interesting and attractive and that it should give pleasure while working on it or reading about it. That is why the pace of the presentation is rather leisurely, eschewing acronyms and *ad hoc* abbreviations. It is also the reason for the inclusion of many diagrams—even in situations in which a formal argument could have been supplied. It is the author's hope that the reader will find this approach inviting and the appeal to geometric intuition useful and stimulating.

Exercises and Problems 1.1. In the following sections we shall present lots of exercises that deal with configurations. As a warm-up, here are some questions that are only marginally relevant to configurations but have much in common with the spirit that permeates the study of configurations.

1. The *orchard problem* is: For sets of p points in the plane, find $t(p)$, the maximal possible number of lines containing precisely three of these points; see [**37**]. It is known that $t(12) = 19$. Can you find a set of 12 points with this property? Concerning $t(13)$, it is known only that $22 \leq t(13) \leq 24$. Can you improve on this?

2. The following is a particular case of the generalized *Sylvester problem*. Let $s(p)$ denote the minimal possible number of points incident with precisely two of a set of p lines, not all concurrent and no two parallel. It is known only that $7 \leq s(15) \leq 9$. What is the correct value?

3. Assume given n lines in the plane, no three concurrent and no two parallel. Show that among the bounded regions that they determine in the plane, there are at least $n - 2$ triangular regions. This is known as Roberts's theorem (see [**107**, p. 398]).

1.2. An informal history of configurations

Somewhat parallel to the history of Western civilization, the development of the theory of configurations can be assigned to distinct periods. However, in the case of configurations it is possible to assign precise dates to each of these periods.

We begin with the **prehistory**. By this we mean the relevant results developed by various mathematicians prior to the year 1876, at which time the concept of configurations was first formulated by T. Reye [**187**]; see also [**188**], where the notation n_k was introduced[1]. The results in question were formulated as theorems pertaining to certain sets of points and lines; in retrospect we can see that the results can be interpreted as configurations, or as implying the existence of certain configurations. Typical for these "prehistoric" results is the theorem of Pappus (see, for example, [**47**, p. 67], [**52**, p. 231], which is usually formulated as follows (see Figure 1.2.1):

> If alternate vertices of a hexagon [2, 6, 8, 3, 5, 9] lie on two lines, then the three intersections 1, 4, 7 of opposite sides of the hexagon are collinear.

---

[1]This notation, or the slight modification (n_k), is in general use. Since in many cases n is either irrelevant or not known, we shall also use "k-configuration" in such instances.

It should be noted that this formulation requires additional explanation and modifications in some special cases, but this is not really relevant to the present discussion.

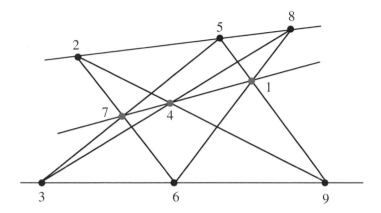

Figure 1.2.1. An illustration of the theorem of Pappus. Solid red dots are the two triplets of collinear vertices of a hexagon, solid green dots are the collinear intersection points of the three pairs of opposite sides of the hexagon.

In our interpretation, Pappus's theorem amounts to the assertion that the incidences indicated in Figures 1.1.4 and 1.1.5 are correct. We used there the same notation as in Figure 1.2.1.

Other results from the prehistoric period can be found in works of Desargues, Steiner, Möbius, Cayley, and others. We shall mention them in due course.

$$* * * * *$$

Next comes the **classical period**, which runs from 1876 to 1910. It starts with the publication of Reye's book [**187**] and ends with the publication of the survey of configurations by Ernst Steinitz [**212**] in the Encyklopädie der mathematischen Wissenschaften. This period covers the formulation of the configuration concept as well as a number of basic results, in particular those related to questions (A) and (B) in Section 1.1. This includes works by such mathematicians as Reye, Kantor, Martinetti, Schröter, Schönflies, Brunel, Burnside, Daublebsky, Steinitz, and others. For example, Kantor [**131**], [**132**] gives answers to question (B) for $n = 8, 9, 10$. Unfortunately, his results are not correct as claimed. One of his errors concerns the enumeration of the combinatorial types of configurations (10_3). Kantor claims that there are precisely ten different types of such configurations and presents diagrams purporting to illustrate these types. In Figure 1.2.2 we reproduce Kantor's drawing of one of these configurations. However, in

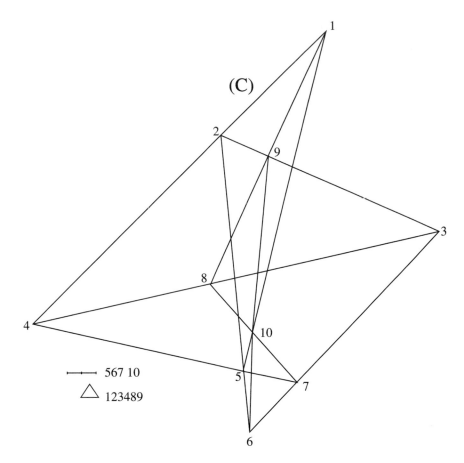

Figure 1.2.2. A diagram from Kantor [**132**] which was supposed to illustrate a configuration (10_3) but which cannot be drawn by straight lines (as we shall see in Section 2.2).

contrast to the situation with the Pappus configuration explained above, a configuration of this type **cannot** be drawn with **straight** lines. The reviews [**193**] of [**132**] by C. Rodenberg and [**203**] of [**131**] and [**132**] by H. Schubert contain several incorrect assertions. The impossibility was first proved by Schröter [**201**] and by other methods more recently by Carver [**40**], Laufer [**143**], Bokowski and Sturmfels [**29**], Glynn [**72**], and Sternfeld et al. [**215**]. Claims by Dolgachev [**61**] that this result is due to Kantor [**131**] (sic) and that "a modern proof" is given in [**18**] are both incorrect. We shall return to this topic in Section 2.2.

While Kantor's error was discovered shortly after its publication, other deficiencies of his paper did not receive attention till very recently (see, for example, Section 2.3). The same late discovery of errors occurred with certain works of Martinetti and Steinitz, which were considered as basic for the theory of configurations throughout the twentieth century; however, the "Ernst Steinitz" article in the *MacTutor History of Mathematics* [**176**] ignores all of Steinitz's work on configurations! We shall discuss Martinetti's and Steinitz's claims, and their corrected versions, in Chapter 2.

It is not clear how the various errors arose, and it is even more mysterious why they remained hidden for close to a century. A possible answer to the latter question is that for a long time apparently nobody cared enough about the topic to immerse oneself in the long and murky expositions of the original papers.

Other investigations during this period dealt with specific types of configurations, such as "mutually inscribed and circumscribed polygons". As relating to k-configurations, with the single exception of a paper by Brunel [**35**], the considerations were limited to $k = 3$. We shall consider these results, as well as some more recent one on these topics, in many of the following pages.

$$* * * * *$$

The classical period, which was characterized more by enthusiasm for configurations than by solid mathematical achievement, was followed by the long-lasting **dark ages** of configuration theory. While configurations did not disappear completely from the mathematical horizon, in the period from 1910 till 1990 there were few significant publications on this topic. It may well have been no fault of configurations that the creative attention of mathematicians was directed to other fields. Rather it is the excitement caused by spectacular development in other mathematical disciplines that attracted most researchers; it led to the general neglect of the intuitively accessible parts of geometry. The interest shifted to more technically intricate fields, leaving configurations dead in the water.

A few exceptions from this gloomy picture deserve to be mentioned. The first is the only book-length serious publication on configurations, by Levi [**145**]. Unfortunately for the theory of configurations, the book had practically no influence on later works on configurations. Possible explanations for this lack of effect may be the very restricted types of questions considered (fifty pages are devoted to the consideration of the lines associated with the theorem of Pascal—a topic that was very popular in the "prehistory" but has practically no relevance to more modern investigations of configurations), together with the dry and pedestrian tone of exposition, the pedantic discussion of topics not connected to configurations (such as the long chapter on regular polyhedra), and the almost complete absence of references to previous work on configurations. Naturally, the fact that the center of gravity of mathematicians' interests shifted to other fields must also be considered as relevant to the book's lack of influence. We shall have occasion to mention Levi's book several times in connection with some of the original results contained in it—but this is meager pickings for a book of more than 300 pages.

Three years after the publication of Levi's work an extremely well-received book was published by Hilbert and Cohn-Vossen [**126**] (see, for example, [**69**]). Twenty years later an English translation was published, but unfortunately the editors of Mathematical Reviews did not feel it deserved more than a listing; a second German edition in 1996 did not get even a listing. This is a grievous mistake, since many later workers (the present author included) became interested in configurations by reading the account in [**126**]. This presentation of the basics of configuration theory is contained in just a part of one chapter of [**126**] but presents an attractive approach to the topic. It has been mentioned often as a justification for studying configurations, by quoting the following sentence from [**126**, English translation p. 95]:

> "... there was a time when the study of configurations was considered the most important branch of geometry."

The author would like to conjecture that this is the greatest exaggeration of the truth that can be found in any of Hilbert's writings. While it is a fact that—as mentioned above—in the "classical period" of the history of configurations there were quite a few people interested in the topic, configurations were never a central topic of mathematical (or geometric) research.

Even so, the influence of these two books can on occasion still be discerned today. For example, the recent work [**179**] by K. Petelczyc mentions only these two sources for its information about configurations—ignoring all earlier and later publications.

Another relevant publication is the book by H. L. Dorwart [**62**].

Other points of light during the "dark ages" of configurations were several papers by Coxeter. Two of his early contribution to the topic are [**45**] and [**46**]. The latter—reproduced in [**51**]—introduced several new ideas and popularized some older ones; we shall mention it frequently in various section of this book. Coxeter's other contributions to configurations are his papers [**48**], [**49**], and [**50**], in which he presents detailed studies of certain specific configurations.

Some other papers on configurations that were published during the "dark ages" will find mention in appropriate places. They did not amount to much, and some of them were wrong.

* * * * *

After the long "dark ages" of the theory of configurations came a **renaissance** that continues to this day. It amounted to posing questions different from the ones considered previously, as well as to application of new tools and methods. For example, the investigation of Euclidean symmetries of a configuration was found to lead to more meaningful presentations of various configurations and also to lead to the construction of previously unimagined ones. This is evident in the illustrations of the 4-configurations in Figure 1.1.2 and will become a leading topic in several later sections. Another difference concerns the more careful attention given to questions of the graphic presentation of configurations, the use of new computational methods and computer graphics in the study of configurations, and—last but not least— to precise definitions, formulation, and proofs of results.

Rather immodestly, the author believes that he played a significant role in the "renaissance" of the theory of configurations. Influenced by the chapter on configurations in the Hilbert and Cohn-Vossen's book [**126**], in the mid-1980s he started studying configurations and presenting some of the results and problems in various seminars at the University of Washington and in some courses that he was giving, for example, [**109**], which is mentioned in [**29**]. The first formal publications resulting from these actions were the publication of the Grünbaum and Rigby paper [**122**] in 1990 and the Sturmfels and White paper [**217**] in the same year. The former contains the first published images of any (n_4) configurations, while the latter answers (affirmatively for $n \leq 12$) the question the author posed earlier (in [**109**]) whether every configuration (n_3) can be realized in the *rational* plane. These and other publications that followed led soon to a revival of interest in configurations, with significant advances in many directions. These will be discussed in Chapters 2 and 3 and seem to justify the use of the term "renaissance".

Readers interested in the history of configurations and the individuals involved in its creation may wish to consult the various papers of H. Gropp ([**81**], [**86**], [**97**], [**99**], [**98**], [**100**], [**101**]). However, it should be borne in mind that Gropp's attitude towards geometric configurations may be inferred from his statement in [**101**]: "From the point of view of pure combinatorics the problems of realizing and drawing configurations may be artificial."

* * * * *

History does not stop today. There is no doubt that the theory of configurations, and related topics, will continue to be studied and to evolve. It is not possible to predict what direction the future investigations will take, but it is possible to hope that deeper connections will be established

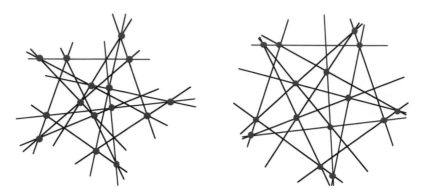

Figure 1.2.3. Two configurations (15_3). Are they "essentially the same" according to your definition?

with algebraic geometry—a rather natural home to configurations—and to topology.

Exercises and Problems 1.2.

1. Define what you understand by, say, that "two configurations are essentially the same". Use your definition to decide whether the configurations in Figure 1.2.3 fit your definition of "sameness".

2. Can you find a configuration that is "essentially the same" according to your definition as the ones in Figure 1.2.3 but has 3-fold rotational symmetry?

3. Find (in a book or Google or ...?) a formulation of the Desargues's theorem that leads to the Desargues configuration in Figure 1.1.1.

1.3. Basic concepts and definitions

In this section we shall clarify the fundamental concepts involved in the study of configurations. We shall start with very general definitions that will enable us to specialize and particularize the concepts as we find appropriate. Considerable care is required in distinguishing the various related concepts; neglect to do so led to many of the problems we mentioned in the historical account in Section 1.2. The reader who finds the plethora of words confusing should skip the details and return to this section when the text uses terms that need explanation. This author's position in connection with the abundance of terms is that Johann Wolfgang Goethe had it all wrong when (in "Faust", lines 1995–1996) he has Mephistopheles say:

> "Denn eben wo Begriffe fehlen,
> Da stellt ein Wort zur rechten Zeit sich ein."[2]

---

[2]Freely translated: "Where concepts are missing, a word soon appears."

On the contrary, words are needed (even if they have to be invented) when concepts appear that need to be distinguished from other concepts. So, here we go.

A *configuration* C is a family of "points" (sometimes called *vertices*) and a family of "lines" such that, for positive integers p, q, n, k each of the p "points" that constitute the family is "incident" with precisely q of the n "lines", while each of these "lines" is "incident" with precisely k of the "points". The use of quotation marks is meant to indicate that these objects can be of any nature whatsoever, as soon as the "incidence" relation satisfies what can be considered the natural conditions:

- It is a symmetric (that is, mutual) relation;
- an "incidence" can involve only a "point" and a "line", never two "points" or two "lines"; and
- two "points" (or "lines") can be incident with at most one "line" ("point").

Moreover, it is assumed throughout the book that use of the word *configuration* implies that there are a "point" and a "line" that are not "incident" with each other. The totality of "points" and "lines" of a configuration will be called its **elements**.

A configuration C with **parameters** p, q, n, k will in general be denoted by (p_q, n_k). The concept of (p_q, n_k) configurations and the notation for it were introduced by de Vries [58] in 1888. If the particular values of p and n are not important, we shall say that we have a **[q, k]-configuration**. If $q = k$, we shall simplify the notation by dispensing with the brackets and write **k-configuration**. In the terminology of a parallel universe, what we call a [q, k]-configuration is known as a "geometry of order $(q-1, k-1)$"; see, for example, van Maldeghem [220]. Note that a [q, k]-configuration is called a "slim geometry" in the terminology of that universe, and a 3-configuration is a "bislim geometry".

By counting the total number of incidences in a configuration (p_q, n_k), it follows that a necessary condition for the existence of a configuration with parameters p, q, n, k is the equation $pq = nk$.

There are additional necessary conditions. Each "point" of (p_q, n_k) is "incident" with q "lines", each of which is "incident" with $k-1$ other "points". Hence there are at least $p \geq q(k-1) + 1$ "points". A similar argument shows that $n \geq k(q-1) + 1$. To avoid trivialities, we shall generally assume that $q \geq 2$ and $k \geq 2$, which implies that $p \geq 3$ and $n \geq 3$. (Exceptions will be signaled explicitly.) Although these necessary conditions are in many cases sufficient for the existence of some configuration (p_q, n_k), we shall see in the following sections that this is not always the case.

Much of the time we are interested in configurations (p_q, n_k) with $p = n$ (and therefore $q = k$). As already mentioned, it is customary to simplify the notation for such configurations and designate them as (n_k) **configurations**, or **k-configurations**. The (n_k) notation (or, more precisely, the very similar n_k notation) was introduced by Reye [**188**] in 1882; it is convenient in many contexts, but in cases where the number of points and lines is not known or is nor relevant, it seems illogical to insert the letter n that has no meaning in such a situation; hence the alternative "k-configuration" notation.

In the literature, k-configurations are often called "symmetric"; but this term is highly unsuitable since the configurations in question may fail to have any symmetry whatsoever, geometric or combinatorial. (The ill-advised use of "symmetric" in this context seems to go back to erroneous understandings in some late nineteenth-century writings on configurations and possibly to its use in the theory of BIBDs—balanced incomplete block designs; the latter are totally irrelevant to the theory of configurations. More about this will follow in Section 1.5.) The main objection to the use of "symmetric" in this context is that, as we shall see throughout the text, configurations exhibiting certain genuine symmetries—combinatorial or geometric—are very important. With much justification it may be claimed that configurations with *geometric symmetries* have been the motivating factor in the recent great expansion of knowledge about configurations. Hence bestowing the descriptor "symmetric" to configurations that may be totally devoid of symmetries is downright misleading. In the present book we shall say that k-configurations are **balanced**; it is hoped that this term will become the accepted designation for k-configurations. (The use of "balanced" in the BIBD theory is not compromised by this use for configurations, just as the use of "symmetric" for BIBDs raises no problems for configurations.)

We shall be concerned with configurations at three levels of generality. In the most restricted sense, "points" and "lines" are interpreted as being the points and lines in some space in which these concepts are defined, so as to satisfy the first two groups of Hilbert's axioms. (For these axioms see, for example, Hilbert [**125**], Noronha [**174**], Sibley [**207**], Stahl [**209**].) In particular, this interpretation includes the traditional Euclidean plane and higher-dimensional Euclidean spaces, as well as the real projective plane and higher-dimensional projective spaces. Unless specifically stated otherwise, we shall consider configurations at this level of generality to be in the real Euclidean plane, or in the real projective plane, and call such configurations **geometric**. In the Appendix the necessary facts about the Euclidean and projective planes are collected. The usual way of presenting geometric configurations is by **diagrams**, in which "points" are represented by solid

Table 1.3.1. A configuration table for a set-configuration $(16_3, 12_4)$.

a	b	c	d	e	f	g	h	i	j	k	l
A	A	A	H	B	B	B	I	C	C	J	D
H	N	G	N	H	N	I	P	O	J	O	K
I	O	L	P	G	M	J	M	M	K	P	L
C	D	E	K	F	E	D	L	F	E	G	F

dots and "lines" by straight lines. Naturally, it must be true (and possible to verify) that the presumed lines are straight and that the incidences actually occur; some diagrams can be misleading, as illustrated by Figure 1.2.2. More details on this topic will be found in Section 2.2. For general discussions of the topic of drawing configurations see [**92**], [**97**].

In the most general sense we shall consider **combinatorial** (or **abstract**) configurations; we shall use the term **set-configurations** as well. In this setting "points" are interpreted as any symbols (usually letters or integers), and "lines" are families of such symbols; "incidence" means that a "point" is an element of a "line". It follows that combinatorial configurations are special kinds of general incidence structures. Occasionally, in order to simplify and clarify the language, for "points" we shall use the term **marks**, and for "lines" we shall use **blocks**. The main property of geometric configurations that is preserved in the generalization to set-configurations (and that characterizes such configurations) is that two marks are incident with at most one block, and two blocks with at most one mark. The usual way of presenting set-configurations is by **configuration tables**. In such a table the marks of each block are listed in a column that represents the block. If necessary or convenient, a label for the block may be indicated at the head of each column, but in other cases this may not be needed. An example of a set-configuration $(16_3, 12_4)$ is shown in Table 1.3.1 (with block labels); in Table 1.3.3 we dispensed with the block labels. Sometimes a combinatorial configuration (p_q, n_k) is presented simply as a family of n k-tuples formed from p marks, with appropriate restrictions. In [**82**] and other papers, H. Gropp faults Hilbert and Cohn-Vossen [**126**] and Coxeter [**46**] for being interested in structures "realized in real geometry" and not in the "more modern 'schematical' configurations". This seems to the present author to be quite inappropriate, especially since in most of his papers Gropp does not warn the reader that his use of "point" and "line" is meant in the purely combinatorial sense.

It is immediate from the definitions that every geometric configuration gives rise to a set-configuration: Just label the points of the geometric configuration, and use these labels as marks to construct a configuration table. As can be checked very easily, the set-configuration in Table 1.3.1 corresponds

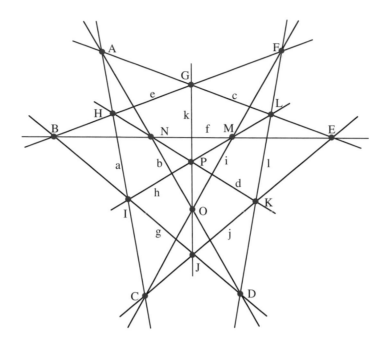

Figure 1.3.1. A geometric configuration $(16_3, 12_4)$ with points labeled in such a way as to yield the configuration Table 1.3.1.

in this way to the $(16_3, 12_4)$ geometric configuration in Figure 1.3.1. However, as we shall illustrate in Section 2.1, the converse is not valid: There are set-configurations that do not correspond to any geometric configuration.

The relationship between the configurations in Table 1.3.1 and Figure 1.3.1 is sometimes formulated by saying that the former **underlies** the latter and that Figure 1.3.1 is a geometric **realization** of the set-configuration in Table 1.3.1.

A remarkable result goes back to Steinitz [**210**], in the classical period of the theory of configurations. In one formulation, it states that every combinatorial k-configuration can be presented in an **orderly** configuration table. A configuration table is orderly if each point (mark) appears in each row of the table once and only once. We shall discuss this result and some of its ramifications in Section 2.5.

If the points and lines of two configurations C' and C'' admit labels such that a 1-to-1 correspondence τ of points to points (and lines to lines) preserves incidences, we shall say that C' and C'' are **isomorphic** (or **combinatorially equivalent**, or of the same **combinatorial type**); sometime we shall also wish to make explicit the correspondence τ. With appropriate interpretation, this terminology applies to set-configurations as well. For example, the set-configuration in Table 1.3.1 is isomorphic to the geometric configuration in Figure 1.3.1 with τ the identity transformation ι of the labels. If an incidence-preserving correspondence maps points to lines (and vice versa), it is said to be a **duality**, and the configurations are said to be **dual** to each other.[3] A configuration table of the configuration dual to the one in Table 1.3.1 is shown in Table 1.3.2 and in Figure 1.3.2 is a geometric configuration dual to the one in Figure 1.3.1.

Here again a warning seems necessary. Some authors (for example, van Maldeghem [**220**]) use the word "realization" with a different meaning. In particular, they allow (geometric) realizations of set-configurations to have additional incidences, not among the ones in the underlying set-configurations. Said differently, they use "realization" for a different concept, which we shall encounter below under the designation "representation". Thus, for us, a geometric realization of a set-configuration is isomorphic to it, while a representation that is not a realization is not isomorphic to it.

Table 1.3.2. A configuration table of the dual of the configuration in Table 1.3.1 and in Figure 1.3.1.

A	B	C	D	E	F	G	H	I	J	K	L	M	N	O	P
a	e	a	b	c	e	c	a	a	g	d	c	f	b	b	d
b	f	i	g	f	i	e	d	g	j	j	h	h	d	i	h
c	g	j	l	j	l	k	e	h	k	l	l	i	f	k	k

Some configurations are isomorphic to configurations dual to them. In such cases we call the configuration **selfdual**. In other words, a configuration C is selfdual if there is an incidence-preserving correspondence τ that maps the points of C onto its lines and vice versa; such a correspondence τ is called a **selfduality**, and it is obvious that in this case the inverse τ^{-1} is a selfduality as well. In many cases (but not always) τ^2 is the identity map ι. An example of a selfdual configuration with $\tau^2 = \iota$ is shown in Figure 1.3.3. We shall discuss this topic in much more detail in Sections 2.10 and 5.8.

There is often a need to consider families of points and lines (or marks and blocks) that fail—by a "few" incidences—to be configurations in the sense we use here. We say that such a family is a **prefiguration**. We shall often encounter two types of prefigurations, although other types occur at times.

A **superfiguration** (p_q, n_k) is a family of "points" and "lines" with "incidences" as in the definition of configurations, such that each of the

---

[3]It is unfortunate that Coxeter [**46**] uses "dual configurations" to mean any pair consisting of a (p_q, n_k) configuration and an (n_k, p_q) configuration; in this terminology any (n_k) configuration is selfdual. This error has been copied by Evans [**66**].

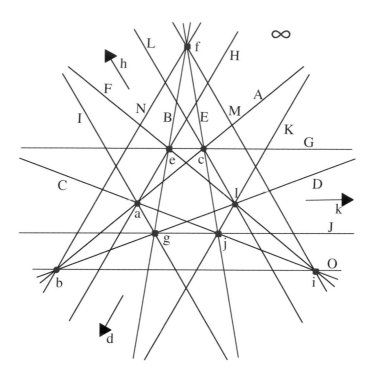

Figure 1.3.2. A geometric configuration that is a realization of the set-configuration in Table 1.3.2. It is dual to the configuration in Figure 1.3.1. The arrows indicate "points-at-infinity", and the "line-at-infinity", indicated by the infinity symbol, is also part of the configuration. Explanations of this kind of diagram in the projective (or extended Euclidean) plane are given in the Appendix.

p "points" is "incident" with **at least** q of the n "lines", and each "line" is "incident" with at least k of the "points". If the number of incidences exceeding that in a $[q, k]$-configuration is s, we shall sometimes say that it is an **#s-superfiguration**. An example of a remarkable (and quite useful—see Section 2.11) superfiguration is shown in Figure 1.3.4; it is a #2-superfiguration of a 3-configuration, since there is one line incident with four points and one point incident with four lines. (This is taken from [**107**]; it also appears in [**29**] and [**169**] and probably in other places as well.) Superfigurations often arise through "accidental" incidences in configurations that have realizations depending on variable parameters.

Another way of looking at this situation is that sometimes it is necessary or convenient to consider certain superfigurations as *representing* combinatorial *configurations*. As already mentioned, by a **representation** we mean a family of points and lines such that all the combinatorial incidences are satisfied but some points may be on lines with which they are not incident

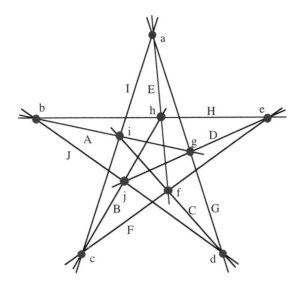

Figure 1.3.3. An example of a selfdual configuration (10_3). The self-duality mapping τ interchanges the upper and lowercase letters. Hence clearly $\tau^2 = \iota$, the identity.

in the combinatorial configuration, or some pairs of distinct points (or lines) of the combinatorial configuration may be represented by single points (or lines). A typical example of such a superfiguration is shown in Figure 1.3.5, which is a *representation* of the set-configuration specified in Table 1.3.3. The point 1 in Figure 1.3.5 lies on the line 089 but is not incident with it according to Table 1.3.3.

Table 1.3.3. A configuration table of a set-configuration (10_3). We shall encounter this configuration again in Section 2.2.

1	1	1	2	2	3	3	0	0	0
2	4	6	4	7	5	6	4	5	8
3	5	7	8	9	9	8	6	7	9

In the same vein we call **subfiguration** (or **#s-subfiguration**) a family that is short by a number of incidences (specifically, s incidences), to be $[q, k]$-configuration. An illustration of this concept is provided in Figure 1.3.6. Other examples will become very important in Section 2.5.

* * * * *

Intermediate in generality between set-configurations and geometric configurations are **topological configurations**, also known as **configurations of pseudolines**. A family of simple curves in the (projective or Euclidean) plane is a family of **pseudolines** provided each curve differs from a straight

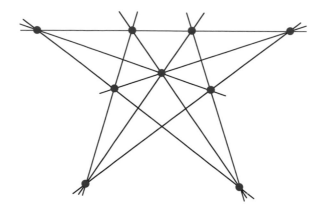

Figure 1.3.4. An example of a geometric #2-superfiguration (9_3) in which one point is incident with four lines and one line is incident with four points.

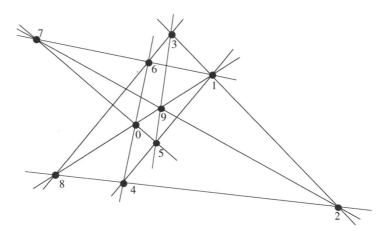

Figure 1.3.5. A #2-superfiguration that is a representation (but not a realization) of the set-configuration specified by the configuration Table 1.3.3.

line in at most one segment of the Euclidean (or projective) line and any two curves have at most one point in common, at which they cross each other. We shall discuss pseudolines in more detail in several of the following sections. The definition implies that any two points are incident with at most one pseudoline.

Configurations of pseudolines are defined in complete analogy to configurations of lines, with pseudolines taking the place of "lines". As an example of a topological configuration we may take Kantor's presumed configuration shown in Figure 1.2.2, which—as we mentioned in Section 1.2 and will prove in Section 2.2—is not realizable as a configuration of lines. Other examples are shown in Figure 1.3.7. Clearly, any geometric configuration

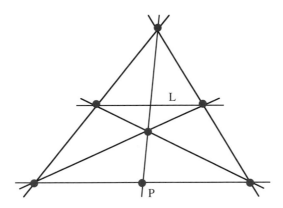

Figure 1.3.6. A subfiguration of a (7_3) configuration. If the line L were incident with the point P, this would be a configuration. However, as we shall see in Section 2.1, there exists no geometric configuration (7_3), although a set-configuration of this type is possible.

can be understood as a topological configuration, and each of the latter has an underlying set-configuration. It is easily seen that each topological configuration is isomorphic to a configuration in which each pseudoline consists of a finite number of (straight) segments (including rays, considered as segments). Several problems of combinatorial geometry that involve pseudolines are discussed in [**108**].

It is clear that if a set-configuration is geometrically representable by a superfiguration C, it is also *realizable* by a topological configuration. The pseudolines may be "bent" to avoid offending incidences. However, not every superfiguration can be realized by a topological configuration. For example, the superfiguration in Figure 1.3.4 is not a representation of any set-configuration and cannot be realized by a topological configuration.

$$* * * * *$$

The terminology of the theory of configurations is very unsettled. Different writers and "schools" use terms that are often quite unrelated to each other and sometimes even carry a different meaning while using the same words. An example of the former is the use of terms like *slim* and *bislim geometries* to indicate what we call configurations. On the other hand, some writers discuss at length "configurations" without bothering to mention that they have **combinatorial**—and *not* geometric—configurations in mind. This was to some extent excusable in the nineteenth century, when the essential difference between the concepts had not yet been recognized. A hundred year later this is exemplary carelessness or, at the least, total disregard for traditional terminology and the work of earlier authors. Still other writers state that configurations are "partial linear spaces with constant and

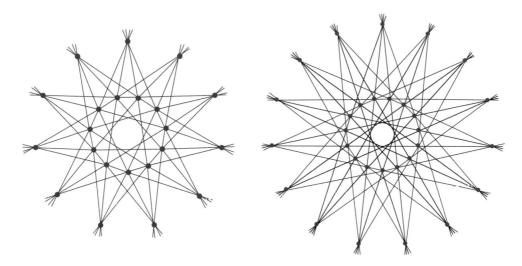

Figure 1.3.7. A (22_4) and a (30_4) configuration of pseudolines. Note how small the departure from straight lines is in these examples. It may be conjectured that these configurations are not isomorphic to any geometric configuration; however, this has not been established.

equal point rank and line rank"; despite the use of explicit geometric terms (including "n-gons") the configurations they consider are only combinatorial (see, for example, [**179**], [**129**]).

The internet also abounds with vague, misleading, or wrong entries. This applies, in particular, to the frequently consulted Wikipedia (see [**225**], as modified November 9, 2007) and Mathworld (see [**223**], quoted from the version dated November 30, 2007). In the latter one finds, among other inaccurate assertions, that the (7_3) and (8_3) configurations are realizable with a "point-at-infinity".

* * * * *

It is easy to see that every realization of a set-configuration by points and lines in any Euclidean or projective space (of any dimension) can lead by suitable projection to a geometric configuration in the Euclidean plane. However, the converse question—can a given geometric configuration be realized in a higher-dimensional space in such a way that it is not contained in a subspace—has a negative answer in some cases. For example, each of the three configurations (9_3) in Figure 1.1.6 is easily seen to be contained in the plane spanned by some three of its points. With this example in view, it is meaningful to define the **dimension** of a configuration as the largest dimension of a space that is spanned by the configuration. We shall discuss this topic in Section 5.6.

In many questions about configurations one is concerned with what is often called "polygons" in the literature. However, this is a misnomer since in most cases it is not segments that are relevant as "sides" of the "polygons"; instead, the intention is to deal with the lines of the configuration. We shall call any sequence of points and lines of a configuration that can be written as $P_0, L_0, P_1, L_1, \ldots, P_{r-1}, L_{r-1}, P_r = P_0$, with each L_i incident with P_i and P_{i+1} (all subscripts understood mod r) **multilateral**. For example, in the prefiguration shown in Figure 1.3.5, the sequence of points 1, 2, 8, 0, 4, 5, 7, 9, 3, 6, 1 (and the lines determined by adjacent pairs) determines a multilateral that involves all points and all lines. A **multilateral path** satisfies the same conditions except the coincidence of the first and last elements. A **Hamiltonian multilateral** passes through all points and uses all lines, each precisely once; hence the example just given is a Hamiltonian multilateral. We shall encounter multilaterals in several sections, and in particular Section 5.2 is devoted to Hamiltonian multilaterals.

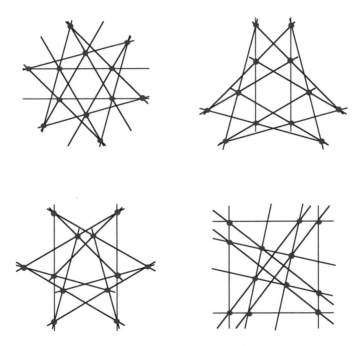

Figure 1.3.8. Four configurations (12_3). Are any isomorphic?

Exercises and Problems 1.3.

1. Show that the superfiguration in Figure 1.3.4 is selfdual. Find a selfduality map τ such that $\tau^2 = \iota$.

2. Decide whether any of the (12_3) configurations in Figure 1.3.8 are isomorphic. (As a practical matter, to show that two configurations are

isomorphic, it is sufficient to find an isomorphism. To show that they are not isomorphic, it is often simplest to find a property that is invariant under all isomorphisms but regarding which the two configurations behave differently. Neither one is all that simple to actually carry out, even for rather small configurations.)

3. What is the smallest n required for the existence of a combinatorial configuration (n_4)? Can you find a configuration table for it? Can you decide whether it is unique (up to isomorphism)?

4. Do the topological configurations in Figure 1.3.7 admit Hamiltonian multilaterals? What about the configurations in Figure 1.3.8?

5. Determine the dimension of each of the configurations in Figure 1.3.8 and of the configuration in Figure 1.3.3.

1.4. Tools for the study of configurations

There are many different ways to relate configurations of points and lines to other mathematical objects; these are often useful in investigating configurations. In Section 1.3 we encountered one example of such a tool—the underlying set-configuration and its *configuration table*. A related concept, which is helpful in the context of more general combinatorial structures as well, is the *incidence matrix* of the configuration. Specifically for configurations, incidence matrices seem to have been introduced by Levi [**145**], and we shall usually call them "Levi incidence matrices". Levi makes the rows of a matrix correspond to the points of the configuration, and the columns correspond to the lines (or vice versa). An incidence between a point and a line is indicated by a marking of the corresponding element of the matrix. This can be done by assigning to such matrix elements the value 1 and to the others 0 or by some other specification. Levi's own preference is to use an array of small squares and an X marked in each square that represents an incidence. As an illustration, consider the set-configuration we shall encounter in Section 2.2 and denote there by $(10_3)_4$. A configuration table is shown in Table 1.4.1, and a Levi incidence matrix is in Figure 1.4.1(a). By a suitable permutation of the rows and columns of that matrix, we obtain the Levi incidence matrix in Figure 1.4.1(b). This latter form demonstrates one of the uses of incidence matrices: It shows at a glance that the configuration in question is selfdual, since the matrix is symmetric with respect to the main diagonal.

$$* \; * \; * \; * \; *$$

Several graphs have been attached to configurations; it is probably simplest to describe them in terms of geometric configurations, although the

Table 1.4.1. A configuration table for a configuration discussed in Section 2.2 and denoted there by $(10_3)_4$. The notation is adapted from Schröter [201]. Two versions of the Levi incidence matrix of this configuration are shown in Figure 1.4.1, and its configuration graph is in Figure 1.4.2. A realization with pseudolines appears in Figure 1.2.2.

a	b	c	d	e	f	g	h	i	j
1	1	1	8	2	3	2	3	4	5
2	4	6	9	4	6	5	7	6	7
3	5	7	0	8	8	9	9	0	0

concepts apply to topological and combinatorial configurations (with appropriate changes in wording), and even to more general combinatorial systems.

The earliest of these graphs was proposed by K. Menger in a course on projective geometry at Notre Dame University in 1945. It seems that he never published on the topic; the first publication discussing it is a paper (doctoral thesis under Menger's supervision) of M. P. van Straten [221] in 1949. The name "Menger graph" appears to have been introduced by Coxeter in [45] and [46]; it has been used in other works as well, for example in van Maldeghem [220]. It will also appear later in this book, in Section 5.1. The **Menger graph** $M(C)$ of a configuration C is the graph with vertices corresponding to those of C; an edge of $M(C)$ connects two of its vertices if and only if the corresponding points of the configuration are collinear on a line of the configuration. There seem to have been only a few uses of this kind of graph. One mention of the Menger graph of a

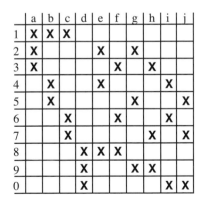

	a	b	c	d	e	f	g	h	i	j
1	X	X	X							
2	X				X		X			
3	X					X		X		
4		X			X				X	
5		X					X			X
6			X			X			X	
7			X					X		X
8				X	X	X				
9				X			X	X		
0				X					X	X

(a)

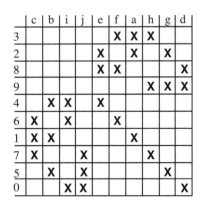

	c	b	i	j	e	f	a	h	g	d
3						X	X	X		
2					X		X		X	
8					X	X				X
9								X	X	X
4		X	X		X					
6	X		X			X				
1	X	X					X			
7	X			X				X		
5		X		X					X	
0			X	X						X

(b)

Figure 1.4.1. Two versions of the Levi incidence matrix of the configuration Table 1.4.1. In (a) the matrix is formed in the obvious way, while the rearranged form in (b) illustrates the selfduality of the configuration. Part (b) is adapted from Laufer [143].

configuration is in the paper by Di Paola and Gropp [**59**] in connection with their definition of the "configuration graph" of a configuration. It will also appear later in this book, in Section 5.1. For a configuration C, the **configuration graph** is the graph with the same vertices as C and with an edge connecting two vertices of the graph if and only if the corresponding points are not on any line of C; Gropp [**103**] calls it the "Martinetti graph", and Mendelsohn *et al.* [**156**] call it the **deficiency graph**. We shall use the latter term. Obviously, in graph-theoretic terminology the deficiency graph of a configuration is the complement of its Menger graph. As we shall see in Section 1.7 and in Chapter 2, constructs related to deficiency graphs have been used in some of the earliest papers on configurations, under the name "Restfigur" ("remainder figure") in enumerations of configurations (n_3) for small values of n. We shall enlarge on these remainder figures below. An interesting recent application is also given in Section 1.7.

The major shortcoming of both the Menger and the deficiency graphs is that they do not uniquely determine the configuration. This was noted already in [**156**] and [**59**]. The simplest example of distinct configurations having the same configuration graphs is that of the two (10_3) configurations which we shall denote (in Section 2.2) by $(10_3)_1$ and $(10_3)_4$; this is illustrated in Figure 1.4.2. In [**156**, p. 96] a family of such pairs is indicated.

Of far greater importance is the third graph associated with a configuration, its "Levi graph". This concept was introduced by Levi in [**146**], and the name was first used by Coxeter in [**46**]; see also van Maldeghem [**220**]. The **Levi graph** $L(C)$ of a configuration C is the bipartite graph, with "black" points corresponding to the points of C and "white" points corresponding to the lines of C; two points of the Levi graph determine an edge if and only if one represents a point and the other a line incident with that point. As an illustration we show in Figure 1.4.3 the Levi graph $L(C)$ of the (combinatorial and topological) configuration $(10_3)_4$.

The importance of Levi graphs derives from their 1-to-1 correspondence with combinatorial configurations. More specifically, we have the following widely used result:

A bipartite graph G is the Levi graph $L(C)$ of a $[q, k]$-configuration C if and only if

- all black vertices are q-valent and all white vertices are k-valent;

- G has girth at least 6, that is, all circuits in G have length at least 6. In particular, G has no loops or digons.

The correspondence between $G = L(C)$ and C is 1-to-1, up to isomorphism in each class of objects.

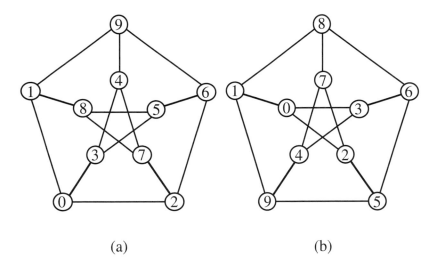

Figure 1.4.2. The configuration graph of the configuration $(10_3)_4$ specified by the configuration Table 1.4.1 and by the Levi incidence matrices in Figure 1.4.1 is shown in (a). In (b) is the isomorphic configuration graph of the Desargues configuration, denoted by $(10_3)_1$ in Section 2.2, and using the labels shown there.

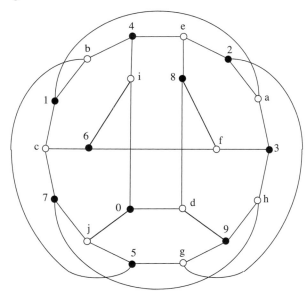

Figure 1.4.3. The Levi graph $L(C)$ of the topological configuration $C = (10_3)_4$. The color-reversing mirror symmetry of the graph shows that the configuration $(10_3)_4$ is selfdual under the correspondence implied by the symmetry.

Various graph-theoretic concepts can be transferred to configurations by the use of Levi graphs. For example, we say that a configuration is **connected** or *c*-**connected** for some integer *c* if and only if its Levi graph is connected, resp. *c*-connected. Translated into the terminology of configurations, connectedness means that any two elements can be included in a multilateral path; a configuration C is *c*-connected if on selecting any $c - 1$ elements of C, any pair of the remaining elements can be connected by a multilateral path that does not use any of the selected elements.

A not entirely trivial result, easily provable using Levi graphs, is that any connected *k*-configuration is 2-connected. We shall see this in Section 5.1, together with a discussion of some related questions.

As mentioned earlier, a concept related to configuration graphs is that of "remainder figures". We shall put it in a more general setting, and we define:

Definition 1.4.1. Given a configuration C and a point P of C, the **complementary graph of** C **at** P consists of the points of C that are not collinear with P on any line of C and the segments connecting pairs of these points. The **complementary graph complex** of C is the family consisting of complementary graphs of C at all its points P. It is usually more convenient to take, instead of the family, the union of the members of the family.

Table 1.4.2. The configuration table of a set configuration (8_3).

$$
\begin{array}{cccccccc}
1 & 2 & 3 & 4 & 5 & 6 & 7 & 8 \\
2 & 3 & 4 & 5 & 6 & 7 & 8 & 1 \\
4 & 5 & 6 & 7 & 8 & 1 & 2 & 3
\end{array}
$$

For example, for the configuration (8_3) given by the configuration Table 1.4.2, the complementary graph at each point is a singleton vertex, and the complementary graph complex consists of eight isolated vertices. Analogously, the complementary graph complex of the combinatorial configuration (14_4) shown in Table 1.4.3 consists of 14 isolated vertices. For each of the combinatorial configurations (9_3) (see Figure 1.1.5) and (15_4) (see Section 3.1) the complementary graph complex consists of one or more circuits that cover all the vertices of these configurations; we will have more to say about these cases in Sections 2.2 and 3.1. It should be noted that in general, the isomorphism of the complementary graph complexes of two configurations does not imply the isomorphism of the configurations; this is illustrated in Section 2.3.

$$* \ * \ * \ * \ *$$

Table 1.4.3. A configuration table (adapted from [**159**]) of a combinatorial configuration (14_4).

A	A	A	A	B	B	B	C	C	C	D	D	D	E
B	F	G	H	G	H	E	H	E	F	E	F	G	F
C	L	N	P	L	M	P	L	M	N	L	Q	M	G
D	M	R	Q	Q	N	R	R	Q	P	N	R	P	H

Each given geometric configuration C determines an **arrangement** $A(C)$ of the Euclidean or projective plane. By this is meant the 2-complex consisting of the intersection points of the lines of C (**vertices** of $A(C)$), of the open segments (**edges** of $A(C)$) of each line constituting the complement of the points of the line, and the 2-dimensional open convex polygons (**cells** of $A(C)$) that constitute the connected components of the complement of the union of the lines. For example, in Figure 1.4.4 we show another drawing of the (10_3) configuration C from Figure 1.3.3, in which we made visible all the intersection points of its lines; the points that are not configuration points are shown by hollow circles. It is easy to count that, in the Euclidean plane, $A(C)$ has 25 points, 70 edges (20 unbounded), and 46 cells (20 unbounded). If considered in the projective plane (that is, the extended Euclidean plane), then it has 60 edges and 36 cells. In either case, one can apply the appropriate Euler relation or other results that have been established for arrangements to deduce properties of configurations. The concept of arrangement associated with a configuration can be applied to topological configurations as well. This will be useful in Sections 2.1 and 3.2.

$$* \; * \; * \; * \; * \; * \; *$$

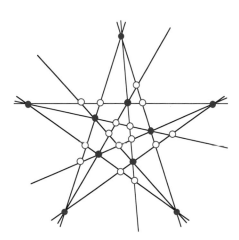

Figure 1.4.4. The arrangement associated with the (10_3) configuration from Figure 1.3.3.

Other tools that have been found very useful in some questions concerning configurations are of a more algebraic character. They have been explained and applied in several widely quoted and studied works, in particular those by Bokowski and Sturmfels [**29**] and Bokowski [**25**]. We shall not have any occasion to use them here, so we advise the reader interested in seeing how this approach works to consult with the literature.

Another topic we are not going into in this book is programs for computers and computer graphics. Many of the enumeration results (especially those of a purely combinatorial character) have been obtained by using computers. We shall mention a number of such cases, but we do not find it appropriate to enter into details beyond giving credit and references to the original works.

Much of the work presented in this book could not have been done at all without the use of widely available software. The author's main tools were the various consecutive versions of Mathematica®, Geometer's Sketchpad®, and ClarisDraw®, as implemented on several generations of Apple® computers. Other computer-related programs and graphics were generated by various collaborators in the joint papers we shall mention in due course, and the author sometimes adapted them to the formats used here.

Exercises and Problems 1.4.

1. Find a "configuration" table for the prefiguration in Figure 1.3.3 and a Levi incidence matrix for it. Find a version of the incidence matrix that exhibits the selfduality of the prefiguration.

2. Consider the graph determined by the vertices and (finite) segments in Figure 1.4.4. Does it admit a Hamiltonian circuit?

3. For each mark of the set-configuration (14_4) in Table 1.4.3 find its complementary graph. Does the complementary graph complex determine any configuration?

1.5. Symmetry

By a **symmetry** of an object we generally mean a mapping of the object onto itself that preserves some relevant features of the object. For configurations, by **symmetry** we shall understand that the incidence relations are preserved, but we will also impose other requirements that will depend on the kind of configuration considered and on other aspects of the discussion.

More specifically, for combinatorial configurations a **symmetry** is just an incidence preserving 1-to-1 mapping (permutation) of the elements of the configuration onto themselves. We find it convenient to distinguish between **automorphisms**, that is, symmetries that map marks to marks and blocks to blocks, and **dualities**, which map marks to blocks and vice versa.

By an **automorphism** of a geometric or topological configuration we shall understand an automorphism of the underlying combinatorial configuration.

For topological configurations a **symmetry** is a *homeomorphism* of the plane onto itself that maps the configuration onto itself. However, in different contexts, this definition should be understood in one of three ways, depending on the **plane** we are considering. This can be either the Euclidean plane E^2 or the extended Euclidean plane E^{2+} or the projective plane P^2. Although the projective plane is homeomorphic with the extended Euclidean plane, when considering symmetries of E^{2+}, we require that the line-at-infinity be mapped onto itself. It follows that the symmetries of a topological configuration in E^{2+} can be a proper subset of the symmetries of such a configuration in P^2.

Analogously, **symmetries** of geometric configurations in E^2 are *isometries* of the plane that map the configuration onto itself. For geometric configurations in E^{2+} we need an isometry of E^2 that maps the finite part of the configuration onto itself and permutes the points-at-infinity.

For both topological and geometric configurations it is sometimes useful to include dualities among their symmetries. In particular, for geometric configurations a special type of duality is called **polarity** or **reciprocation**, since it arises by the polarity (also called reciprocation by some) in a circle.

As is obvious, in each of the interpretations of the term "symmetry", all symmetries of a configuration form a group, the **symmetry group** of the configuration (in the appropriate sense). Quite often it is convenient to consider only a subgroup of the symmetry group of a configuration. In such a case we shall say that the group in question is a **group of symmetries** of the configuration.

We shall soon see examples of these various interpretations of "symmetry". But first we should discuss two aspects of symmetries of geometric configurations (that apply in some cases to the other kinds of configurations as well) that lead to classifications of the appropriate configurations.

Our first concern is the collection of orbits of the configuration under *the* symmetry group of the configuration. If a configuration has h_1 orbits of points and h_2 orbits of lines, we shall occasionally say that it is of **orbit type** $[h_1, h_2]$ or $[h_1, h_2]$-**orbital**. We note that no *geometric or topological* (n_k) configuration with $k \geq 3$ can have a single orbit of points or a single orbit of lines; in contrast, there are many such *combinatorial* configurations of this type and even of $[1, 1]$ orbit type. More generally, if a geometric $[q, k]$-configuration (that is, a (p_q, n_k) configuration) is $[h_1, h_2]$-orbital, then $h_1 \geq [(k+1)/2]$ and $h_2 \geq [(q+1)/2]$. This is a consequence of the fact that no isometric symmetry can map the middle one of three collinear points onto one of the other points of the triplet, and analogously for lines. In case equality holds in both inequalities, we shall say that the configuration is **astral**. The most interesting seem to be $[h_1, h_2]$-orbital configurations with $h = h_1 = h_2$; in that case we shall simplify the language by calling the configuration **h-orbital** or, more often, **h-astral**. If the values of h_1, h_2, or h are not relevant or are unknown, we shall speak of **multiastral** configurations. (We give more on this terminology at the end of the present section.)

In many situations we shall be dealing with configurations in which all orbits have the same number of elements; however, some cases in which this condition is not fulfilled do have interesting features and lead to various questions. In any case, this is not a requirement included in the definition.

The other aspect of symmetry considerations for a geometric configuration is the determination of its **symmetry group**. From the well-known classification of isometries of the Euclidean plane it follows that the symmetry group of a geometric or topological configuration is either a **cyclic group** c_r or a **dihedral group** d_r, where r is a positive integer. The group c_r consists of rotations about a **center** through integer multiples of $2p/r$, the zero multiple being the identity. The group d_r consists of the same rotations as its subgroup c_r, together with r **mirrors**, that is, lines of reflective isometry.

For example, the configuration in Figure 1.5.1(a) has symmetry group d_{10}, and the one in (b) has symmetry group d_5. Other illustrations are given in Figures 1.5.2 and 1.5.3.

Although configurations with non-trivial symmetry group occurred in the literature from time to time, it is the recent—last twenty years or so—systematic concern with very symmetric configurations that led to the revival of interest in the whole topic of configurations. We shall investigate symmetric configurations of various kinds in several of the following sections.

As an example of the use of the different notions of symmetry, we reproduce Figure 1.3.3 once more as Figure 1.5.4 and show its Levi graph in Figure 1.5.5. Although the symmetry group of this configuration is c_5, the symmetries of its Levi graph show that the automorphism group of this configuration is c_{10} and the group of automorphisms and selfdualities is d_{10}.

We give one urgent note of caution.

As is the case in many rapidly developing fields, the terminology of configurations with varying degrees of symmetry is still unsettled. One could almost claim that each author introduces separate concepts and often even changes them from paper to paper. This has certainly been the case with the present writer—naturally, on each occasion there was some good reason for the terms introduced and used.

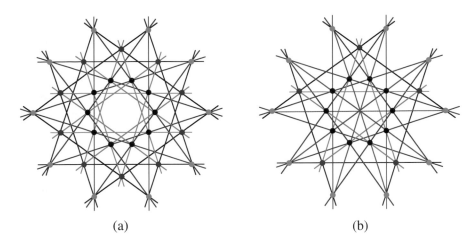

(a)   (b)

Figure 1.5.1. Geometric configurations (30_4) of orbit type $[3,3]$ (that is, 3-astral) and (25_4) of orbit type $[3,4]$; the orbits are color-coded. The configuration in (a) has symmetry group d_{10} and all (point and line) orbits of size 10. The configuration in (b) has symmetry group d_5, two orbits of size 10 and one of size 5 for points, and one orbit of size 10 and three of size 5 for lines.

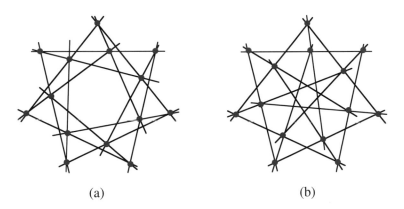

(a)   (b)

Figure 1.5.2. Two geometric configurations (14_3) of the orbit type $[2,2]$ and with symmetry group c_7. Both are astral.

The astral, h-astral, and $[h_1, h_2]$-astral terminology we shall use in this book is a development of the various similar concepts introduced in [**110**], [**111**], and [**117**]. It should be stressed that although astral configurations are visually attractive and theoretically most easily investigated, for many kinds of configurations they are not the smallest possible.

In [**24**] M. Boben and T. Pisanski introduced related terminology dealing with *polycyclic* configurations in the Euclidean plane. They call a configuration C k-cyclic provided there exists an automorphism α of order k of the underlying abstract configuration such that all orbits of points and lines of

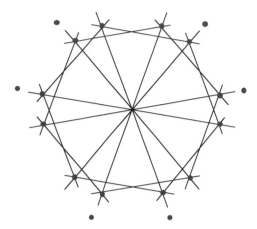

Figure 1.5.3. A geometric configuration (18_3) of the orbit type $[2, 2]$ with symmetry group d_6 in the extended Euclidean plane. This configuration cannot be represented with high symmetry in the Euclidean plane, but it is astral in the extended Euclidean plane E^{2+}.

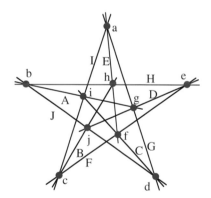

Figure 1.5.4. This (10_3) configuration has symmetry group c_5 and orbit type $[2, 2]$; hence it is astral.

C under α have the same size k (that is, the number of elements in each is k). In this terminology the configuration in Figure 1.5.1(a) is 10-cyclic (with three orbits of points and three orbits of lines), while the configuration in Figure 1.5.1(b) is 5-cyclic (with five orbits each of points and lines). The two configurations in Figure 1.5.2 are 7-cyclic.

Another related concept is that of celestial configurations, considered by L. Berman in [**10**]. We shall discuss it in Section 3.5.

Exercises and Problems 1.5.

1. Decide whether the two (14_3) configurations in Figure 1.5.2 are isomorphic.

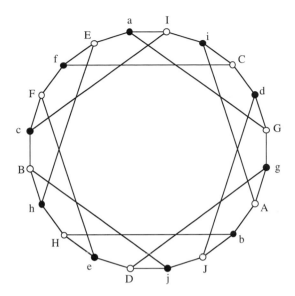

Figure 1.5.5. The Levi graph of the (10_3) configuration shown in Figure 1.5.4.

2. Find the symmetry group of each of the configurations (12_3) in Figure 1.3.8.

3. Consider the different realizations of the Pappus configuration $(9_3)_1$ in Figure 1.5.6. Maps between the different realizations establish automorphisms of the underlying combinatorial configuration. Find permutation representations for each of these mappings. Which of them correspond to geometric (isometric) symmetries?

4. Show that all points of the combinatorial configuration underlying the Pappus configuration $(9_3)_1$ form a single orbit (under automorphisms). What about the lines? What about the other two configurations (9_3) (see Figure 1.1.6)?

5. Show that there exist combinatorial configurations such that all the points are in one orbit (under automorphisms) but the lines belong to more than one orbit.

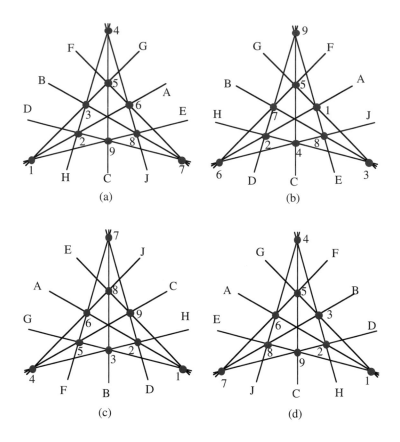

Figure 1.5.6. Four realizations of the Pappus configuration $(9_3)_1$.

1.6. Reduced Levi graphs

We introduced Levi graphs of configurations in Section 1.4. Now, with the symmetry concepts available, we can modify Levi graphs in such a way that for symmetric geometric configurations the information appears in a much more condensed form. We call these graphs "reduced Levi graphs" and describe them separately for cyclic and for dihedral symmetry groups[4].

The **reduced Levi graph** $R(C)$ of a geometric (or topological or combinatorial) configuration C with a **cyclic** group c_r and of orbit type $[h_1, h_2]$ is a bipartite graph that consists of h_1 black vertices and h_2 white vertices corresponding to the **orbits** of points and of lines of C. An edge connects two vertices (of different colors, naturally) if and only if points of the corresponding orbit are incident with the lines of the corresponding line orbit.

---

[4]The term "reduced Levi graph" has been used with a different meaning by R. Artzy in [**2**]. We shall discuss this in Section 2.11.

For the (12_3) configuration C in Figure 1.6.1 this first step leads to the graph in Figure 1.6.2(a).

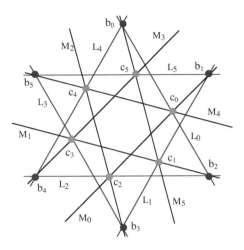

Figure 1.6.1. An astral (12_3) configuration C, with cyclic symmetry group c_6 and with labels and colors convenient in the construction of its reduced Levi graph $R(C)$.

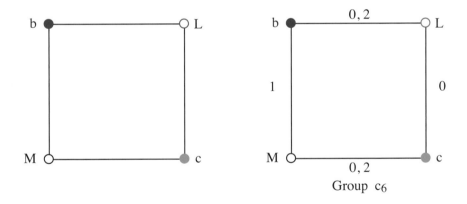

Figure 1.6.2. The formation of the reduced Levi graph $R(C)$ of the configuration C in Figure 1.6.1. The first step is shown on the left and the complete graph $R(C)$ on the right. All subscripts are understood mod 6.

The reduced Levi graph $R(C)$ relies on the labeling of the points and vertices in a way that corresponds to the action of (a generator of) the symmetry group. With such a labeling, an edge of the $R(C)$ carries as labels the differences between the labels of the points that led to the particular edge. Thus in the case illustrated in Figures 1.6.1 and 1.6.2, each line L_j is incident with points b_j and b_{j+2} and c_j, leading to the labels shown on the edges connecting L to b and c in the reduced Levi graph, and similarly for the lines M_j. An indication of the relevant symmetry group completes the reduced Levi graph.

As we shall see in Section 2.8, changing only the symmetry group in this graph helps specify a whole infinite family of graphs of astral configurations.

A different example is presented in Figure 1.6.3. It is a slightly more symmetric version of a configuration shown in Figure 2 of Daublebski [55] and specified combinatorially in his list as #88. It is interesting to note that (as stated by Daublebski) the *automorphisms* of the underlying set-configuration act transitively on the points; in fact the automorphisms are transitive on the lines, and even on the flags, see Exercise 6 at the end of this section. This difference between the symmetry group of a geometric configuration and its underlying set-configuration is a frequent phenomenon; we noted it in connection with the configuration in Figure 1.5.4 and its Levi graph in Figure 1.5.5.

It is quite obvious that the reduced Levi graph $R(C)$ of a configuration C can be used to find the configuration table or its equivalent Levi graph $L(C)$ (as originally defined in Section 1.4). It is also clear that the construction of a reduced Levi graph can be carried out for combinatorial configurations that have a cyclic symmetry group.

A slight modification of the construction outlined above is necessary in case some lines M_j of a configurations C are mapped onto themselves by a c_2 subgroup of the cyclic group of symmetries of C. Then the order of the cyclic group c_k must be even, $k = 2t$, and for each j we have $M_j = M_{j+t}$. In the reduced Levi graph $R(C)$ we indicate this fact by writing $M^\sim$. An illustration is given in Figure 1.6.4, in the case of a configuration $(4_3, 6_2)$ with cyclic symmetry group c_4. Similar to this is the modification required if the configuration includes points-at-infinity; such points are mapped onto themselves by a $180°$ rotation, that is, by a c_2 symmetry. This is illustrated by the configuration and its reduced Levi diagram in Figure 1.6.5.

A concept closely related to reduced Levi graphs of configurations with cyclic symmetry group was introduced by T. Pisanski under the name "voltage graphs" and was first published in [24]. The description presented above differs from the one used in [24] in that our starting point is a given symmetric (geometric or topological) configuration, while Pisanski starts with a combinatorial configuration that admits a non-trivial cyclic automorphism. For general graphs the concept of "voltage" was introduced by Gross [104] and presented in detail in several chapters of [105] and [106].

For configurations with dihedral symmetry the construction of reduced Levi graphs is slightly more complicated—but in some cases it allows for a commensurately more compressed encoding of the configuration. Naturally, a configuration with dihedral symmetry group d_j can also be considered

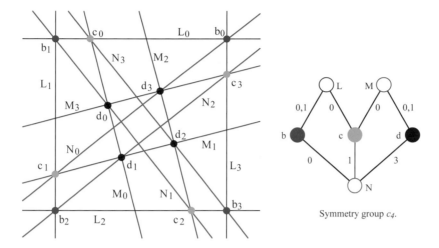

Figure 1.6.3. The (12_3) configuration listed as #88 in the enumeration of Daublebski [**55**] and its reduced Levi graph.

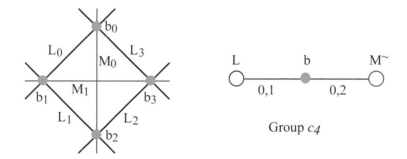

Figure 1.6.4. An example illustrating the formation of the reduced Levi diagram in the presence of lines that are mapped onto themselves by a subgroup c_2 of the symmetry group. The cyclic symmetry group of the $(3, 2)$-configuration shown on the left is c_4.

under the action of the cyclic symmetry group c_j. However, this disregard of reflections as symmetry operations leads in many cases to an increase in the *number* of orbits (and decrease in the *size* of some orbits). For example, if the configuration in Figure 1.5.1(b) is considered under the symmetry group c_5, all orbits have size 5 and the orbit type is $[5, 5]$.

The main steps to consider in the construction of a reduced Levi graph of a configuration with dihedral symmetry follow.

- One reflection and its images under the cyclic part of the group are selected and kept throughout. In each orbit a representative pair of elements or a singleton is chosen. The labels of mirror-image chosen pairs carry + or − superscripts, the superscript is ± if the

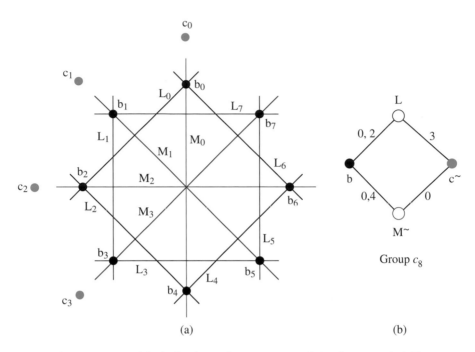

Figure 1.6.5. (a) A (12_3) configuration astral in the extended Euclidean plane, with cyclic symmetry group c_8; the points at infinity (indicated by the detached dots) and the lines M_j are individually invariant under a c_2 subgroup. (b) The reduced Levi diagram of this configuration.

chosen element is mapped onto itself by the mirror. In the latter case it is often convenient to drop the superscript entirely.

- Each black point of the graph corresponds either to a pair of configuration points related by the appropriate reflection or to a point on a mirror; white points (empty circles) correspond analogously to lines—pairs related by reflection in the mirror or single lines invariant under the reflection chosen. Rotations carry the labels to all points and lines of the configuration.

- The edges of the graph carry the information regarding which signed labels of lines lead to which signed labels of points, as well as the actual difference in label subscripts. The positive or negative superscripts are indicated by p and n, respectively.

Details of the construction and labeling are best understood by examples. In Figure 1.6.6 we show a labeled configuration (12_3) with symmetry group d_3. Its reduced Levi diagram is shown in Figure 1.6.7. More complicated examples are shown in Figure 1.6.8 to 1.6.10. In cases where the configuration is part of a family with varying numbers of points, it is sometimes convenient to use negative integers to indicate the difference in the labels of points and lines. This is illustrated by the examples in Figure 1.6.10 that belong to families we shall discuss in Section 3.3.

As in the case of cyclic symmetry groups, for configurations with dihedral symmetry group that contain lines or points that are mapped onto themselves by halfturns, a few special conventions are needed for the labeling.

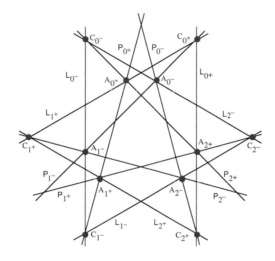

Figure 1.6.6. An astral configuration (12_3) with symmetry group d_3.

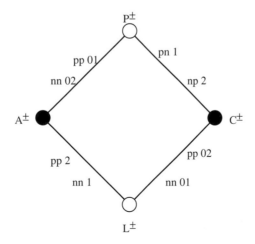

Figure 1.6.7. The reduced Levi graph of the configuration in Figure 1.6.6.

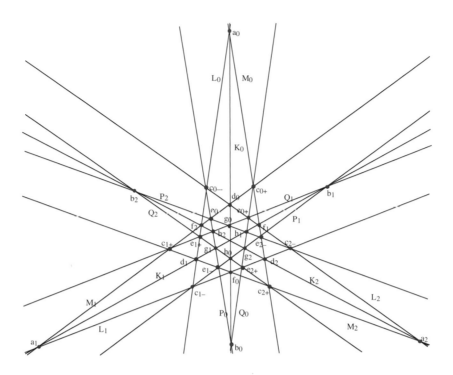

Figure 1.6.8. A $(30_3, 15_6)$ configuration with symmetry group d_3.

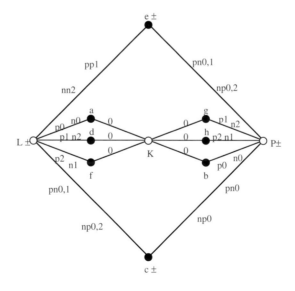

Figure 1.6.9. The reduced Levi graph of the configuration in Figure 1.6.8.

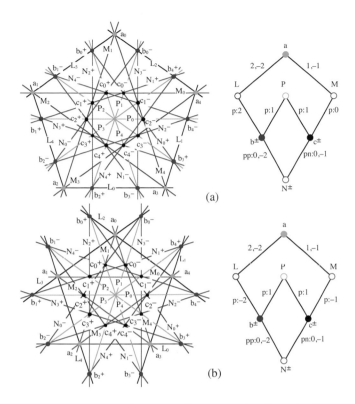

Figure 1.6.10. On the left, two configurations (25_4), each of orbit type $[3, 3]$ and with symmetry group d_5. The second one is the same as the configuration in Figure 1.5.1(b). On the right, we have the reduced Levi graphs of these configurations. All subscripts are mod 5. The configuration in (a) is due to J. Bokowski; see Section 3.3.

Exercises and Problems 1.6.

1. Find the reduced Levi graph of the (14_3) configuration in Figure 1.5.2(a), and compare it with the reduced Levi graph of the (12_3) configuration shown in Figure 1.6.1.

2. Find labels for the configuration in Figure 1.5.2(b) that yield the reduced Levi graph shown in Figure 1.6.11.

3. Show that the (10_3) configuration in Figure 1.6.12 is isomorphic to the Desargues configuration (10_3) shown in Figure 1.1.1.

4. Find a reduced Levi graph for the (10_3) configuration in Figure 1.6.12.

5. Find the reduced Levi graphs of the configurations in Figure 1.3.8.

6. (i) Use the labels in Figure 1.6.3 to label the Levi graph in Figure 1.6.12.

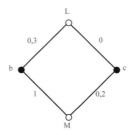

Figure 1.6.11. A reduced Levi graph used in Exercise 2.

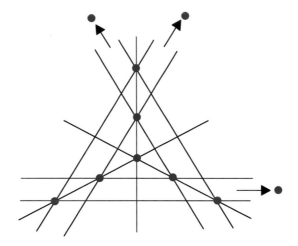

Figure 1.6.12. A configuration used in Exercises 3 and 4.

(ii) Use the Levi graph in Figure 1.6.12 to show that all points of the (12_3) configuration in Figure 1.6.3 form one orbit under its group of automorphisms.

(iii) Use the Levi graph in Figure 1.6.12 to show that the (12_3) configuration in Figure 1.6.3 is selfdual.

(iv) Use the Levi graph in Figure 1.6.13 to show that all flags of the (12_3) configuration in Figure 1.6.3 form one orbit under its group of automorphisms.

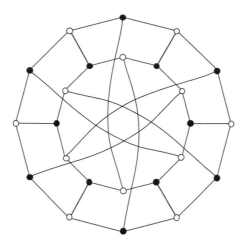

Figure 1.6.13. The Levi graph of the (12_3) configuration in Figure 1.6.3.

1.7. Derived figures and other tools

Testing whether a **given** mapping between the elements of two configurations (p_q, n_k) is an isomorphism is straightforward, though tedious and somewhat time consuming; its complexity is polynomial in p (and n) for given q and k. However, deciding whether **there exists** an isomorphism is much harder if only brute force is used. That is why, since the beginning of the study of configurations in the nineteenth century, variants of the idea of **derived figures** have been useful in finding isomorphisms between configurations or establishing that they are not isomorphic. The method is, in essence, a sort of "preprocessing" and is particularly time saving if many configurations are to be considered simultaneously or if they are known to have only a few transitivity or symmetry classes. It is also very helpful if one aims at determining the automorphism group of a configuration.

The idea underlying "derived figures" is to associate with each point (or line) of a given configuration C a small "figure", determined by the point (or line) and the incidences in C. The associated ("**derived**") figure should be easy to determine, and it should be easy to see whether two such figures are isomorphic. The vagueness of the above description should not bother you too much: it is not supposed to be an algorithm, just a heuristic approach which has been found convenient and which can be best explained by examples.

Consider the (10_3) configuration indicated in Figure 1.7.1(a), which we shall call $(10_3)_9$ as in Section 2.2. We would like to determine whether it is isomorphic to the Desargues (10_3) configuration shown in Figure 1.7.1(b). We would also like to determine its automorphism group, the transitivity

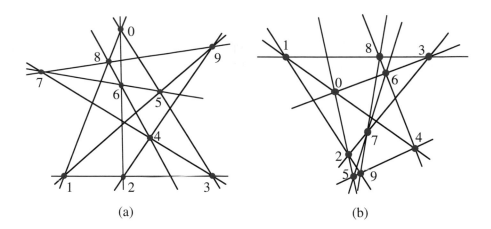

Figure 1.7.1. Two configurations (10_3).

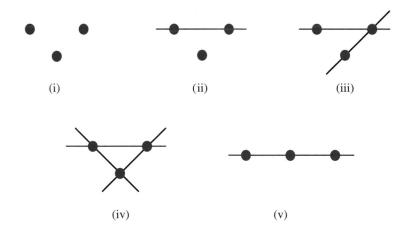

Figure 1.7.2. The possible "derived figures" for points of configurations (10_3).

classes of its elements, and the possible symmetry groups that isomorphic configurations can have.

We start by observing that since in any (10_3) configuration C each point lies on three lines which together contain six other points of C, for each point of C there are three points of C to which it is not connected by any line of C. (For example, the point 2 in Figure 1.7.1(a) is connected by no line of the configuration to any of 5, 7, 8.) One kind of "derived figure" associates to each point of C the set of the three points of C that are not connected to it, together with any lines of C that contain two of these points or all three of them. (In the literature, this is often called the "**remainder** figure", or

"Restfigur" in the German literature.) Clearly, the derived figure could be any of the five schematically indicated in Figure 1.7.2. It is easily checked that for the configuration $(10_3)_9$ of Figure 1.7.1(a) the derived figure is of type (iv) at points 0, 1, 4, 7 and is of type (iii) for the other six points. Thus the points of $(10_3)_9$ form at least two transitivity classes; since the points of the Desargues configuration form one transitivity class, these two configurations cannot be isomorphic. The same conclusion can be reached without appealing to the automorphisms group of the Desargues configuration: We note that the derived figure at each vertex of the Desargues configuration is of type (v)—and even a single such derived figure shows the impossibility of an isomorphism to $(10_3)_9$.

A closer examination of the derived figures of $(10_3)_9$ shows that the points 3 and 8, which lie on lines determined by 0, 1, 4, 7, cannot be in the same transitivity class as 2, 5, 6, 9; hence there are at least three transitivity classes. Each of the sets $\{0, 1, 4, 7\}$, $\{2, 5, 6, 9\}$, $\{3, 8\}$ is either an equivalence class of points of $(10_3)_9$ or a union of such classes. We shall try to determine which is the case. We start by looking for a permutation of the vertices that maps 0 to 1. Since they determine a line, 1 must be mapped onto 0 (since no line connects 1 to 4 or 7). Hence the permutation we are trying to find has the cycle $(0, 1)$ and also the singleton cycle (8). The points 4 and 7 are either invariant or else interchanged. In the former case, we would have 9 invariant as well; but then the line through 1 and 9 would be mapped on the line through 0 and 9—which is not a line of the configuration. On the other hand, if we assume that the permutation contains the cycle $(4, 7)$, then we find, successively, that it must contain the cycles (3), $(2, 5)$, and $(6, 9)$ as well. Hence the only candidate for an automorphism that maps 0 to 1 is the permutation $s = (0, 1)(2, 5)(3)(4, 7)(6, 9)(8)$. A check reveals that s is indeed an automorphism of $(10_3)_9$. A similar analysis shows that there is no automorphism that maps 0 to 4, or 2 to 6, or 3 to 8. Hence the decomposition of the vertices of $(10_3)_9$ into transitivity classes is $\{0, 1\}\{2, 5\}\{3\}\{4, 7\}\{6, 9\}\{8\}$. Moreover, the *automorphism* group consists of two elements, the identity and s. Thus a labeling of the points of $(10_3)_9$, more rational than the one in Figure 1.7.1(a), is as shown in Figure 1.7.3; now s interchanges the starred and double-starred versions of each letter, while keeping those without stars invariant. Concerning the *symmetry* groups of geometric configurations isomorphic to $(10_3)_9$, we can say that they either have the trivial symmetry group c_1 as in Figure 1.7.3 or else d_1 or c_2. However, since two of the points of the configuration remain invariant under s, it cannot have symmetry group c_2. By an elementary but slightly longer argument it can be shown that d_1 is impossible as well; hence no geometric realization has any nontrivial symmetry.

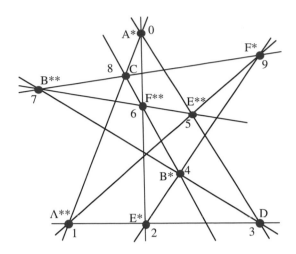

Figure 1.7.3. A revised labeling of the configuration $(10_3)_9$, making visible its automorphism group.

As an illustration of a second variant of the method of derived figures we investigate the four $(24_3, 18_4)$ configurations shown in Figure 1.7.4.

We shall call them C_1, C_2, C_3, and C_4. In this case we shall associate a "derived figure" with every line L of the configuration, by taking the 9 lines of the configuration that do not meet L in a point of the configuration, as well as the points of the configuration incident with these 9 lines. Since the configurations in Figure 1.7.4 have only two symmetry classes of lines, it is easy to determine all the derived figures; the isomorphism types that occur are **schematically** indicated in Figure 1.7.5. (Since we are interested only in isomorphisms, the relative positions of the points and lines are not relevant; only the incidences that occur in the configuration matter.) The results are

C_1 has 6 derived figures of type (i) and 12 of type (ii), and so does C_4;

C_2 has 6 derived figures of type (iii) and 12 of type (iv);

C_3 has 18 derived figures of type (v); note that these are isomorphic to the ones of type (iii).

It follows that the lines of C_1 form two transitivity classes, and so do the lines of C_2 and those of C_4. As far as the derived figures are concerned, the lines of C_3 could all belong to the same transitivity class. They, in fact, do so, as is shown, for example, by the permutation $t = (1)(2, 18)(3, 11, 9, 5)(4, 14, 10, 17)(6, 13)(7)(8, 15)(12, 16)$.

It also follows that C_2 and C_3 are isomorphic neither to each other nor to either of C_1 and C_4. However, as far as the derived figures are concerned, C_1 and C_4 could be isomorphic. The labeling of these configurations in

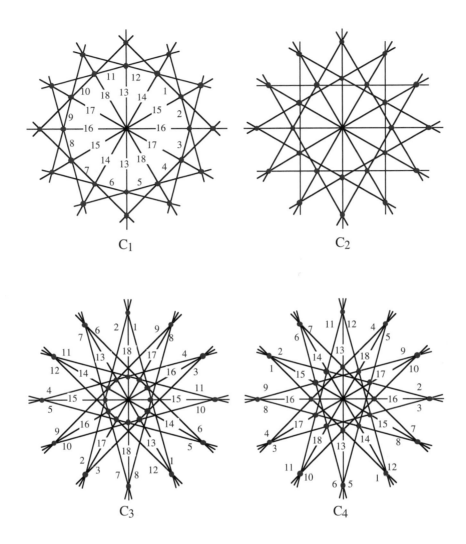

Figure 1.7.4. Four configurations $(24_3, 18_4)$.

Figure 1.7.4, which was obtained with the help of the derived figures, is easily checked to represent an isomorphism between the two configurations.

As the next example, we consider the three (9_3) configurations in Figure 1.1.6 and again find the derived figures for the various points. (We will consider these three configurations again in Section 2.2. There we will use a slightly different method.) It is clear that for all points, in all three configurations, the derived figure consists of two points. In $(9_3)_1$ none of these pairs is incident with a line of the configuration; in $(9_3)_2$ each such pair of points is incident with a line of the configuration, while in $(9_3)_3$ in six of the pairs the two points are incident with a line of the configuration and in the three remaining pairs this is not the case. Therefore no two of

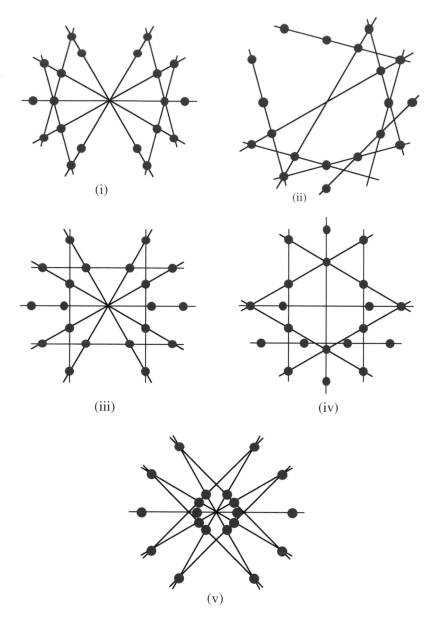

(i)

(ii)

(iii)

(iv)

(v)

Figure 1.7.5. Derived figures that are possible for the configurations $(24_3, 18_4)$ in Figure 1.7.4.

the configurations in Figure 1.1.6 are isomorphic. Moreover, the points in $(9_3)_3$ form (at least) two transitivity classes, while it is possible that in each of $(9_3)_1$ and $(9_3)_2$ they are in a single transitivity class. This is indeed the situation, as is easy to verify and as we will see in Section 2.2. (Note that $(9_3)_1$ is the Pappus configuration, which we encountered in Section 1.1.)

So far all the derived figures we considered were of a "local" type, related to features of the configuration that depended on the relations of each individual point with the other elements of the configuration. We shall now consider a "global" derived figure, which was used in several works of H. Gropp.

In Section 1.4 we briefly mentioned the Menger graph of a configuration. Its nodes are all the vertices of the configuration, and edges connect pairs of nodes that correspond to vertices incident with a line of the configuration. As is easy to see, this graph is almost always rather unwieldy and has had very little use. However, in many instances its **complement**, that is, the graph on the same nodes but with edges connecting precisely those pairs which are not endpoints of an edge, in the Menger graph, is more useful. This graph has been called by Gropp the **configuration graph**; in Section 1.4, following [**156**], we called it the **deficiency graph**. The deficiency graphs of the three (9_3) configurations in Figures 1.1.6 and 2.2.1 are shown in Figure 2.2.2 and used in Section 2.2 to distinguish between the three possible configurations (9_3).

The deficiency graph can be used to quickly decide whether the vertices of the ($12_4, 16_3$) configuration in Figure 1.7.6 form one transitivity class. We consider its deficiency graph shown in Figure 1.7.7. The nodes P, Q, X are not in any 3-circuit, while the other nine nodes are in such circuits. Together with the obvious d_3 symmetry of the geometric realization in Figure 1.7.6, this means that P, Q, X are in one orbit and that the other vertices form either one orbit or they are in two orbits. It is easy to verify that the permutation $(NMO)(RTV)(SWU)(P)(Q)(X)$, together with the geometric symmetries, establishes the single transitivity class of the nine points.

Another example of the use of deficiency graphs is given in Di Paola-Gropp [**59**], where combinatorial configurations (21_4) are studied. Using various combinatorial techniques, they produce 200 non-isomorphic configurations of this kind, 12 of which are selfdual. They are recognized as non-isomorphic by the use of their deficiency graphs—except that one pair of non-isomorphic configurations has the same graph. They do not list the other configurations but present configuration tables for these two (and their configuration graphs). They also do not make any statements regarding geometric realizability. The author was curious about whether the selfdual geometric configuration (21_4) from [**122**] (shown in Figure 1.7.8) is isomorphic to one of these—and using the given configuration graph of the Di Paola-Gropp paper, it is easy to verify that our configuration is not isomorphic to either of these. In Table 1.7.1 we show a corrected copy of their tables. Since the vertices of our (21_4) form one orbit under automorphisms, in order to show that it is not isomorphic to the Di Paola-Gropp configurations, it is

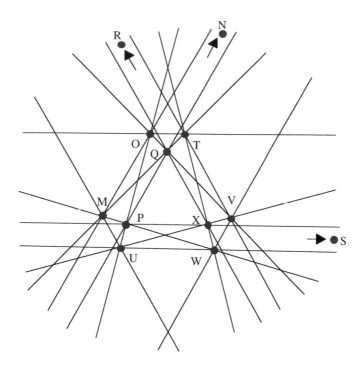

Figure 1.7.6. A $(12_4, 16_3)$ configuration.

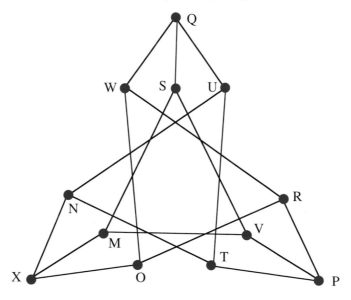

Figure 1.7.7. The deficiency graph of the $(12_4, 16_3)$ configuration in Figure 1.7.6.

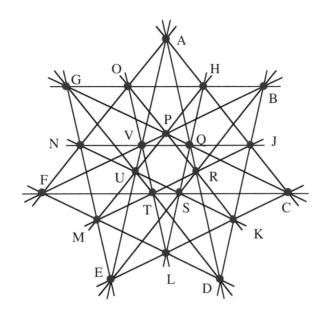

Figure 1.7.8. A geometric configuration (21_4) from [**122**].

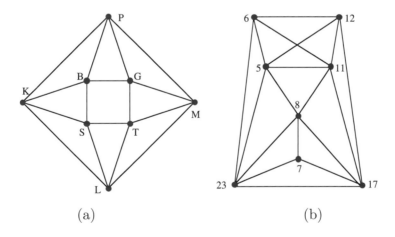

(a)   (b)

Figure 1.7.9. (a) The remainder figure of vertex A in Figure 1.7.8. All other vertex figures are isomorphic to this. (b) The remainder figure of vertex 1 in the configuration (74_9) of Table 1.7.1.

enough to compare one of its remainder figures with any one of the latter. The remainder figure of the vertex A in Figure 1.7.8 is shown in Figure 1.7.9(a). The remainder figure of node 1 of the first graph in Table 1.7.1 is shown in Figure 1.7.9(b); since it has a 5-valent vertex, the geometric configuration is not isomorphic to either of the two combinatorial ones.

Table 1.7.1. The configuration tables of two non-isomorphic combinatorial configurations (21_4) and their isomorphic configuration graphs (from [**59**]). The isomorphism is established by the permutation mapping Graph (749) onto Graph (799): (1, 13, 2, 14, 3, 15)(4, 8, 6, 7, 5, 9)(10, 17, 11, 18, 12, 16)(22, 23). The peculiar names of the marks (19, 20, 21 not used) are from [**59**]. The two red entries are not correct in the original.

Configuration (749)   Graph (749)

1	2	3	22	(1, 5)	(1, 6)	(1, 7)	(1, 8)
1	4	13	24	(1, 11)	(1, 12)	(1, 17)	(1, 23)
1	9	16	18	(2, 4)	(2, 6)	(2, 8)	(2, 9)
1	10	14	15	(2, 10)	(2, 12)	(2, 18)	(2, 23)
2	5	14	24	(3, 4)	(3, 5)	(3, 7)	(3, 9)
2	7	16	17	(3, 10)	(3, 11)	(3, 16)	(3, 23)
2	11	13	15	(4, 8)	(4, 9)	(4, 14)	(4, 15)
3	6	15	24	(4, 16)	(4, 18)	(5, 7)	(5, 9)
3	81	7	18	(5, 13)	(5, 15)	(4, 16)	(5, 17)
3	12	13	14	(6, 7)	(6, 8)	(6, 13)	(6, 14)
4	5	6	23	(6, 17)	(6, 18)	(7, 11)	(7, 12)
4	7	10	22	(7, 14)	(7, 15)	(8, 10)	(8, 12)
4	11	12	17	(8, 13)	(8, 15)	(9, 10)	(9, 11)
5	8	11	22	(9, 13)	(9, 14)	(10, 13)	(10, 17)
5	10	12	18	(10, 21)	(10, 23)	(11, 14)	(11, 18)
6	9	12	22	(11, 21)	(11, 23)	(12, 15)	(12, 16)
6	10	11	16	(12, 21)	(12, 23)	(13, 16)	(13, 17)
7	8	9	24	(13, 22)	(14, 17)	(14, 18)	(14, 22)
7	13	18	23	(15, 15)	(15, 18)	(15, 22)	(16, 22)
8	14	16	23	(16, 24)	(17, 22)	(17, 23)	(18, 22)
9	15	17	23	(18, 24)	(22, 23)	(22, 24)	(23, 24)

Configuration (799)   Graph (799)

1	2	3	22	(1, 5)	(1, 6)	(1, 8)	(1, 9)
1	4	13	24	(1, 10)	(1, 12)	(1, 16)	(1, 23)
1	7	17	18	(2, 4)	(2, 6)	(2, 7)	(2, 9)
1	11	14	15	(2, 10)	(2, 11)	(2, 17)	(2, 23)
2	5	14	24	(3, 4)	(3, 5)	(3, 7)	(3, 8)
2	8	16	18	(3, 11)	(3, 12)	(3, 18)	(3, 23)
2	12	13	15	(4, 8)	(4, 9)	(4, 14)	(4, 15)
3	6	15	24	(4, 17)	(4, 18)	(5, 7)	(5, 9)
3	9	16	17	(5, 13)	(5, 15)	(5, 16)	(5, 18)
3	10	13	14	(6, 7)	(6, 8)	(6, 13)	(6, 14)
4	5	6	23	(6, 16)	(6, 17)	(7, 11)	(7, 12)
4	7	10	22	(7, 13)	(7, 14)	(8, 10)	(8, 12)
4	11	12	16	(8, 14)	(8, 15)	(9, 10)	(9, 11)
5	8	11	22	(9, 13)	(9, 15)	(10, 15)	(10, 16)
5	10	12	17	(10, 23)	(10, 24)	(11, 13)	(11, 17)
6	9	12	22	(11, 23)	(11, 24)	(12, 14)	(12, 18)
6	10	11	18	(12, 23)	(12, 24)	(13, 16)	(13, 18)
7	8	9	24	(13, 22)	(14, 16)	(14, 17)	(14, 22)
7	15	16	23	(15, 17)	(15, 18)	(15, 22)	(16, 22)
8	13	17	23	(16, 24)	(17, 22)	(17, 24)	(18, 22)
9	14	18	23	(18, 24)	(22, 23)	(22, 24)	(23, 24)

Exercises and Problems 1.7.

1. For the $(16_3, 12_4)$ configuration of Figure 1.7.4(C_3) consider the derived figures of lines of each of the two symmetry classes. Find an automorphism that maps one line in one symmetry class to a line in the other class, and use that to show that all the lines belong to a single transitivity class. Show that all points belong to a single transitivity class and that, in fact, all flags are in one transitivity class.

2. Use derived figures of lines to show that the two $(20_3, 15_4)$ configurations in Figures 4.3.3 and 4.3.4 are not isomorphic. Show also that in each of these configurations the lines (as well as the points) form two transitivity classes. How many transitivity classes of flags are there?

3. For the configuration that we denote by $(10_3)_3$ in Section 2.2, conduct an analysis of its automorphisms and symmetries analogous to the one we did above for $(10_3)_9$.

4. In continuation of our discussion concerning the configuration $(12_4, 16_3)$ shown in Figure 1.7.6, decide how many orbits of lines there are under the group of automorphisms.

5. Investigate the orbits, automorphisms, and symmetries of the configuration $(12_4, 16_3)$ shown in Figure 4.3.7(b).

6. Are any of the three configurations in Figure 4.3.7 isomorphic?

7. Show that the vertices of the (21_4) configuration in Figure 1.7.8 are in a single orbit under automorphisms.

8. Find the automorphisms of the configuration denoted (749) in Table 1.7.1.

Chapter 2

3-Configurations

2.0. Overview

This is the longest of our chapters, reflecting not only the long history of the topic of 3-configurations to which the overwhelming majority of the early works was devoted, but also the much more recent discovery of the deficiencies in the original works of a century or more ago, and their corrections. We also present the recent developments that center on the study of configurations with a high degree of symmetry.

Section 2.1 investigates the existence of combinatorial, topological, and geometric configurations (n_3) for various values of n.

In Sections 2.2 and 2.3 this is made more detailed by describing the efforts to determine the numbers of distinct configurations (n_3) for specific values of n. These investigations started in the first period of the study of configurations more than a century ago but have been resumed and advanced only in the recent past.

Section 2.4 is devoted to the attempts to construct all combinatorial configurations (n_3) recursively. This goal seemed to have been achieved by V. Martinetti some 125 years ago—but a few years ago his result was shown to be incomplete. The corrected result was obtained in the doctoral thesis of M. Boben!

Sections 2.5 and 2.6 present the result of the 1894 doctoral thesis of E. Steinitz. This is a remarkable work, even though it has a significant blemish in its geometric part. It is interesting that the last part of this work has remained undeciphered ever since Steinitz wrote it—nobody claims to understand what he is claiming! However, it is clear that at least some parts of the claim are not true.

The next three sections represent recent developments. These are investigations of 3-configurations with remarkably large cyclic or dihedral symmetry groups.

Section 2.10 deals with some unexpected aspects of duality and polarity of these configurations.

Finally, Section 2.11 makes explicit a few of the most intriguing problems about 3-configurations.

2.1. Existence of 3-configurations

Among the questions attacked and solved in the early years of the study of configurations is the one we formulated as question (A) in Section 1.1. We can now formulate it more specifically as follows:

Determine all values of n such that there exists a combinatorial or a topological or a geometric configuration (n_3).

The complete answer to this problem is given by

Theorem 2.1.1. *Combinatorial configurations (n_3) exist if and only if $n \geq 7$.*

Theorem 2.1.2. *Topological configurations (n_3) exist if and only if $n \geq 9$.*

Theorem 2.1.3. *Geometric configurations (n_3) exist if and only if $n \geq 9$.*

To prove Theorem 2.1.1, we note that the inequalities of Section 1.3 imply, for $k = 3$, that $n \geq 7$. Hence we only have to show that for each $n \geq 7$ there exist a combinatorial configuration (n_3). Of the various ways of fulfilling this task, probably simplest is the listing of a configuration table for a *cyclic* (n_3), as illustrated in Table 2.1.1. We shall encounter this configuration repeatedly, and we reserve the symbol $\mathscr{C}_3(n)$ for it. Besides, the existence of topological and geometric configurations (n_3) for $n \geq 9$ implies the existence of the corresponding combinatorial configurations.

Table 2.1.1. A configuration table for the cyclic combinatorial configuration $\mathscr{C}_3(n)$ for $n \geq 7$. It also shows that for $n \leq 6$ this would not be a configuration, since some pairs of points would belong to two different lines.

$$
\begin{array}{cccccccccc}
1 & 2 & 3 & 4 & \cdots & n-3 & n-2 & n-1 & n \\
2 & 3 & 4 & 5 & \cdots & n-2 & n-1 & n & 1 \\
4 & 5 & 6 & 7 & \cdots & n & 1 & 2 & 3
\end{array}
$$

This completes the proof of Theorem 2.1.1.

In the exercises at the end of this section we shall enlarge upon the configurations $\mathscr{C}_3(n)$ and other cyclic configurations.

The configuration $\mathscr{C}_3(7)$ is known as the **Fano configuration**; it was described by Gino Fano in 1891 [**68**, p. 111] in connection with his axiomatic studies of projective geometries. In fact, it was found earlier (in 1888) by Schönflies [**195**], who dismissed it by saying that "a configuration 7_3 with all points distinct does not exist", as well as by Schroeter [**199**], also in 1888. Schönflies' assertion makes a limited sort of sense when one realizes that he was thinking of geometric configurations—albeit in the complex projective plane! Schroeter was the first to stress the distinction between combinatorial and geometric configurations and between geometric configurations in the real plane as distinct from the ones in the complex projective plane.

The configuration $\mathscr{C}_3(8)$ is known as the Möbius-Kantor configuration. In the prehistory of configurations it was described by Möbius in 1828 [**172**], who proved that it cannot be realized geometrically in the real Euclidean plane. The configuration was later described by Kantor in 1881 [**131**], although not as a combinatorial configuration but as a configuration *geometric in the complex plane*. Reye noted in 1882 [**188**] that $\mathscr{C}_3(8)$ does not exist as a geometric configuration in the real plane. In the same paper [**131**] the three configurations (9_3) are described by Kantor for the first time, as geometric configurations in the real plane.

To prove Theorems 2.1.2 and 2.1.3, it is sufficient to show that *geometric* configurations (n_3) exist for each $n \geq 9$ and that *topological* configurations (n_3) do not exist for $n = 7, 8$.

To establish the latter, we shall first prove a lemma.

Lemma 2.1.1. *Let C be a family of pseudolines in the real projective plane such that no point is incident with all members of C. Then there is a point that is contained in precisely two of the pseudolines in C.*

Such a point is customarily called an *ordinary point* of the family C.

Proof of the lemma. If all intersection points of pseudolines in C are ordinary, there is nothing to prove. Otherwise, there exist "triangles" (that is, regions of the plane) bounded by three pseudolines and such that, at one of the vertices of this "triangle", one of the pseudolines L incident with that vertex V enters the interior of the "triangle". Indeed, start with any non-ordinary point of C and three arbitrarily chosen pseudolines through V. Then any pseudoline L^* not through V will determine a triangle with the required properties. See Figure 2.1.1. Call such a triangle a good triangle. Among the (possibly many) good triangles of C find the (or one) that has a minimal area. Then the pseudoline L that enters the triangle at V must meet L^* at some point P on the boundary of the "triangle". Now P has to be an ordinary point of C, since any pseudoline through P different from L and L^* would belong to a "good" triangle with smaller area.

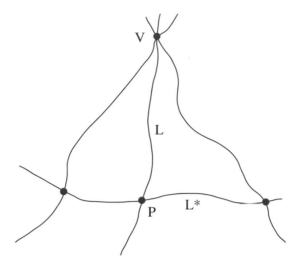

Figure 2.1.1. Any vertex incident with three or more pseudolines of a family C as in Lemma 2.1.1 can serve as one of the vertices of a "good" triangle.

Resuming now the proof of the non-realizability by pseudolines of any combinatorial configuration (7_3), we note that if a realization were possible, then the seven points of such a configuration would account for 21 pairwise incidences of points and pseudolines. On the other hand, seven pseudolines can have at most $7(7-1)/2 = 21$ pairwise intersections, that is, at most 21 pairwise incidences with points of the configuration. Thus all intersection points of the pseudolines would be at points of the configuration, hence triple points, and there would be no ordinary points—contradicting Lemma 2.1.1. It follows that there is no realization of combinatorial configurations (7_3) by pseudolines. This completes the proof of this part of our assertion.

Concerning the case of topological configurations (8_3), let us assume we have a realization C of such a configuration by pseudolines. We begin by selecting one of them, say L. It is met by six other pseudolines in the three vertices of C that are on L. Hence there is one pseudoline L^* of C that meets L in an ordinary point, which is not a vertex of C. Choosing the line-at-infinity to pass through that point, we can represent L, L^* and the vertices of C incident with them as shown in Figure 2.1.2.

The remaining six pseudolines must all pass through the three vertices on L and through the three vertices of L^*, as well as through the two remaining vertices of C. Let V be one of these two vertices; without loss of generality we can assume that it is in the "strip" between L and L^*; then, since at the point of intersection the pseudolines must cross each other, the three pseudolines pass through V as schematically shown in Figure 2.1.3.

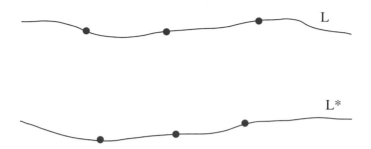

Figure 2.1.2. Two of the pseudolines and the six vertices discussed in the proof.

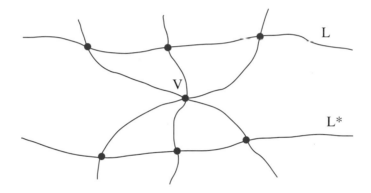

Figure 2.1.3. The arrangement of the pseudolines incident with the vertex V.

Now the last three pseudolines, which are incident with the last vertex of C, must be connected either by the scheme represented in Figure 2.1.4 by the dotted connections or by the dashed connections. Since in either case two of them meet in a point (which therefore must be the last vertex) that is inaccessible to the third, it follows that the realization of (8_3) by pseudolines is impossible.

The proof of the weaker result that *geometric* configurations (7_3) and (8_3) do not exist is much older and simpler. For example, Schroeter [**199**] argues that, in the notation used in Table 2.2.2, if a geometric realization of the configuration (7_3) were possible, the points 2, 3, 4, 5 would generate a complete quadrangle (in the sense of projective geometry), with diagonal points 1, 6, 7. But these points are collinear in (7_3) while diagonal points of a complete triangle cannot be collinear unless the starting points are collinear; hence there is no geometric configuration. A different proof of the impossibility of geometric realization of the (7_3) configuration appears in Bokowski-Sturmfels [**29**, p. 39]; it relies on the method of "final polynomials". Essentially the same proof is used by Levi [**145**, p. 95]. It shows

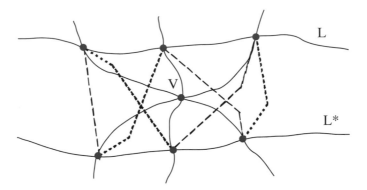

Figure 2.1.4. The three dotted connections or the three dashed connections are schematic representations of the relative positions of the three pseudolines that should be incident with the last vertex of C.

that (7_3) can be "realized" only in projective planes with characteristic 2. Sidorov's statement in [**208**] that (7_3) is realizable in the complex plane is plain wrong.

The method of "final polynomials" is used by Bokowski-Sturmfels [**29**, p. 35] to show that (8_3) cannot be geometrically realized in the *real* plane, although it can be realized in the *complex* plane. That result itself belongs to the "prehistory" of configurations; it appears (in somewhat different formulation) in Möbius [**172**, p. 445], as described by Schroeter [**199**, p. 239]. An explicit calculation of feasible coordinates of points of a realization of (8_3) by Möbius [**172**], as well as by Levi [**145**, p. 99], shows that such coordinatization is possible only using complex numbers.

The only publication the author is aware of in which the non-existence of topological configurations (7_3) and (8_3) has been considered (and proved) is Levi's book [**145**, pp. 95, 100]. However, Levi's proofs are quite laborious, and part of the argumentation in the case of (8_3) is left to the reader to complete. Instead of our Lemma 2.1.1, Levi relies on a lemma (Satz 21, [**145**, p. 85]) that we may formulate as follows: No topological configuration (n_k) with $k \geq 3$ contains as vertices all points determined by its pseudolines. This is clearly a weaker version of Lemma 2.1.1. On the other hand, Kelly and Rottenberg prove in [**135**] the stronger result that every family of n pseudolines, not all incident with one point, must determine at least $3n/7$ ordinary points. That result is a generalization of the well-known result of Kelly and Moser [**134**] for families of straight lines. This topic has had an interesting history and is still the subject of widespread interest. It is not possible to enlarge upon it here; the interested reader should consult [**32**], where Section 7.2 presents details and gives a large number of references.

In order to complete the proof of Theorem 2.1.3 (hence also of Theorem 2.1.2), we shall describe the construction of a suitable geometric configuration (n_3).

We shall show that for $n \geq 9$, the cyclic combinatorial $\mathscr{C}_3(n)$ configuration of Table 2.1.1 can be realized as a geometric configuration of points and lines (see Figure 2.1.5). We begin by placing the first triplet on the x-axis in a coordinate system, with point 2 at the origin and point 4 at $x = 2$; we shall specify the location of the point 1 shortly. We draw a line through 2 with positive slope and place 3 on it near 2, so that the line $3, 4$ has negative slope small in absolute value, and place 5 on the same line sufficiently far to the right so that $4, 5$ has positive slope. On line $3, 4$ we locate 6 so that its x-coordinate is larger than the x-coordinate of 5. Then on $4, 5$ we locate 7 so that its x-coordinate is larger than that of 6, and so on up to and including the line through $n - 5$ and $n - 4$ on which we locate $n - 2$ so that its x-coordinate is larger than that of $n - 3$. Clearly, all these steps can be carried out. Now, the choice of location for vertex 1 determines the only possible position of vertex $n - 1$ (since it is on the already determined lines 1, $n - 2$ and $n - 4$, $n - 3$), as well as the position of vertex n (which must be on the by now determined lines 2, $n - 1$ and $n - 3$, $n - 2$). The only remaining question is whether the last triplet $1, 3, n$ is collinear—and this depends on the choice of 1 (see Figure 2.1.5). It is easy to check that if 1 is chosen to be on the x-axis between points 2 and 4 and near 2 (see part (a) of Figure 2.1.5), then the halfplane determined by the line $1, n$ and containing the positive x-axis contains the point 3 in its interior. On the other hand, if 1 is chosen between 2 and 4 but near 4 (see part (b) of Figure 2.1.5), then 3 in not in the interior of that half-plane determined by $1, n$ that contains the positive x-axis. Due to the continuity of all construction steps, it follows that there must exist a position of vertex 1 for which the line $1, n$ passes through the point 3—thus yielding the desired geometric realization of the combinatorial configuration in question. $\square$

It should be noted that the construction fails unless $n - 5 > 3$; this provides an explanation for why this construction requires $n \geq 9$.

This proof is quite analogous to the first published proof by Schroeter [199]. The main difference is that in the last part, instead of the continuity argument used in our proof, Schroeter gives a purely geometric proof which utilizes properties of sets of points on cubic curves. He published a book on this topic in the same year [200]. The advantage of Schroeter's proof over ours is that it shows that the cyclic configuration of Table 2.1.1 can be geometrically realized, for every $n > 9$, by a linear construction—that is, with just a straightedge. This implies that all these configurations can also be geometrically realized in the rational plane. In their combinatorial guise

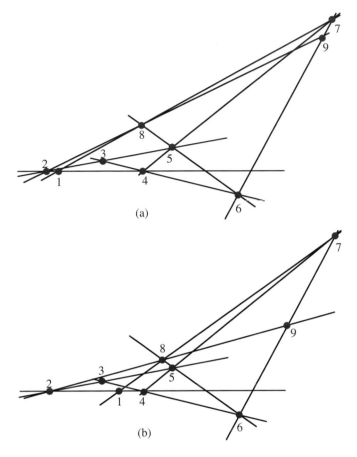

Figure 2.1.5. An illustration (for $n = 9$) of the construction of a geometric realization of the cyclic combinatorial configuration $\mathscr{C}_3(n)$ from Table 2.1.1.

these configurations were studied, more or less simultaneously, by Schönflies [**195**] and Martinetti [**152**].

✳ ✳ ✳ ✳ ✳ ✳

A few remarks seem appropriate at this time.

Early papers on configurations often considered configurations in the complex plane. Obviously, all geometric configurations as considered here (that is, in the real Euclidean plane) can also be considered as being in the complex plane. The interesting point is that in the complex plane Theorems 2.1.2 and 2.1.3 require modification: There exists a geometric configuration (n_3) in the complex plane if and only if $n \geq 8$. The fact that a *configuration* (8_3) exists in the complex plane was first announced by Kantor [**131**], although in "prehistoric" formulation it goes back at least to Möbius [**172**, p. 445].

Like many other writers on configurations in the last quarter of the nineteenth century, Kantor [**131**] did not make explicit what kind of configurations (n_3) he is investigating. This is particularly amusing in connection with the configuration (8_3), which he describes as two quadrangles, each inscribed to and circumscribed about the other. Only later does he make an off-hand remark that at most four of the eight vertices of the configurations are real!!!

The cyclic configuration $\mathscr{C}_3(7)$ is the only configuration (7_3); this will be proved explicitly in Section 2.2. The configuration (7_3) does not appear in the paper by Kantor [**131**] in which he considers configurations (n_3) for $n \leq$ 9. Although he relies on some combinatorial arguments, the combinatorial configuration (7_3) was probably invisible to him since it cannot be realized in the complex plane; it seems that he was considering only configurations that have realizations in the complex plane, although he is not explicit about that. On the other hand, (7_3) appears in many other publications and guises—for example, as a Steiner triple system on 7 elements, as the projective plane of order 2, and several others. A Levi incidence matrix of the (7_3) configuration is shown in Figure 2.1.7.

Levi [**145**] established Theorem 2.1.1 by considering generalizations of the cyclic configuration $\mathscr{C}_3(n)$ in Table 2.1.1. The same idea appeared earlier, most explicitly in Schönflies [**195**]. A generalization of this is the *cyclic configuration* $\mathscr{C}_3(n, m)$, which consists of triples $\{j, j+1, j+m\}$, for $1 \leq j \leq n$, all entries taken mod n. Such configurations were studied (with slightly different notation) by Levi [**145**, p. 91]. Levi proved that $\mathscr{C}_3(n, m)$ is a combinatorial configuration whenever $3 \leq m < n/2$. He does not discuss their geometric realizability and mentions no earlier works on any cyclic configurations.

There are familiar diagrams intended to illustrate the Fano (7_3) and Möbius-Kantor (8_3) configurations, shown in Figure 2.1.6. They are not topological configurations, since they involve one "line" that is a circle. If one of the incidences is not insisted upon, then this line can be "opened up" and a geometric realization of the resulting subfiguration is obtained.

Exercises and Problems 2.1.

1. Find the isomorphism between the labeling of the points of the configurations (7_3) and (8_3) in Figure 2.1.6 and the cyclic configuration $\mathscr{C}_3(7)$ and $\mathscr{C}_3(8)$.

2. Use the illustration of the Möbius-Kantor (8_3) configuration given in Figure 2.1.6 to find a Levi incidence matrix for it. Can you use it to show that the configuration is selfdual?

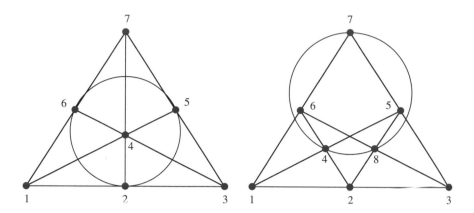

Figure 2.1.6. Diagrams often used to illustrate the combinatorial Fano (7_3) and Möbius-Kantor (8_3) configurations. The labeling shown is the "greedy" one: it uses a new mark only if unavoidable.

	L_1	L_2	L_3	L_4	L_5	L_6	L_7
Q_1	X		X				X
Q_2		X				X	X
Q_3	X				X	X	
Q_4				X	X		X
Q_5			X	X		X	
Q_6		X	X		X		
Q_7	X	X		X			

Figure 2.1.7. A Levi incidence matrix for the Fano (7_3) configuration, which shows that it is cyclic and selfdual.

3. A *general cyclic configuration* $\mathscr{C}_3(n, a, b)$ consists of triples $\{j, a + j, b + j\}$, for given a, b with $0 < a < b < n$ and for $1 \leq j \leq n$, all entries taken mod n. The configuration $\mathscr{C}_3(n)$ in Table 2.1.1 is, in this notation, $\mathscr{C}_3(n, 1, 3)$. For which n, a, b is $\mathscr{C}_3(n, a, b)$ a combinatorial configuration? As a first step, investigate configurations $\mathscr{C}_3(n, 1, m)$. (This was done as early as 1895 by G. Brunel in [**34**].)

4. Is $\mathscr{C}_3(n, 1, 4)$ geometrically realizable for some n? Generalize.

2.2. Enumeration of 3-configurations (Part 1)

We turn now to the presentation of the results known about the number of non-isomorphic configurations (n_3) of the three kinds, for each value of n—as far as these numbers are known. As we shall see, this is not very far. Moreover, we have to discuss some other questions concerning these enumerations.

For the purposes of this section, we shall denote by $\#_c(n), \#_t(n)$, and $\#_g(n)$ the number of non-isomorphic combinatorial, topological, or geometric configurations (n_3), respectively. We begin with

Theorem 2.2.1. *The complete list of known numbers $\#_c(n)$, $\#_t(n)$, and $\#_g(n)$ is given in Table 2.2.1.*

Table 2.2.1. The number of non-isomorphic configurations (n_3) of the three kinds, for each n. All known values are shown.

n	$\#_c(n)$	$\#_t(n)$	$\#_g(n)$
≤ 6	0	0	0
7	1	0	0
8	1	0	0
9	3	3	3
10	10	10	9
11	31	31	31
12	229	229	229
13	2,036		
14	21,399		
15	245,342		
16	3,004,881		
17	38,904,499		
18	530,452,205		
19	7,640,941,062		

We start by presenting the proofs of the enumerations for $n \leq 8$. Following this, we shall first discuss the case $n = 9$, then the rather unexpected situation for $n = 10$, and finally the cases of $n \geq 11$. Some general considerations will be explained next, with exercises and problems to follow.

From Section 2.1 we already know that all three numbers are 0 for $n \leq 6$. Now we first show that each of the (combinatorial) configurations (7_3) and (8_3) is unique. This follows easily from the consideration of the formation of their configuration tables. For (7_3), starting with the three lines that contain 1 and then continuing by using the freedom of assigning labels to previously uncommitted points in the only possible way, we obtain the unique configuration table shown in Table 2.2.2. For (8_3) we first note that since each

Table 2.2.2. A configuration table of the unique combinatorial configuration (7_3).

$$
\begin{array}{ccccccc}
1 & 1 & 1 & 2 & 2 & 3 & 3 \\
2 & 4 & 6 & 4 & 5 & 4 & 5 \\
3 & 5 & 7 & 6 & 7 & 7 & 6
\end{array}
$$

Table 2.2.3. A configuration table of the unique combinatorial configuration (8_3).

$$
\begin{array}{cccccccc}
1 & 1 & 1 & 2 & 2 & 3 & 3 & 5 \\
2 & 4 & 7 & 4 & 5 & 4 & 6 & 6 \\
3 & 6 & 8 & 7 & 8 & 5 & 7 & 8
\end{array}
$$

point is connected (by a line of the configuration) to six other points, it fails to be connected to a unique point. Designating the unconnected pairs by $\{1,5\}, \{2,6\}, \{3,7\},$ and $\{4,8\}$, a similar procedure leads to the unique configuration table shown in Table 2.2.3. The uniqueness of these configurations has been known since early in the study of configurations. The uniqueness of (8_3) seems to have been established first by Kantor [**131**], while $\#_c(7) = 1$ was proved by Martinetti [**152**, pp. 3–4]; for other proofs see, for example, Levi [**145**, pp. 94, 98] and Hilbert and Cohn-Vossen [**126**]).

Similar arguments can be applied to the determination of the different combinatorial configurations (9_3). Easier to carry out is an application of the "remainder figures" method described in Section 1.4. First comes the observation that each point fails to be connected to precisely two other points. Drawing an edge (segment) between any two unconnected points, we see that the unconnected pairs form one or more circuits. (This is the *deficiency graph* of this configuration introduced in Section 1.4.) Since a circuit has to have at least three points, there are four potential sets of circuits: a single 9-circuit, a 6-circuit and a 3-circuit, a 5-circuit and a 4-circuit, and three 3-circuits. It is obvious that different sets of circuits imply that the configurations are not isomorphic, since any isomorphism preserves connected pairs of points, hence also disconnected pairs. Similarly to the earlier cases, it is possible to show that each case corresponds to a (unique) configuration, except that the case of one 5-sided and one 4-sided circuits corresponds to no configuration. The reason for this is the following: Assume that it is possible, and consider the lines incident with the vertices of the 4-circuit. Two lines correspond to the "diagonals" of the 4-circuit, while each of the four vertices has to be on two additional lines, all distinct and different from the earlier two; this would require the existence of at least 10 lines. Hence such a possibility cannot lead to a configuration. The result, using these or other arguments, appears in Kantor [**131**], Martinetti [**152**],

Table 2.2.4. A configuration table for the configuration $(9_3)_1$.

$$
\begin{array}{ccccccccc}
1 & 1 & 1 & 2 & 2 & 2 & 3 & 3 & 3 \\
4 & 5 & 6 & 4 & 5 & 6 & 4 & 5 & 6 \\
7 & 8 & 9 & 8 & 9 & 7 & 9 & 7 & 8
\end{array}
$$

Table 2.2.5. A configuration table for the configuration $(9_3)_2$.

$$
\begin{array}{ccccccccc}
1 & 1 & 1 & 2 & 2 & 2 & 3 & 3 & 4 \\
3 & 4 & 5 & 4 & 5 & 6 & 5 & 6 & 7 \\
7 & 6 & 8 & 8 & 7 & 9 & 9 & 8 & 9
\end{array}
$$

Table 2.2.6. A configuration table for the configuration $(9_3)_3$.

$$
\begin{array}{ccccccccc}
1 & 1 & 1 & 2 & 2 & 2 & 3 & 3 & 3 \\
4 & 5 & 8 & 4 & 5 & 7 & 4 & 6 & 7 \\
7 & 6 & 9 & 6 & 8 & 9 & 5 & 9 & 8
\end{array}
$$

Schröter [**199**], and again in Levi [**145**, p. 103], Hilbert and Cohn-Vossen [**126**], and Gropp [**94**].

Configuration tables of the three combinatorial configurations (9_3) are shown in Tables 2.2.4, 2.2.5, and 2.2.6. All three of these configurations can be geometrically realized; this was first proved by Kantor [**131**] and more thoroughly analyzed by Schröter [**199**]. Representative examples of such realizations are shown in Figure 2.2.1, and the same representatives are shown in Figure 2.2.2 with the circuits formed by non-connected pairs of points.

Concerning the (10_3) configurations, we start by presenting in Table 2.2.7 configuration tables for all ten combinatorial configurations. The existence of precisely ten non-isomorphic combinatorial configurations (10_3) has been established repeatedly, by more or less brute force enumeration; historical details and references will be given below. In order to prove that these configurations are distinct, we shall use the concept of "remainder figures" which was introduced in Section 1.4: For each vertex of the combinatorial configuration (10_3) consider the three vertices that are not on any of the lines passing through the given vertex. There are three possibilities concerning these three points: Either all three are on one configuration line or they determine two lines of the configuration or they determine three such lines (a *trilateral* or "triangle"). We shall denote the three possibilities by I, V, Δ. No other situations are possible. Indeed, if the three points were collinear, but on a line that is not in the configuration, there would be nine configuration lines through them and three additional lines through the original vertex—while only ten lines are available. But if the three points are not

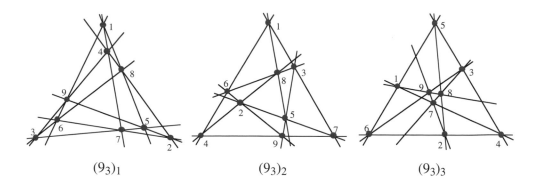

$(9_3)_1$ $\qquad\qquad\qquad$ $(9_3)_2$ $\qquad\qquad\qquad$ $(9_3)_3$

Figure 2.2.1. Examples of the three types of geometric configurations (9_3). The claim by Steinitz [**212**, p. 489] that H. A. Schwarz [**205**] found the form of the configuration $(9_3)_3$ with 3-fold rotational symmetry is not correct.

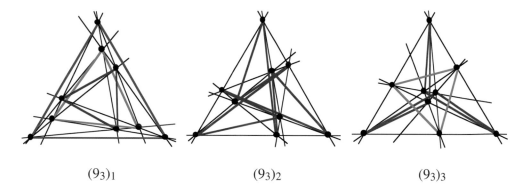

$(9_3)_1$ $\qquad\qquad\qquad$ $(9_3)_2$ $\qquad\qquad\qquad$ $(9_3)_3$

Figure 2.2.2. The circuits formed by the non-connected pairs in the three types of geometric configurations (9_3).

collinear, they determine a triangle. If none of the three lines determined by the points were a configuration line, the configuration would again have to have at least 12 lines. On the other hand, if just one of the lines determined by the sides of the triangle were a line of the configuration, then there would have to be present in the configuration at least $1 + 2 + 2 + 3 + 3 = 11$ lines. Thus, the three cases listed earlier are the only ones possible.

The above arguments that the remainder figure in the case of (10_3) must be one of I, V, Δ are taken from Schröter [**201**], together with his notation. By very exhaustive and exhausting argumentation one can show that only ten combinations of the different remainder figures (listed in Table 2.2.8) can occur in a combinatorial configuration (10_3) and that each corresponds to a unique isomorphism type of combinatorial configurations, represented by one of the ten configurations in Table 2.2.7. The detailed discussion of the possible combinatorial configurations depending on the kind and number

Table 2.2.7. The ten non-isomorphic combinatorial configurations (10_3), in the notation of Schröter [201]. They were first determined by Kantor [132], using other methods and different notation and labeling.

$(10_3)_1$

1	1	1	8	2	3	2	3	4	5
2	4	6	9	4	5	6	7	6	7
3	5	7	0	8	8	9	9	0	0

$(10_3)_2$

1	1	1	8	2	3	2	3	4	5
2	4	6	9	4	7	6	5	6	7
3	5	7	0	8	8	9	9	0	0

$(10_3)_3$

1	1	1	8	2	3	2	3	4	5
2	4	6	9	4	6	7	5	6	7
3	5	7	0	8	8	9	9	0	0

$(10_3)_4$

1	1	1	8	2	3	2	3	4	5
2	4	6	9	4	6	5	7	6	7
3	5	7	0	8	8	9	9	0	0

$(10_3)_5$

1	1	1	8	2	3	2	4	3	5
2	4	6	9	4	7	5	6	6	7
3	5	7	0	8	8	9	9	0	0

$(10_3)_6$

1	1	1	8	2	3	2	5	3	4
2	4	6	9	4	7	6	7	5	6
3	5	7	0	8	8	9	9	0	0

$(10_3)_7$

1	1	1	2	4	6	5	3	7	2
2	4	6	8	8	9	7	5	3	4
3	5	7	9	0	0	8	9	0	6

$(10_3)_8$

1	1	1	3	5	7	2	6	4	2
2	4	6	8	8	9	7	5	3	4
3	5	7	9	0	0	8	9	0	6

$(10_3)_9$

1	1	1	2	4	6	5	3	2	3
2	4	6	8	8	9	7	5	7	4
3	5	7	9	0	0	8	9	0	6

$(10_3)_{10}$

1	1	1	3	2	7	5	6	4	2
2	4	6	8	8	9	7	5	3	4
3	5	7	9	0	0	8	9	0	6

of the remainder figures is spread over 22 pages in [201]. It leads to the conclusion that each column in Table 2.2.8 corresponds to one and only one combinatorial configuration (10_3).

As far as geometric realizations go, sketches similar to the ones in the first enumeration of the (10_3) configurations by Kantor [132] are shown in Figure 2.2.3; our Figure 1.2.2 is a copy of one of the Kantor diagrams. The diagram of $(10_3)_1$ is easily checked to be an illustration of the Desargues configuration. We shall encounter $(10_3)_{10}$ in Section 2.6 as the astral configuration denoted $5\#(2,2;1)$.

One of the most striking features exhibited by Table 2.2.1 is the inequality $\#_t(10) = 10 > \#_g(10) = 9$. This arises because the diagram in Figure 2.2.4, which *appears* to show a (10_3) geometric configuration (which is a

Table 2.2.8. The number of occurrences of the different remainder figures in the ten combinatorial configurations (10_3). The last two rows give the notation by Martinetti [**152**] and Kantor [**132**].

	$(10_3)_1$	$(10_3)_2$	$(10_3)_3$	$(10_3)_4$	$(10_3)_5$	$(10_3)_6$	$(10_3)_7$	$(10_3)_8$	$(10_3)_9$	$(10_3)_{10}$
I	10	4	2	6	1	1	0	0	0	0
V	0	6	6	0	3	9	9	3	6	0
Δ	0	0	2	4	6	0	1	7	4	10
	X	VIII	V	II	I	IX	III	VI	IV	VII
	B	G	D	C	H	F	E	J	K	A

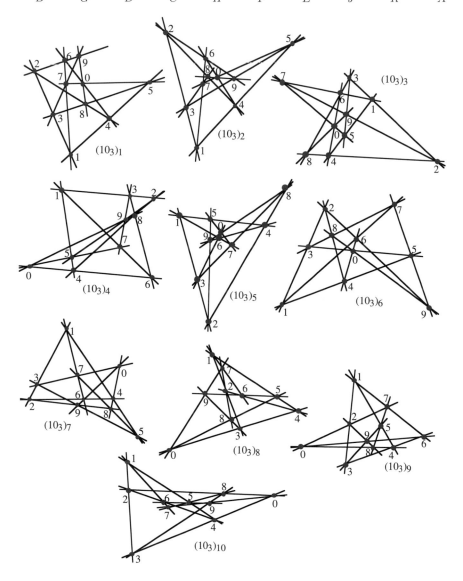

Figure 2.2.3. Sketches of configurations (10_3), analogous to the ones presented in Kantor's paper [**132**].

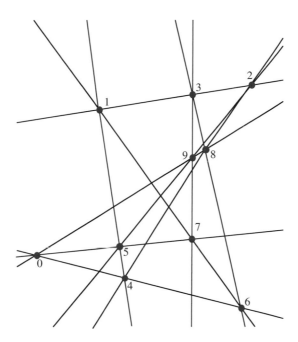

Figure 2.2.4. An apparent geometric realization of the combinatorial configuration $(10_3)_4$, which was also shown in Figure 1.2.2. However, both diagrams are misleading. This configuration is not isomorphic to any configuration of points and (straight) lines. On the other hand, the "lines" are (very mildly curved) pseudolines; hence a topological realization of this configuration is possible.

cleaner drawing of the configuration $(10_3)_4$ in Figure 2.2.3), cannot in fact be realized by straight lines—although the diagram clearly indicates that a topological realization is possible.

The impossibility of a geometric realization of $(10_3)_4$ can be established as follows.

The complete quadrangle $2, 3, 8, 9$ contains the three pairs of opposite sides

$$23 - 1 \qquad 28 - 4 \qquad 29 - 5$$
$$89 - 0 \qquad 39 - 7 \qquad 38 - 6$$

while the complete quadrangle $6, 7, 9, 0$ contains the three pairs of opposite sides

$$67 - 1 \qquad 60 - 4 \qquad 70 - 5$$
$$90 - 8 \qquad 97 - 3 \qquad 96 -^*$$

By a basic theorem of projective geometry, the three pairs of lines of each quadrangle intersect the line 145 in three pairs of points of an involution. But these involutions must coincide, since two of the pairs coincide: 1 and $890 \cap 145$, and 4 and $379 \cap 145$. Then the point paired with 5 in the involution must be the intersection point of the three lines $145, 368, 96^*$. This cannot be 6, since then 145 would contain four points; the only alternative is that 368 and 96* coincide—but then 368 would contain the fourth point 9. Hence the configuration $(10_3)_4$ cannot be realized geometrically.

This proof of the impossibility of a geometric realization of the configuration $(10_3)_4$ is due to Schröter [**201**]. The difference between topological and geometric realizability seems to have been taken as a challenge by many people, leading to a variety of proofs of geometric non-realizability, or at least mention of it; see Carver [**40**], Laufer [**143**], van de Craats [**219**], Glynn [**72**], Killgrove et al. [**138**], Sternfeld et al. [**215**], and others. Zacharias [**235**] is not aware of the earlier works and attempts to enumerate all the (10_3) configurations. There are several errors in [**235**] (as well as in the review [**218**] by Togliatti); corrections appear in [**237**]. Some other publications discuss just the enumeration of combinatorial (10_3) configurations; for example, we may mention Betten and Schumacher [**18**].

This situation makes it even more important to make sure that the remaining nine combinatorial configurations (10_3) are geometrically realizable. Following Schröter [**201**], we present here a method of stepwise construction for each of the nine *geometric* configurations (10_3). The method leads to several important results; among them are the number of parameters needed to determine each of these configurations (that is, the number of "degrees of freedom"), the possibility of constructing each of them using only an unmarked ruler, and the possibility of realizing each in the rational plane (or, equivalently, with all vertices at points of the integer lattice). Since these are quite non-trivial results, which can be found in few of the more recent publications, Schröter's constructions are shown in Figure 2.2.5. Naturally, each of these constructions requires justification, which is given in the paper; examples follow.

From the configuration table for $(10_3)_1$ we see that the intersections $(24, 35) = 8$, $(26, 37) = 9$, and $(46, 57) = 0$ are collinear; hence by the Desargues theorem the triangles 246 and 357 are in perspective from a point—which in the table is identified as 1. This justifies the construction in Figure 2.2.5. Moreover, it enables one to find out how many degrees of freedom there are in the construction (precisely 11) and that the construction is linear. It is meant by this that only systems of linear equations need to be solved and hence that it can be carried out in the rational plane (the plane in which only points with rational coordinates are considered).

For the combinatorial configuration $(10_3)_2$ we see that the triangles 246 and 357 are again in perspective from point 1, but the sides of the triangles 246 and 375 (in that order!) intersect in the collinear points 8, 9, and 0; hence by Desargues they must be in perspective from some point. This is

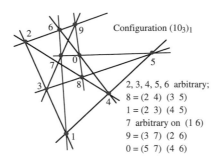

Configuration $(10_3)_1$

2, 3, 4, 5, 6 arbitrary;
8 = (2 4) (3 5)
1 = (2 3) (4 5)
7 arbitrary on (1 6)
9 = (3 7) (2 6)
0 = (5 7) (4 6)

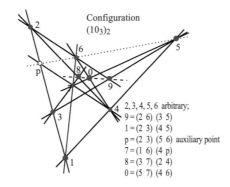

Configuration $(10_3)_2$

2, 3, 4, 5, 6 arbitrary;
9 = (2 6) (3 5)
1 = (2 3) (4 5)
p = (2 3) (5 6) auxiliary point
7 = (1 6) (4 p)
8 = (3 7) (2 4)
0 = (5 7) (4 6)

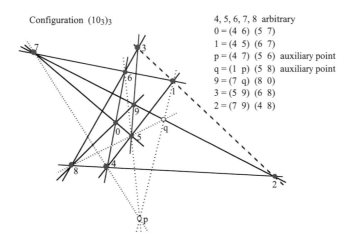

Configuration $(10_3)_3$

4, 5, 6, 7, 8 arbitrary
0 = (4 6) (5 7)
1 = (4 5) (6 7)
p = (4 7) (5 6) auxiliary point
q = (1 p) (5 8) auxiliary point
9 = (7 q) (8 0)
3 = (5 9) (6 8)
2 = (7 9) (4 8)

Figure 2.2.5. Part 1. The construction of the nine geometric configurations (10_3) following Schröter [**201**] (continued on the next page).

a point p that is not a point of the configuration. The construction now follows. Note that in this case there are only 10 degrees of freedom.

Arguments of similar kinds can be made in the seven remaining cases. They are explained in detail in Schröter [**201**]. The steps outlined with each

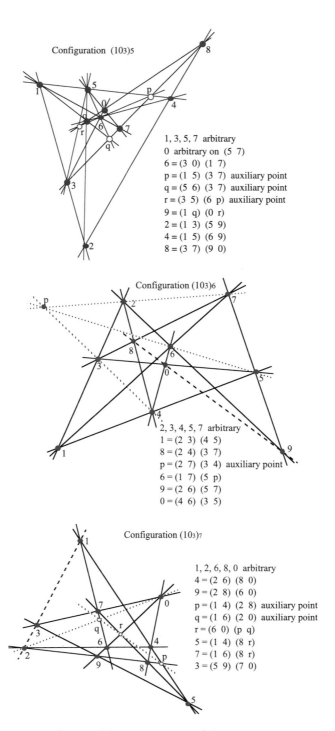

Configuration $(10_3)_5$

1, 3, 5, 7 arbitrary
0 arbitrary on (5 7)
6 = (3 0) (1 7)
p = (1 5) (3 7) auxiliary point
q = (5 6) (3 7) auxiliary point
r = (3 5) (6 p) auxiliary point
9 = (1 q) (0 r)
2 = (1 3) (5 9)
4 = (1 5) (6 9)
8 = (3 7) (9 0)

Configuration $(10_3)_6$

2, 3, 4, 5, 7 arbitrary
1 = (2 3) (4 5)
8 = (2 4) (3 7)
p = (2 7) (3 4) auxiliary point
6 = (1 7) (5 p)
9 = (2 6) (5 7)
0 = (4 6) (3 5)

Configuration $(10_3)_7$

1, 2, 6, 8, 0 arbitrary
4 = (2 6) (8 0)
9 = (2 8) (6 0)
p = (1 4) (2 8) auxiliary point
q = (1 6) (2 0) auxiliary point
r = (6 0) (p q)
5 = (1 4) (8 r)
7 = (1 6) (8 r)
3 = (5 9) (7 0)

Figure 2.2.5. Part 2. The construction of the nine geometric configurations (10_3) following Schröter [**201**] (continued from the previous page).

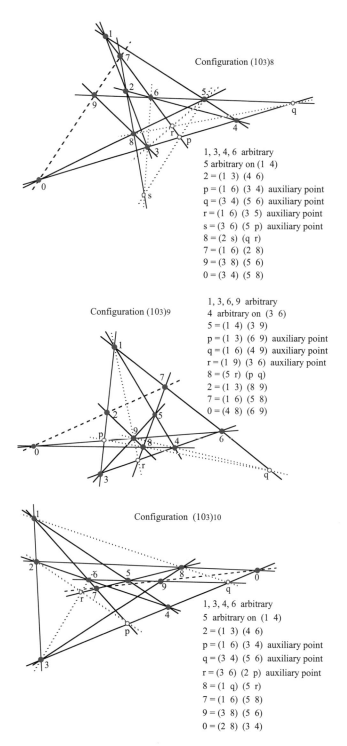

Configuration (10₃)8

1, 3, 4, 6 arbitrary
5 arbitrary on (1 4)
2 = (1 3) (4 6)
p = (1 6) (3 4) auxiliary point
q = (3 4) (5 6) auxiliary point
r = (1 6) (3 5) auxiliary point
s = (3 6) (5 p) auxiliary point
8 = (2 s) (q r)
7 = (1 6) (2 8)
9 = (3 8) (5 6)
0 = (3 4) (5 8)

1, 3, 6, 9 arbitrary
4 arbitrary on (3 6)
5 = (1 4) (3 9)
p = (1 3) (6 9) auxiliary point
q = (1 6) (4 9) auxiliary point
r = (1 9) (3 6) auxiliary point
8 = (5 r) (p q)
2 = (1 3) (8 9)
7 = (1 6) (5 8)
0 = (4 8) (6 9)

Configuration (10₃)9

Configuration (10₃)10

1, 3, 4, 6 arbitrary
5 arbitrary on (1 4)
2 = (1 3) (4 6)
p = (1 6) (3 4) auxiliary point
q = (3 4) (5 6) auxiliary point
r = (3 6) (2 p) auxiliary point
8 = (1 q) (5 r)
7 = (1 6) (5 8)
9 = (3 8) (5 6)
0 = (2 8) (3 4)

Figure 2.2.5. Part 3. The construction of the nine geometric configurations (10₃) following Schröter [**201**] (continued from the previous page).

Table 2.2.9. The number of degrees of freedom beyond projective transformations, for each of the geometric configurations (10_3).

$(10_3)_1$	$(10_3)_2$	$(10_3)_3$	$(10_3)_4$	$(10_3)_5$	$(10_3)_6$	$(10_3)_7$	$(10_3)_8$	$(10_3)_9$	$(10_3)_{10}$
3	2	2	—	1	2	2	1	1	1

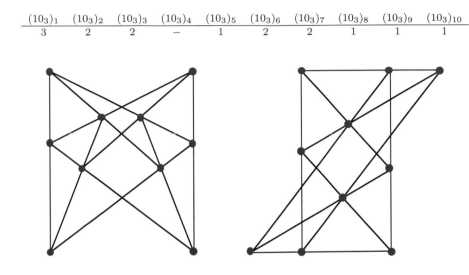

Figure 2.2.6. Two (10_3) configurations.

construction enable the determination of the degree of freedom of each configuration. The result—after taking into account that projective transformations account for eight degrees of freedom, which are not deemed essential in the present context—is shown in Table 2.2.9.

Exercises and Problems 2.2.

1. Since the configuration (7_3) is combinatorially unique, the configuration in Table 2.1.1 for $n = 7$ must be isomorphic to the configuration in Table 2.2.2. Find the mapping that transforms one into the other. Perform the same task for Table 2.1.1 for $n = 8$ and Table 2.2.3.

2. Verify the entries for $(10_3)_2$, $(10_3)_4$, and $(10_3)_9$ in Table 2.2.8.

3. Justify the numbers in Table 2.2.9.

4. Explain and justify Schröter's construction of the $(10_3)_5$ and $(10_3)_6$ configurations.

5. For each of the two configurations in Figure 2.2.6 decide whether it is a "fake". If not, find the coordinates of its points and determine to which of the configurations in Figure 2.2.5 it is isomorphic.

6. Use the configuration tables of the (10_3) configurations to find the automorphism group of each.

7. Is there a topological realization of the $(10_3)_4$ configuration that has a nontrivial symmetry?

8. In Section 1.7 we demonstrated that the configuration $(10_3)_9$ has no geometric realization with nontrivial symmetry. What about the configuration $(10_3)_5$?

2.3. Enumeration of 3-configurations (Part 2)

Combinatorial configurations (11_3) were first enumerated by Martinetti [152] in 1887; using the method we shall describe in the next section, he found that $\#_c(11) = 31$. The enumeration of these configurations was independently carried out by Daublebsky [54] in 1894; he used a variant of the remainders method. Diagrams supposed to show geometric realizations of all 31 combinatorial configurations (that is, $\#_g(n) = 31$) were provided by Daublebsky in an appendix to [55] in 1895 (shown in Figure 2.3.1 below; see also Figure 2.3.2).

Daublebsky states that all these combinatorial configurations can be realized as geometric configurations (that is, with points and straight lines) given by his diagrams but does not give any justification beyond the intimation that he followed the method of Schröter [201]. An independent verification of the geometric realizability of all 31 configurations (11_3) was provided only nearly a century later, by Sturmfels and White [216], [217] in 1988 and 1990, with a different method. Sturmfels and White also proved that each of these configurations can be realized in the rational plane; in other words, one can always draw the configurations so that the vertices are at points of the integer lattice. The value of $\#_c(11) = 31$ was independently confirmed by Gropp (see [79]) and by Betten et al. [17], among others.

$$* \; * \; * \; * \; *$$

The first enumeration of the combinatorial configurations (12_3) was carried out by Daublebsky [55] in 1895, again using the method of remainder figures. He found that only 18 different remainder figures could possibly occur in such a configuration. Through various arguments (described only in general terms) Daublebsky arrived at the conclusion that these remainder figures could be combined to yield something like 1,600 configurations (12_3). Then he "...drew a schematic diagram of each configuration on a separate piece of paper..." and determined for each the "remainder system", that is, a list of the different remainder figures occurring in the configuration. Finally, configurations with the same remainder system were investigated to see whether they are isomorphic. This turned out to be true in most—but not all—cases (see Exercises 3 and 4 at the end of this section). Daublebsky presented the resulting 228 combinatorial configurations by their configurations tables (in the form he gave them, they take 23 pages!!!). He also gave some other data and provided drawings for geometric realizations of a few of the configurations. In a later paper [56], Daublebsky gave results of his investigations of the groups of automorphisms of each of the 228 combinatorial configurations (12_3). However, not all of these are correct. The first independent enumeration of the combinatorial (12_3) configurations was carried out only in 1990, by Gropp (see [79]). It showed that Daublebsky missed one, so that there are in fact $\#_c(12) = 229$ such configurations. Gropp published the configuration table of this additional configuration in [84] and communicated it to the author; the table can also be read off from the illustrations in the more readily available [63] and [96]. As with configurations (11_3), the 229 combinatorial configurations (12_3) have been independently enumerated (by two different methods) in [17]. Even so, Dolgachev [61] in 2004 still quotes $\#_c(12) = 228$.

The only published proof that all 228 combinatorial configurations (12_3) found by Daublebsky are geometrically realizable was given only recently, by Sturmfels and White [216], [217]. Sturmfels and White also proved that all these (12_3) configurations are realizable in the rational plane. In a private communication, B. Sturmfels showed that the "new" combinatorial configuration found by Gropp is also geometrically realizable, even in the rational plane; a diagram is shown in Dorwart–Grünbaum [63] in 1992.

The numbers $\#_c(n)$ of combinatorial configurations for $13 \leq n \leq 19$ were determined by various computer programs. For $12 \leq n \leq 14$ these values were first found by Gropp [79] and for $n = 15$ by Betten and Betten, [16]; the values for $16 \leq n \leq 18$ in Table 2.2.1 are from Betten, Brinkmann, and Pisanski [17]. The value $\#_c(19) = 7,640,941,062$ was determined by these authors and published in [22] and [117]. However, there is no information available about the possibilities of realization of the combinatorial configurations (n_3) for $n \geq 13$ by topological or geometric configurations, beyond individual examples—these will be discussed in the following sections. This is not very surprising in view of the number of combinatorial configurations. As shown in Table 2.2.1, this number is well above 2,000 for $n = 13$ and increases by factors exceeding 10 for larger n.

This completes the discussion of the data in Table 2.2.1. The only additional information that is available is that $\#_c(n) > \#_t(n)$ for all $n \geq 14$ and that $\#_t(n) > \#_g(n)$ for all $n \geq 16$. The former happens due to the existence of **disconnected configurations**—that is, configurations that are disjoint unions of two or more configurations, between the elements of which there are no incidences.

As examples, consider the (14_3) configuration which consists of two disjoint copies of the Fano configuration (7_3) or the (15_3) configuration formed by disjoint copies of (7_3) and (8_3); the latter was implicitly recognized as disconnected by Betten and Betten [16], the former is explicitly mentioned

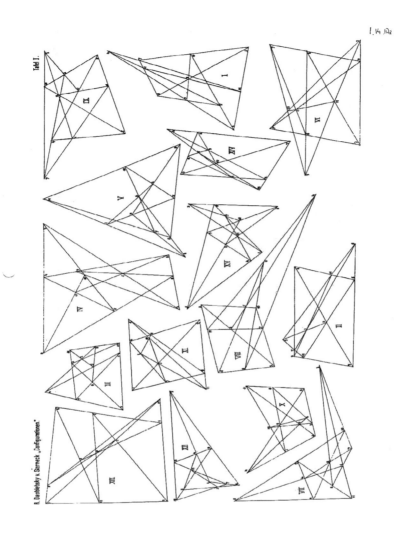

Figure 2.3.1. (first half). The diagrams of the (11_3) configurations, from Daublebsky [**55**].

by Gropp [**78**]. Since disconnected configurations arise as unions of smaller configurations, it is easy to determine the number of such configurations for all $n \leq 19$. Since the (7_3) and (8_3) set-configurations are not geometrically realizable, the smallest geometrically realizable disconnected configurations are the six arising as unions of two configurations (9_3). The same is true for topological configurations.

On the other hand, the inequality between the numbers $\#_t(n)$ and $\#_g(n)$ of topological and geometric configurations for $n \geq 16$ is a consequence of the existence of topological configurations of the kind illustrated by the scheme

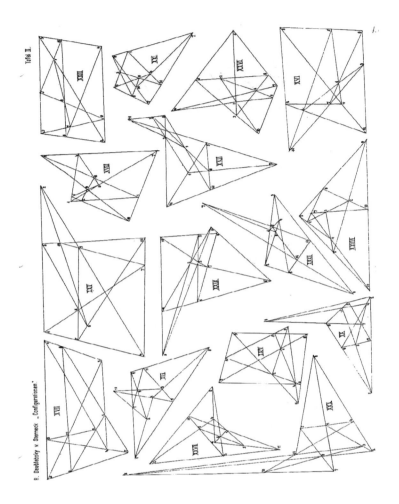

Figure 2.3.1. (second half). The diagrams of the (11_3) configurations, from Daublebsky [**55**].

in Figure 2.3.3. Due to the theorem of Pappus, if this configuration scheme is rendered with straight lines instead of line segments, the points A_2, B_2, C_2, and F_3 are seen to be collinear. Hence this is a superfiguration and not a geometric configuration; clearly, this is not a problem if pseudolines are used. This example can be understood as arising from a "melding" of the Pappus configuration $(9_3)_1$ and the Fano configuration (7_3). (Note that the Fano part is missing one incidence, and this subfiguration is realizable by straight lines.) This construction can be modified in various ways. For example, instead of the Fano configuration one could use any (n_3) configuration, and instead of the Pappus configuration one could use the Desargues configuration $(10_3)_1$. This completes the proof of $\#_t(n) > \#_g(n)$ for all $n \geq 16$. It is not known whether $\#_t(n) = \#_g(n)$ for $n = 13, 14, 15$.

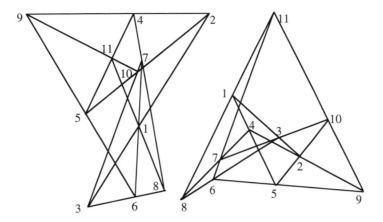

Figure 2.3.2. Diagrams of Daublebsky's configurations $(11_3)_4$ and $(11_3)_5$ redrawn for better visibility.

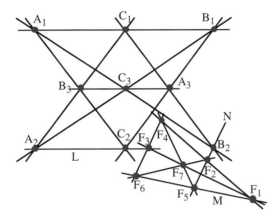

Figure 2.3.3. Pappus's theorem implies that the points A_2, B_2, C_2, and F_3 are collinear; hence this does not realize a configuration (16_3). It is obvious that by using pseudolines the unwanted incidence can be avoided.

This lack of knowledge is part of a larger open question. The single example establishing $\#_t(10) > \#_g(10)$ differs in one important respect from the examples just given with $n \geq 16$: The latter are only 2-connected, while the combinatorial and topological $(10_3)_4$ is 3-connected. The lack of any other 3-connected examples leads to the following conjecture.

Conjecture 2.3.1. *Every 3-connected topological configuration (n_3) with $n \geq 11$ is geometrically realizable.*

$* \ * \ * \ * \ *$

The Schröter constructions explained and illustrated above would nowa-
days be said to be *generic* constructions, the terminology supposing to indicate that it applies to run-of-the-mill situations. In fact, if understood
literally—that all the choices can be made arbitrarily, with only the stated
restrictions—the constructions may fail to lead to the *configurations* they are
supposed to yield. Instead, *superfigurations* may result due to "accidental"
incidences. This is illustrated in Figure 2.3.4.

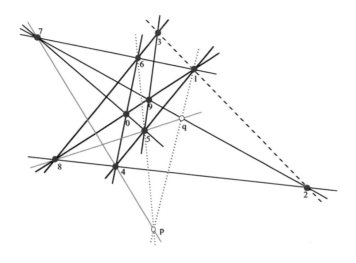

Figure 2.3.4. Failure of the Schröder construction of the configuration
$(10_3)_3$: The line 890 contains the point 1. Notation is the same as in
Figure 2.2.5.

It is hard to understand that no publication on configurations during the
classical period even mentioned the possibility of superfigurations arising in
the construction of geometric configurations. This is astonishing since the
study of accidental incidences in the Desargues configuration was already
old hat at that time. In Figure 2.3.5 we show a Desargues superfiguration,
with a line on four points and a point on four lines. The exceptional point
and line are shown in contrasting color. In the paper [**215**] Sternfeld et al.
study possible superfigurations of (10_3) configurations (and more general
incidence systems), both combinatorial and geometric. We conjecture:

Conjecture 2.3.2. *Every geometric configuration (n_3) with $n \geq 10$ admits
superfigurations with at least one pair of "accidental" incidences.*

It is worth mentioning that the three (9_3) configurations do not have
limiting positions that are superfigurations. On the other hand, the Pappus
configuration $(9_3)_1$ has representatives in which an additional point incident
with three lines or a line incident with three of the points or both can be

found. The last alternative is illustrated in Figure 2.3.6. It is not known whether many other configurations have this property.

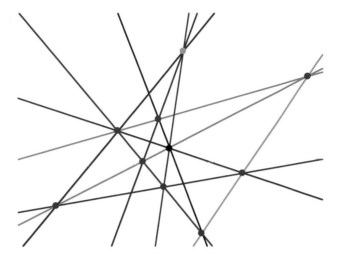

Figure 2.3.5. A superfiguration arising from the Desargues configuration $(10_3)_1$ through multiple incidences. The point and line of perspectivity are shown in teal.

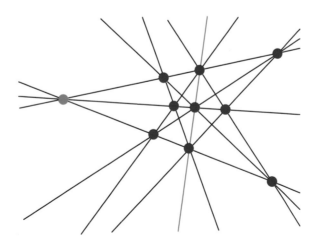

Figure 2.3.6. A superfiguration arising from the Pappus configuration with the addition of a point and a line (green) to create elements incident with four other elements.

Exercises and Problems 2.3.

1. Find the remainder systems of Daublebsky's configurations $(11_3)_4$ and $(11_3)_5$ shown in Figure 2.3.2, and use them to show that these are distinct configurations.

2. Determine the remainder systems of the (12_3) configurations in Figures 1.6.1 and 1.6.3.

3. Find a superfigure of the Desargues configuration that has three points and three lines incident with four elements of the other kind—or prove that such a configuration cannot exist.

4. Consider the two (12_3) configurations from Daublebski's paper [**55**] shown in Figure 2.3.7, with their labels as given by Daublebski. Although they are tantalizingly similar, show that they are not isomorphic.

5. Determine whether the two (12_3) configurations in Figure 2.3.8 are isomorphic and whether any is isomorphic to either of the configurations in Figure 2.3.7.

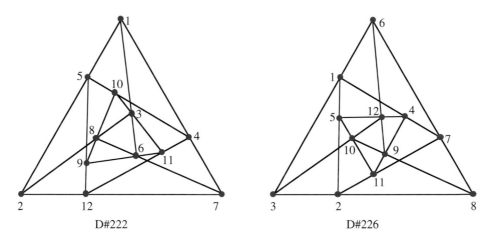

Figure 2.3.7. Two configurations (12_3) from [**55**].

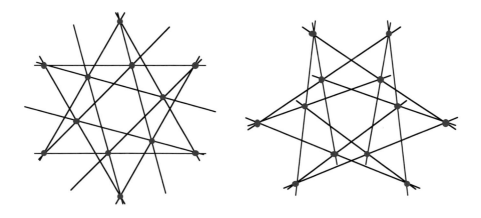

Figure 2.3.8. Two (12_3) configurations.

2.4. General constructions for combinatorial 3-configurations

From the start of the investigations of configurations, the problem of constructing all configurations (n_3) for each value of n attracted considerable attention. Many of the results on that topic have been presented in Sections 2.2 and 2.3 for specific small values of n. However, already in 1887 Martinetti [152] described an inductive procedure that can be used to generate the (n_3) configurations if all configurations with fewer points are known. He illustrated his method by determining all configurations (n_3) with $n \leq 11$, starting from the Fano configuration (7_3). As mentioned in Section 2.3, his enumeration of the 31 configurations (11_3) was correct. However, one of his claims was unfounded: He considered the enumeration of geometric configurations (n_3) to be the same as the enumeration of the combinatorial (n_3). This claim was also stressed in the review [142] of [152] by E. K. Lampe. As we have seen, $\#_c(n) = \#_g(n)$ for $n = 11$ and 12 but not for $n = 10$, and certainly $\#_c(n) \neq \#_g(n)$ for all $n \geq 14$. In the remaining part of this section, we consider *only* combinatorial configurations, even though we speak of "points" and "lines".

The central idea of Martinetti's construction is the following: Assume that in a combinatorial (n_3) configuration we have two "parallel" lines (that is, lines of the configuration that have no point of the configuration in common). If $[A, A', A'']$ and $[B, B', B'']$ are such lines and if A and B are on no line of the configuration, then we delete the two parallel lines and introduce a new point C, together with the three lines $[A, B, C]$, $[A', A'', C]$, $[B', B'', C]$. This is illustrated in Figure 2.4.1. Clearly, this leads to a combinatorial configuration $((n+1)_3)$. A configuration is called **reducible** if it can be obtained from a smaller configuration by the process just described; otherwise, it is **irreducible**. Martinetti's main result is the claim that for each n there are very few irreducible (n_3) configurations, and he purports to give a complete description of all irreducible configurations. More precisely:

Martinetti's claim. A connected (n_3) combinatorial configuration is irreducible if and only if one of the following holds:

(i) It is the cyclic configuration $\mathscr{C}_3(n)$ with lines $[j, j+1, j+3] \pmod n$, for $n \geq 8$.

(ii) $n = 10m$ for some $m \geq 1$, and the configuration is the one described below and denoted $\mathbf{M}(m)$; $\mathbf{M}(1)$ is the Desargues configuration $(10_3)_1$.

(iii) $n = 9$, and the configuration is the Pappus configuration $(9_3)_1$.

(iv) $n = 10$, and the configuration is $(10_3)_2$ or $(10_3)_6$.

Martinetti's combinatorial configuration $\mathbf{M}(m)$ can best be explained as consisting of m copies of the family of the ten points indicated by solid dots in Figure 2.4.2, and the ten solid lines shown there. The jth copy is joined to the $(j+1)^{\text{st}}$ by identifying A_j''', B_j''', C_j''' with A_{j+1}, B_{j+1}, C_{j+1}, respectively; all subscripts taken $(\bmod\, n)$.

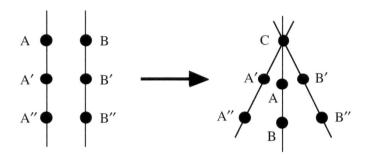

Figure 2.4.1. Martinetti's addition of a point and a line to a combinatorial configuration (n_3).

Martinetti's proof is, not surprisingly, involved and long. The result was quoted or mentioned many times over the next century; see, for example, Steinitz [**212**, pp. 486–487], Steinitz-Merlin [**214**, pp. 153–154], Gropp [**78**], [**79**], [**96**], [**101**], Carstens et al. [**39**]. In some of these it was noticed that Martinetti should have included connectedness among the requirements of his claim. Moreover, in lecture notes for the author's configurations courses in 1999 and 2002 he wrote:

> I have not checked the details, and I do not know it as a fact that anybody has. The statement has been accepted as true for these 115 years, and it may well be true. On the other hand, Daublebski's enumeration of the (12_3) configurations was also considered true for a comparable length of time. . . .

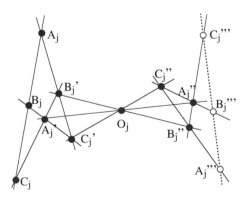

Figure 2.4.2. The "module" used in the Martinetti construction. Only the ten solid dots and the ten solid lines form one module.

As it turned out, the author's suspicion has been vindicated by Boben [19], [20]. He showed that Martinetti's list of irreducible configurations is incomplete. The problem in Martinetti's proof arises as follows. When constructing $\mathbf{M}(m)$, we attach to each other m copies of the "module" in Figure 2.4.2 as indicated above but stop before attaching $\mathbf{M}(m)$ to $\mathbf{M}(1)$. Martinetti formed that attachment "straight", by identifying A_n''' with A_1, and similarly for the B's and C's, thus obtaining $\mathbf{M}(m)$. However, as shown by Boben, that attachment can be done in "twisted" ways as well; two such attachments yield irreducible configurations which we may denote by $\mathbf{M}^*(m)$ and $\mathbf{M}^{**}(m)$. These are obtained by identifying A_n''' with C_1, B_n''' with B_1, and C_n''' with A_1 for the former and A_n''' with C_1, B_n''' with A_1, and C_n''' with B_1 for the latter. A separate argument shows that the three resulting configurations are non-isomorphic for every m. With this modification, we have the following corrected version of Martinetti's result:

Theorem 2.4.1 (Boben [19], [20]). *A connected (n_3) combinatorial configuration is irreducible if and only if one of the following holds:*

(i) *It is the cyclic configuration $\mathscr{C}_3(n)$ with lines $[j, j+1, j+3] \pmod{n}$, for $n \geq 7$.*

(ii) *$n = 10m$ for some $m \geq 1$, and the configuration is one of $\mathbf{M}(m)$, $\mathbf{M}^*(m)$, or $\mathbf{M}^{**}(m)$ described above. For $m = 1$ these are the configurations $(10_3)_1$, $(10_3)_2$, and $(10_3)_6$ (in the notation used in Section 2.2).*

(iii) *$n = 9$, and the configuration is the Pappus configuration $(9_3)_1$.*

A remarkable aspect of the situation is that all the irreducible configurations (n_3) with $n \geq 9$ are geometrically realizable by straight lines in the Euclidean plane. For the cyclic configurations, we have seen this in the proof of Theorem 2.1.3. A different construction, involving cubic curves, was given by Schröter [199] in 1888. For the configurations $\mathbf{M}(m)$ the realizability is almost obvious from Figure 2.4.2 and can be proved in general.

Concerning configurations (n_3) for particular values of $n \geq 13$, there is very little specific information available in print. Gropp [84] applied Martinetti's theorem to enumerate the combinatorial configurations with up to 14 points. He reports that there are 2,036 combinatorial configurations (13_3) and 21,399 combinatorial configurations (14_3). These numbers were confirmed by [17]; this paper reports the numbers $\#_c(n)$ of combinatorial configurations (n_3) for $n \leq 18$ (see Table 2.2.1). The number $\#_c(19)$ was reported in [22] and [117].

One of the combinatorial configurations (14_3) consists of two disjoint copies of the (7_3) configuration and is therefore not geometrically realizable. It is not known whether the other (13_3) and (14_3) combinatorial configurations are geometrically realizable. Clearly, analogous disconnected and non-realizable configurations exist for all $n \geq 14$. However, even if considering only connected configurations, the statement in Steinitz [**212**, p. 490] and Steinitz-Merlin [**214**, p. 158] that for $n \geq 11$ all (n_3) combinatorial configurations are "probably realizable" is contradicted by the example in Figure 2.2.8 (from [**63**]; see also Gropp [**96**]), which shows that the statement is invalid even if restricted to configurations that are "connected" and "realizable by pseudolines". (In the example in Figure 2.2.8, the left part of the figure has all but one of the incidences of the Pappus configuration, and therefore by Pappus's theorem the line L must be incident with the point P.) It is not known whether the (16_3) configuration in Figure 2.3.3 is the smallest configuration with these properties. We shall discuss this and related question in Sections 2.5 and 2.6 dealing with a remarkable result of Steinitz.

$$* \ * \ * \ * \ *$$

Levi [**145**, p. 93] mentions the possibility of obtaining a combinatorial configurations $((n+1)_3)$ from the configurations (n_3). He achieves this by manipulating Levi incidence matrices in a way that is equivalent to the Martinetti method illustrated in Figure 2.4.1. However, Levi does not mention Martinetti or irreducible configurations—nor does he claim that all $((n+1)_3)$ configurations are obtainable in this way.

Exercises and Problems 2.4.

1. Prove that all the irreducible configurations with at least nine points specified in Theorem 2.4.1 are geometrically realizable by points and straight lines.

2. Decide whether the (12_3) configurations in Figures 2.3.7 and 2.3.8 are reducible or irreducible. If any is reducible, to which irreducible configuration does it ultimately reduce? Is it possible for one configuration to reduce to different irreducible configurations?

3. Investigate the reducibility of the cyclic configurations $\mathscr{C}_3(n, 1, 4)$.

4. Give a formulation of Theorem 2.4.1 that is valid for all 3-configurations.

2.5. Steinitz's theorem—the combinatorial part

As we have seen in previous sections, the question whether a given combinatorial (or topological) 3-configuration can be geometrically realized is very hard. This is the reason why the 1894 Ph.D. thesis of Ernst Steinitz [**210**] is remarkable in its generality: Although Steinitz fails in completely characterizing the realizability of combinatorial or topological 3-configurations, he come as close to doing so as anybody since then.

Steinitz's claim (in our terminology). Every connected combinatorial 3-configuration (n_3) has a geometric realization by points and lines as a #1-subfiguration in the Euclidean plane; moreover, the point and line of the ignored incidence can be arbitrarily chosen.

Recall from Section 1.3 that a *#1-subfiguration* of a combinatorial configuration is a family of points and (straight) lines that satisfies all the incidence requirements except possibly one that is ignored and has no additional incidences.

In the next section we shall see that this claim is not correct. However, even the weaker result that Steinitz's arguments actually establish (see Theorem 2.6.1) is remarkable in several ways. Steinitz's proof has two parts, a combinatorial and a geometric part. The combinatorial part is correct and was much ahead of its time. However, the geometric part is defective; we discuss this in the next section. We start with the combinatorial part of Steinitz's theorem and first recall from Section 1.3 a useful definition.

A configuration table for a combinatorial configuration is said to be **orderly** if every row of the table contains all the points (hence contains each precisely once). For example, the configuration Table 2.1.1 is orderly, and the configuration tables in Sections 1.3 and 2.2 are not orderly.

The following is a basic result, due to Steinitz [**210**].

Theorem 2.5.1. *Every combinatorial k-configuration admits an orderly configuration table.*

A statement that Theorem 2.5.1 holds for $k = 3$ appears in Martinetti [**152**], without any justification or hint of proof. The majority of later writers do not mention the result—much less its proof—although many seem to accept it as selfevident. On the other hand, the statement in Page and Dorwart [**178**] regarding this result is incorrect, as are the consequences deduced by them from the erroneous statement. It is interesting, as stressed by Gropp [**97**], that Steinitz's result is very well known in combinatorics, but under a different name and credited to other people. It is considered a part of the branch of combinatorics called *matching theory.* (For this discussion, a *matching* in a graph is a collection of disjoint edges that contain all the nodes.) In this guise Steinitz's result from 1894 was independently discovered by König [**141**] in 1916; in modern terminology König's theorem can be formulated it as: *Every bipartite graph having all nodes of the same valence has a matching.* This statement is completely equivalent to Theorem 2.5.1, although neither König nor many later writers seem to have been aware of Steinitz's theorem. In still another guise, Steinitz's theorem has been generalized by the theorem of P. Hall [**123**] in 1935 concerning the existence of systems of distinct representatives. For details and proofs see, for example, Roberts [**192**, Chapter 12] or Brualdi [**33**, Chapter 9]. None of these authors is aware of Steinitz either, although the idea of Steinitz's proof is central to the topic.

We shall start by presenting a proof of this result and then discuss some of its corollaries. Our proof is modeled after Steinitz's presentation, but using what we hope is a better notation. For easier understanding of the proof, a worked-out example is given later in the section. Except for the names of the points and lines, the steps in the example are precisely parallel to those of the proof. In contrast to most of the proofs of the equivalent results mentioned in the preceding paragraph, Steinitz's proof is constructive; it can be used to find effectively an orderly configuration table, convenient for geometric constructions. We shall see such an application in Section 5.2.

Given a fixed combinatorial configuration (n_k), the first goal is to define a 1-to-1 correspondence between points and lines such that each point is incident with (that is, is contained in) the corresponding line. If we have such a correspondence, the first step in the proof is complete. We can certainly start constructing the correspondence by a *greedy algorithm*: We pick an arbitrary point and pair it up with one of the lines that are incident with it; then we chose a point not on this line and assign it to one of the lines containing it, then a point on neither of these lines, etc. Continuing with such a selection as long as possible, we find ourselves at the end in the following situation (adjusting the notation as appropriate and convenient):

The points in a subset $\mathscr{A} = \{a_1, a_2, \ldots, a_p\}$ of the set of configuration points have been assigned to the lines of the subset $\mathbf{A} = \{A_1, A_2, \ldots, A_p\}$ of the set of configuration lines, so that $a_j \in A_j$ for $j = 1, 2, \ldots, p$. We can assume that $p < n$, since otherwise we would be done with the first part of the proof. Hence there is a set $\mathscr{B} = \{b_1, b_2, \ldots, b_q\}$ of points of the configuration and a set $\mathbf{B} = \{B_1, B_2, \ldots, B_q\}$ of lines of the configuration, such that no point in $\mathscr{B}$ is incident with any line in $\mathbf{B}$; clearly, $q = n - p$. Now we shall describe a procedure by which we shall change some of the assignments between points in $\mathscr{A}$ and lines in $\mathbf{A}$, so that it will be possible to modify and extend the assignment to include one point in $\mathscr{B}$ and one line in $\mathbf{B}$.

Let B be an arbitrarily chosen line in $\mathbf{B}$, and let $\mathscr{A}_0$ be the subset of $\mathscr{A}$ consisting of the points of B. We denote by $\mathbf{A}_0$ the set of lines in $\mathbf{A}$ that are associated with the points of $\mathscr{A}_0$. Let $\mathscr{A}_1 \subseteq \mathscr{A} \setminus \mathscr{A}_0$ be the set of points of $\mathscr{A}$ not in $\mathscr{A}_0$ that are on lines of $\mathbf{A}_0$, and let $\mathbf{A}_1$ be the set of lines associated with the points in $\mathscr{A}_1$. Next, let $\mathscr{A}_2 \subseteq \mathscr{A} \setminus (\mathscr{A}_0 \cup \mathscr{A}_1)$ be the set of points of $\mathscr{A}$ not in $\mathscr{A}_0 \cup \mathscr{A}_1$ that are on lines of $\mathbf{A}_1$, and let $\mathbf{A}_2$ be the set of lines associated with the points in $\mathscr{A}_2$. We continue with assignments of this kind till we reach an r such that $\mathscr{A}_{r+1}$ is empty.

This clearly has to happen due to the finiteness of the configuration. Let now $\mathscr{A}^* = \mathscr{A}_0 \cup \mathscr{A}_1 \cup \cdots \cup \mathscr{A}_r$ and $\mathscr{A}^{**} = \mathscr{A} \setminus \mathscr{A}^*$. Note that $\mathscr{A}^*$ is the **disjoint** union of the sets $\mathscr{A}_0, \mathscr{A}_1, \ldots, \mathscr{A}_r$. Let $\mathbf{A}^*$ and $\mathbf{A}^{**}$ be the sets of lines associated with the points in $\mathscr{A}^*$ and $\mathscr{A}^{**}$, respectively.

We now pick a line $L_0 \in \mathbf{A}_0 \cup \mathbf{A}_1 \cup \cdots \cup \mathbf{A}_r$ such that L_0 is incident with at least one point b of $\mathscr{B}$, so that $b \in L_0$. (Such a line always exists, by a simple counting argument that will be given below.) Let p_0 be the point of $\mathscr{A}^*$ that corresponds to L_0; then p_0 belongs to a well-determined set $\mathscr{A}_s$ for some $s \in \{0, 1, \ldots, r\}$. Then $p_0 \in L_1$ for some $L_1 \in \mathbf{A}_{s-1}$, and let p_1 be the point of $\mathscr{A}_{s-1}$ that corresponds to L_1. Continuing in this way, we reach a line $L_s \in A_0$ and the corresponding point $p_s \in \mathscr{A}_0$. Finally, there is a line $B \in \mathbf{B}$ such that $p_s \in B$. Notice that we have the chain of incidences and correspondences

$$b \in L_0 \leftrightarrow p_0 \in L_1 \leftrightarrow p_1 \in \cdots \in L_s \leftrightarrow p_s \in B.$$

Next, we change *for the points of this chain* the assignments with which we started by making b correspond to L_0, p_0 to L_1, p_{s-1} to L_s, and p_s to B. Thus we now have a new 1-to-1 correspondence which decreased the size of the sets $\mathscr{B}$ and $\mathbf{B}$.

Repeating the procedure a finite number of times leads to an assignment of every point of the configuration to a line that is incident with it; this completes the first step of the proof, except for the demonstration of the assertion that we always can pick a line $L_0 \in \mathbf{A}^*$ which contains a point of $\mathscr{B}$. If this were not the case, then all points of $\mathscr{B}$ would have to belong to $\mathbf{A}^{**}$, since they do not belong to lines in $\mathbf{B}$ either. But this is not possible, since the cardinalities of $\mathscr{A}^{**}$ and $\mathbf{A}^{**}$ are the same due to the correspondence established at the beginning, and all incidences of points in $\mathscr{A}^{**}$ are with lines in $\mathbf{A}^{**}$ and vice versa—implying that no line in $\mathbf{A}^{**}$ can be incident with any point of $\mathscr{B}$.

For the second step we rewrite the configuration table in such a way that for each line (that is, each column) the point assigned to it is in the first row. Then the first row contains all the points, each once. The other rows of the configuration table now form a configuration (n_{k-1}), for which we repeat the steps we just did for the original configuration. Continuing in this way, we clearly reach an orderly configuration table in a finite number of steps. $\square$

It may be mentioned that when we have only two rows to deal with, a simple interchange of the order of the entries in some columns may be used instead of the more complicated procedure used in the general case.

We next illustrate the algorithm used in the proof of Theorem 2.5.1 by an example, the construction of an orderly configuration table for the combinatorial configuration (14_4) given below:

A	B	C	D	E	F	G	H	J	K	L	M	N	P
a	a	a	a	b	b	b	c	c	c	d	d	d	e
b	f	g	h	g	h	e	h	e	f	e	f	g	f
c	k	n	p	k	m	p	k	m	n	k	q	m	g
d	m	r	q	q	n	r	r	q	p	n	r	p	h

We select the starting assignments as follows:

A	B	C	D	E	F	G	H	J	K	L	M	N	$\mathbf{A}$
a	f	g	h	b	m	e	c	q	n	d	r	p	$\mathscr{A}$

and rewrite the table as

A	B	C	D	E	F	G	H	J	K	L	M	N	P
a	f	g	h	b	m	e	c	q	n	d	r	p	e
b	a	a	a	g	b	b	h	c	c	e	d	d	f
c	k	n	p	k	h	p	k	e	f	k	f	g	g
d	m	r	q	q	n	r	r	m	p	n	q	m	h

so that the assigned points are in the first row for better visibility. We are left with

$$\{k\} = \mathscr{B}, \qquad\qquad \{P\} = \mathbf{B}.$$

We put

$\mathscr{A}_0 = \{e, f, g, h\} =$ set of points on P, which happens to be the only line of B. Then

$\mathbf{A}_0 = \{G, B, C, D\} =$ associated set of lines of $\mathbf{A}$,

$\mathscr{A}_1 = \{b, p, r, a, m, n, q\} =$ points of $\mathscr{A} \setminus \mathscr{A}_0$ on lines of $\mathbf{A}_0$,

$\mathbf{A}_1 = \{E, N, M, A, F, K, J\} =$ associated set of lines of $\mathbf{A}$,

$\mathscr{A}_2 = \{d, c\} =$ points of $\mathscr{A} \setminus (\mathscr{A}_0 \cup \mathscr{A}_1)$ on lines of $\mathbf{A}_1$,

$\mathbf{A}_2 = \{L, H\} =$ associated set of lines of $\mathbf{A}$. Finally

$\mathscr{A}_3 =$ empty.

Hence we have

$\mathscr{A}^* = \mathscr{A}_0 \cup \mathscr{A}_1 \cup \mathscr{A}_2 = \{a, b, c, d, e, f, g, h, m, n, p, q, r\}$,

$\mathscr{A}^{**} = \mathscr{A} \setminus \mathscr{A}^*$, in this case empty, but it need not be empty in general.

Now we pick a line of $\mathbf{A}_0 \cup \mathbf{A}_1 \cup \mathbf{A}_2$ that contains an element of $\mathscr{B}$. In our case there is only one such element, k, and we have a choice of lines: B, E, or L. In each case we can form a chain:

$k \in B \leftrightarrow f \in P$ or

$k \in E \leftrightarrow b \in G \leftrightarrow e \in P$ or

$k \in L \leftrightarrow d \in N \leftrightarrow p \in G \leftrightarrow e \in P$, and we use it to change the assignments.

We use the last assignment, and it leads to a rewritten table:

A	B	C	D	E	F	G	H	J	K	L	M	N	P
a	f	g	h	b	m	p	c	q	n	k	r	d	e
b	a	a	a	g	b	e	h	c	d	d	p	f
c	k	n	p	k	h	b	k	e	f	e	f	g	g
d	m	r	q	q	n	r	r	m	p	n	q	m	h

Now we deal in the same way with the last three rows:

A	B	C	D	E	F	G	H	J	K	L	M	N	$\mathbf{A}$
b	a	n	p	g	h	e	k	c	f	d	q	m	$\mathscr{A}$

Then we are left with

$\{r\} = \mathscr{B}$ $\qquad$ $\{P\} = \mathbf{B}$,

This time we put

$\mathscr{A}_0 = \{f, g, h\} = $ set of points on a line (P) of $\mathbf{B}$,

$\mathbf{A}_0 = \{K, E, F\} = $ associated set of lines in $\mathbf{A}$,

$\mathscr{A}_1 = \{c, p, k, q, b, n\} = $ points of $\mathscr{A} \setminus \mathscr{A}_0$ on lines of $\mathbf{A}_0$,

$\mathbf{A}_1 = \{J, D, H, M, A, C\} = $ associated set of lines in $\mathbf{A}$,

$\mathscr{A}_2 = \{e, m, d, a\} = $ points of $\mathscr{A} \setminus (\mathscr{A}_0 \cup \mathscr{A}_1)$ on lines of $\mathbf{A}_1$,

$\mathbf{A}_2 = \{G, N, L, B\} = $ corresponding set of lines in $\mathbf{A}$,

$\mathscr{A}_3 = $ empty.

Then we put

$\mathscr{A}^* = \mathscr{A}_0 \cup \mathscr{A}_1 \cup \mathscr{A}_2 = \{a, b, c, d, e, f, g, h, m, n, p, q, r\}$,

$\mathscr{A}^{**} = \mathscr{A} \setminus \mathscr{A}^*$, in this case empty, but it need not be empty in general.

Now we pick a line of $\mathbf{A}_0 \cup \mathbf{A}_1 \cup \mathbf{A}_2$ that contains an element of $\mathscr{B}$.

In our case there is only one such element, r, and we have a choice of the lines C and G. In each case we can form a chain:

$r \in C \leftrightarrow n \in F \leftrightarrow h \in P$,

$r \in G \leftrightarrow e \in J \leftrightarrow c \in K \leftrightarrow f \in P$.

We shall use the former to change the assignments:

A	B	C	D	E	F	G	H	J	K	L	M	N	P
a	f	g	h	b	m	p	c	q	n	k	r	d	e
b	a	r	p	g	n	e	k	c	f	d	q	m	h
c	k	a	a	k	b	b	h	e	c	e	d	p	f
d	m	n	q	q	h	r	r	m	p	n	f	g	g

Making interchanges in columns C, E, G, J, K, N, we finally reach the orderly table

A	B	C	D	E	F	G	H	J	K	L	M	N	P
a	f	g	h	b	m	p	c	q	n	k	r	d	e
b	a	r	p	g	n	e	k	c	f	d	q	m	h
c	k	n	a	q	b	r	h	m	p	e	d	g	f
d	m	a	q	k	h	b	r	e	c	n	f	p	g

in which each point appears in every row.

$$* \ * \ * \ * \ *$$

Before proceeding with the next step in our study of Steinitz's theorem and its ramifications, we recall from Section 1.3 the concept of "multilaterals". A **multilateral** (often inconsistently called "polygon" in the literature) is any sequence of distinct points and distinct lines of a configuration that can be written as $P_0, L_0, P_1, L_1, \ldots, P_{r-1}, L_{r-1}, P_r \ (= P_0)$, with each L_i incident with P_i and P_{i+1} (all subscripts understood mod r). Some examples of multilaterals are shown in Figure 2.5.1. If the last point is not required to coincide with the first one, we are dealing with a **multilateral path**. A family of multilaterals in a configuration that contains all points and all lines but each just once is called a **multilateral decomposition** of the configuration. We shall return to the topic of multilaterals later (for example, in Chapter 5).

Our next aim is to modify an orderly configuration table in a way that will preserve its orderly character but will be useful for the geometric steps. We assume that a line and one if its points are selected to be ignored in the geometric implementation and that, as before, the configuration is connected. We also assume that we are concerned with a 3-configuration.

First, the rows are permuted so that the selected point of the selected line is in the first row. Note that since the table is orderly, this yields a correspondence (possibly different from the one we started with) in which each point is associated with a line that contains it. As mentioned earlier, and as is easily seen, by possibly interchanging the order of the columns (that is, lines) the orderly configuration table can be rearranged to show the multilateral decomposition in such a way that the lines of each constituent

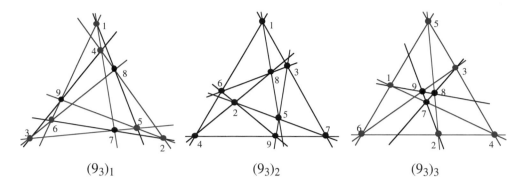

$(9_3)_1$ $(9_3)_2$ $(9_3)_3$

Figure 2.5.1. Some examples of multilaterals, in the three configurations (9_3) shown in Figure 2.2.1. The first is a 6-lateral with sequence of points $1, 4, 2, 5, 3, 6$ (1) in the configuration $(9_3)_1$. The second is 9-lateral, with sequence of points $1, 8, 3, 7, 5, 9, 4, 2, 6$ (1); since this multilateral involves all points (and hence also all lines), it is a Hamiltonian multilateral. The last diagram shows a trilateral $7, 8, 9$ (7) and a 6-lateral $1, 6, 3, 5, 2, 4$ (1). Note that another 6-lateral is $1, 5, 2, 6, 3, 4$ (1). The 3-lateral and either of the 6-laterals taken together form a multilateral decomposition of the configuration $(9_3)_3$.

multilateral occur consecutively. In each of the multilaterals we can assume that the point in the last position in one column is in the middle position in the following column (understood modulo length of the multilateral). We rearrange the columns in such a way that the multilateral that contains the chosen line is placed last, and the selected line is chosen as the last line in the multilateral. If the multilateral is Hamiltonian, this part of the proof is completed. Otherwise, since the configuration is connected, at least one of the points of the last multilateral must be associated to (that is, be in the first row of) a line which is not in the multilateral. Choose the multilateral containing this line to be the next to last and the line in question to be its last line. Then some point of this multilateral must be associated with another multilateral not used so far, and we continue in the same way. At the end we reach what we may call an arranged configuration table. This proves

Theorem 2.5.2. *Every connected 3-configuration has an arranged configuration table.*

As an illustration, we show in Table 2.5.1 an orderly configuration table of a configuration (14_3). Rearranging the columns so as to make the multilateral decomposition visible, we obtain the arranged configuration table, Table 2.5.2.

Table 2.5.1. An orderly configuration table of a connected combinatorial configuration (14_3).

A	B	C	D	E	F	G	H	J	K	L	M	N	P
c	k	n	a	q	b	r	h	m	p	e	d	g	f
d	m	a	q	k	h	b	r	e	c	n	f	p	g
b	a	r	p	g	n	e	k	c	f	d	q	m	h

Table 2.5.2. A rearranged configuration table of the (14_3) configuration of Table 2.5.1, in which the lines of each multilateral appear as consecutive columns. The point e of line G was chosen as the exceptional point, so its row is the first one. The line G is the last line of its multilateral, which is the last multilateral. Each multilateral is specified by rows 2 and 3 of the table. The table is arranged in the sense described earlier.

A	M	P	N	K	B	J	L	C	D	E	F	H	G
b	q	h	m	f	a	c	d	r	p	g	n	k	e
c	d	f	g	p	k	m	e	n	a	q	b	h	r
d	f	g	p	c	m	e	n	a	q	k	h	r	b

We shall see in the next section how such an arranged multilateral decomposition can be used geometrically. Here we shall conclude the section by discussing certain ramifications of the results we have seen so far.

Corollary 2.5.3. *Every connected k-configuration C, with $k \geq 2$, admits multilateral decompositions.*

Indeed, any two rows of an orderly configuration table determine, by the above, a multilateral decomposition of C.

Corollary 2.5.4. *Every connected k-configuration C, with $k \geq 2$, is 2-connected.*

Proof. Assume that C is a connected k-configuration such that, without loss of generality, there is a line L for which for suitable elements R' and R'' there is no R'-to-R'' multilateral path that misses L. By the connectedness of C, there is a multilateral path M that uses L; that is, there are two points Q' and Q'' of L that are part of this path M. In an orderly configuration table of C, permuting the rows if necessary, we may put Q' and Q'' in the last rows of the block L. Let S be a multilateral decomposition of C determined by the last two rows of this orderly configuration table. Then one of the multilaterals of this decomposition uses the points Q' and Q''. But since the multilateral is a circuit, there is a multilateral path (formed by the lines other than L) that connects Q' and Q''. Substituting this path for the one that originally connected R' and R'' eliminates the use of L. Hence the assumption that each path between R' and R'' uses L is incorrect, and so C is 2-connected.   □

We shall discuss additional connectedness results in Section 5.1.

Exercises and Problems 2.5.

1. Use the procedure applied in the proof of Theorem 2.5.1 to replace the configuration Table 2.5.3 by an orderly configuration table.

Table 2.5.3. A (14_4) configuration table.

a	b	c	d	e	f	g	h	i	j	k	l	m	n
1	1	1	1	2	2	2	3	3	3	4	4	6	7
2	5	6	10	3	5	8	4	5	11	5	9	7	8
4	8	9	13	9	6	12	6	7	12	10	11	10	9
7	11	12	14	10	14	13	8	13	14	12	13	11	14

2. Find orderly configuration tables for the two (12_3) configurations in Figure 2.5.2.

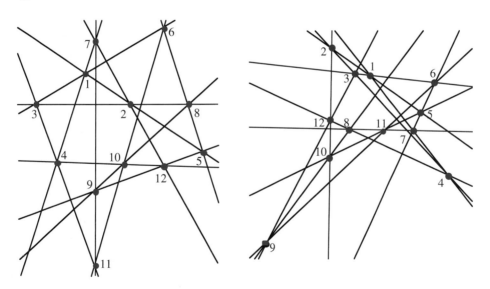

Figure 2.5.2. Two (12_3) configurations.

3. Justify the statement: *Any selection of two rows of an orderly configuration table defines a multilateral decomposition of the configuration.* List all multilateral decompositions resulting from possible choices of the rows in the orderly configuration tables found in Exercise 2 for the configurations (12_3) shown in Figure 2.5.2.

4. Justify the statement: For every k-configuration **C** and every multilateral decomposition of **C**, there is an orderly configuration table in which the multilateral decomposition can be obtained from the first two rows of the table.

5. Modify the proof of Theorem 2.5.1 to establish the following strengthening: *Every combinatorial k-configuration admits an orderly configuration table in which an arbitrarily chosen line is the last line of the table.*

2.6. Steinitz's theorem—the geometric part

We turn now to a consideration of the geometric part of Steinitz's claim. We recall from Section 2.5 the claim from Steinitz's Ph.D. thesis [**210**]:

Steinitz's claim. Every connected combinatorial 3-configuration has a geometric realization by points and lines as a #1-subfiguration in the Euclidean plane; moreover, the point and line of the ignored incidence can be arbitrarily chosen.

Recall from Section 1.3 that a *#1-subfiguration* of a combinatorial configuration is a family of points and (straight) lines that satisfies all the incidence requirements except possibly one that is ignored and has no additional incidences.

By our definition, which coincides with the definition generally used, a **geometric configuration** (n_k) is a family of n points and n (straight) lines such that each point is incident with k lines and each line is incident with k points. The intention of this definition is that each of the points and lines is incident with **precisely** k objects of the other kind. Even though this requirement often was not explicitly stated, in many instances it was stated, and it has been taken as selfunderstood by all nineteenth-century writers on configurations.

However, the following situation does arise: We start with a combinatorial configuration and find a set of points and a set of straight lines which fulfill the incidence requirements of the combinatorial configuration. In other words, every combinatorial incidence corresponds to a geometric incidence. However, it is possible that the points and lines we found have **additional** geometric incidences, not specified in the combinatorial configuration. As mentioned in Section 1.3, in such a case we shall say that the points and lines form a **representation** (**superfiguration** and **weak realization** are terms also used) of the combinatorial configuration. It may happen that a different choice of points and lines will result in a geometric configuration without additional incidences; if it is necessary to stress this fact, we shall

Table 2.6.1. A combinatorial configuration (16_3).

A	B	C	D	E	F	G	H	I	J	K	L	M	N	O	P
a	b	c	d	e	f	g	h	i	j	k	l	m	n	o	p
e	d	h	m	f	a	c	g	j	k	l	p	b	i	n	o
c	f	d	e	g	h	b	i	o	n	m	j	a	p	l	k

say that we have a **realization** (or, if there is need for a more specific expression, a **strong realization**). But, as is easy to see, some combinatorial configurations admit **only** superfigurations.

For example, consider the combinatorial configuration given by Table 2.6.1. A superfiguration is shown in Figure 2.6.1, in which the line H passes through the point m although they are not combinatorially incident according to Table 2.6.1. (This configuration is isomorphic to the one in Figure 2.3.3.) The reason why **every** geometric presentation of this configuration by points and lines is a *representation* (and not a *realization*) lies in the fact that the hexagon $abcdef$ has vertices that alternate on the two lines A and B, and therefore the three points g, h, m are collinear by the Pappus theorem. More complicated examples can have several unintended geometric incidences—in fact, there is no upper bound on the possible number of such incidences. In Figure 2.6.2 we show a topological configuration (18_3) such that each of its representations is a #2-superfiguration. This possibility is also ignored in the Wikipedia article [**227**] on Ernst Steinitz (as of February 4, 2008).

A more detailed analysis of this topic will be presented in [**21**].

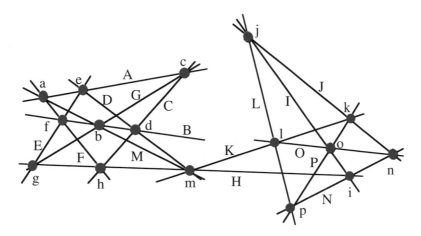

Figure 2.6.1. A **representation** of the combinatorial configuration (16_3) given by Table 2.6.1.

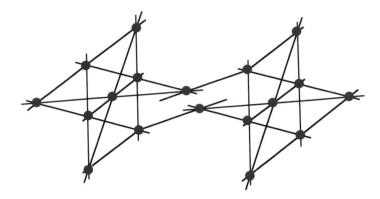

Figure 2.6.2. A topological configuration which can be realized by straight lines only as a #2-superfiguration.

The above preamble to the geometric part of Steinitz's theorem about geometric "realizations" of combinatorial 3-configurations was needed to set the stage for dealing with this rather remarkable result and its history. The following result is what Steinitz actually proved:

Theorem 2.6.1. *For every connected combinatorial 3-configuration and every choice of one incident point-line pair, there is a selection of distinct points and (straight) lines which realize all the incidences except possibly the incidence of the chosen line with the chosen point.*

In other words, every connected 2-configuration has a **near-representation** in the Euclidean plane, in which the point and line of the ignored incidence can be arbitrarily chosen. As shown by the example in Table 2.6.1 (and illustrated in Figure 2.6.1), there are combinatorial configurations for which no near-representation is a near-realization. (The "near" part of these terms is meant to convey that one incidence is disregarded.)

Note that, since the chosen line is incident with only three points (one of which is the chosen point), it is in every case possible to find a curve of degree at most two (even a circle, unless there is a straight line), incident with these three points. This is illustrated in Figure 2.6.3.

There is no indication in any of the writings of Steinitz or other mathematicians during the twentieth century that they were aware of the fact that Steinitz's claim as formulated by him is not valid, since it pretends to prove strong "near-realizability". The first indication (known to me) of the awareness that the theorem actually *established* by Steinitz has to be formulated in terms of weak "near-realizations"—that is, near-representations, as we did above—was in a talk by T. Pisanski [**180**] at a meeting in Ein Gev (Israel) in April 2000. The fact that some combinatorial 3-configurations have only geometric representations and no geometric realization was also

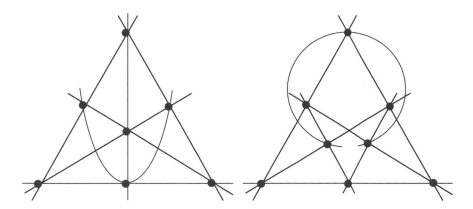

Figure 2.6.3. Frequently shown illustrations of Steinitz's theorem, by 1-subfigurations of the Fano (7_3) and Möbius-Kantor (8_3) configurations.

noticed by W. Kocay and R. Szypowski [**140**] in 1999 and by Glynn [**73**] in 2000; neither work mentions Steinitz.

Proof of Theorem 2.6.1. Starting from a connected combinatorial configuration (n_3) given by a configuration table, the first step is to convert the table into an orderly table. This is possible by Theorem 2.5.1. Next, the rows are permuted so that the exceptional point of the exceptional line is in the first row. Note that since the table was orderly, this yields a correspondence (possibly different from the one we started with) in which each point is associated with a line that contains it. As mentioned earlier, and as is easily seen, by possibly interchanging the order of the columns (that is, lines), the orderly configuration table can be rearranged to show the multilateral decomposition in such a way that the lines of each constituent multilateral occur consecutively. We rearrange the columns in such a way that the multilateral that contains the chosen line is placed last, and the selected line is chosen as the last line in the multilateral. If the multilateral is Hamiltonian, this part of the proof is completed. Otherwise, since the configuration is connected, at least one of the points of this multilateral must be associated to (that is, be in the first row of) a line which is not in the multilateral. Choose the multilateral containing this line to be the next to last and the line in question to be its last line. Then some point of this multilateral must be associated with another multilateral not used so far, and we continue in the same way. At the end we reach what we called an **arranged** configuration table.

As an illustration, we show in Table 2.6.2 an orderly configuration table of a configuration (14_3). Choosing as the exceptional elements the point e

and the line G, and rearranging the columns so as to make the multilateral decomposition visible, we obtain Table 2.6.3.

Table 2.6.2. An orderly configuration table of a connected combinatorial configuration (14_3).

A	B	C	D	E	F	G	H	J	K	L	M	N	P
c	k	n	a	q	b	r	h	m	p	e	d	g	f
d	m	a	q	k	h	b	r	e	c	n	f	p	g
b	a	r	p	g	n	e	k	c	f	d	q	m	h

Table 2.6.3. A rearranged configuration table of the (14_3) configuration of Table 2.6.2, in which the lines of each multilateral appear as consecutive columns. The point e of line G was chosen as the exceptional point, so its row is the first one. The line G is the last line of its multilateral, which is the last multilateral. Each multilateral is specified by rows 2 and 3 of the table.

A	M	P	N	K	B	J	L	C	D	E	F	H	G
b	q	h	m	f	a	c	d	r	p	g	n	k	e
c	d	f	g	p	k	m	e	n	a	q	b	h	r
d	f	g	p	c	m	e	n	a	q	k	h	r	b

In an abbreviated form, using just the names of the vertices (second entries in the list of points in each column) and the lines to which they belong, this multilateral decomposition can be written as

(1) $c\,A\,d\,M\,f\,P\,g\,N\,p\,K \mid k\,B\,m\,J\,e\,L\,n\,C\,a\,D\,q\,E \mid b\,F\,h\,H\,r\,G$

With line G and point e chosen as the exceptional elements, we make the final rearrangement of the columns. The multilateral containing G is placed last, and G is placed as the last line of it. The entry b of the first column in this multilateral is the first point which is associated with a line not in the multilateral. Since b is associated with the line A of the first multilateral, this multilateral is placed next to last, and A is placed at its last column; then M is the first line, and the corresponding first point is d. Therefore the multilateral preceding it has L as its last line (column). The rearranged multilateral decomposition (1) now has the following representation:

(2) $n\,\underline{C}\,\underline{a}\,\underline{D}\,\underline{q}\,\underline{E}\,\underline{k}\,\underline{B}\,m\,\underline{J}\,e\,\underline{L} \mid \underline{d}\,\underline{M}\,f\,\underline{P}\,\underline{g}\,\underline{N}\,\underline{p}\,\underline{K}\,\underline{c}\,\underline{A} \mid \underline{b}\,\underline{F}\,\underline{h}\,\underline{H}\,\underline{r}\,\underline{G}$

From this decomposition (2) it is obvious that each element (point or line) is incident with at most two elements that come before it, except that the last line (G in the present case) is incident with three of the preceding points. (For the other elements, the situation is indicated by the single or

Table 2.6.4. The multilateral decomposition (2) in configuration table form.

C	D	E	B	J	L	M	P	N	K	A	F	H	G
r	p	g	a	c	d	q	h	m	f	b	n	k	e
n	a	q	k	m	e	d	f	g	p	c	b	h	r
a	q	k	m	e	n	f	g	p	c	d	h	r	b

double underline of the symbols.) This means that elements incident with no previous element can be chosen completely arbitrarily in the plane, those incident with one previous element can be chosen freely as a point on a line or as a line through a point, while those incident with two earlier ones are determined without any freedom of choice. The last triplet may be collinear but need not be—in which case a second degree curve can be passed through it.   □

For clarity, the final rearranged configuration table is shown in Table 2.6.4.

The geometric subfiguration that resulted from a set of particular choices is shown in Figure 2.6.4. Figure 2.6.5 illustrates the possibility of making choices which happen to satisfy the last incidence as well and hence yield a proper geometric realization of this configuration (14_3).

As is clear from the above proof, at no point has there been made any claim that the lines and/or points we introduce have no additional incidences. It has been tacitly assumed that the construction is not undermined by (mis)using the freedom of choice to intentionally place points on already-present points or lines they are not supposed to be incident with, and analogously for the selection of lines. However, as shown at the beginning of this section, there are circumstances in which unintended incidences cannot be avoided. The remarkable fact that this possibility has been ignored for a century is nearly incomprehensible.

One mystery still remaining in this context is the fact that all known instances in which unwanted incidences in geometric realizations of combinatorial 3-configurations are unavoidable deal with configurations that are connected (hence 2-connected) but not 3-connected. It is possible to guess that the following holds:

Conjecture 2.6.1. *Every 3-connected combinatorial 3-configuration admits geometric realizations by points and straight lines with no incidences except the required ones.*

Steinitz's Theorem 2.6.1 was proved in [**210**] in 1894. In 1999 it was independently discovered by Kocay and Szypowski [**140**] in a different setting

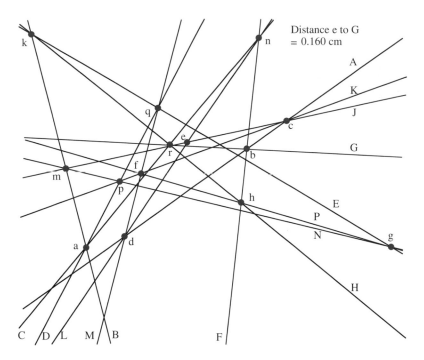

Figure 2.6.4. A "near-realization" following Table 2.6.4 of the configuration (14_3) of Table 2.6.1, in which all the incidences except the one of point e and line G are satisfied.

* * * * *

and in 2000 by Glynn [**73**]. A presentation of the above material and other aspects of Steinitz's theorem appears in [**117**].

* * * * *

The realization in Figure 2.6.5 was obtained by utilizing continuity: In the near-realization shown in Figure 2.6.4 the point e is above the line G, while by choosing some other appropriate positions for the points used in the construction, a near-realization can be obtained in which the point e is below the line G. This is a situation that *seems* to be quite general. Steinitz made the same observation in [**210**]. Steinitz devoted more than half the dissertation [**210**] (24 pages) to a consideration of ways in which one could *guarantee* that the final step in the above proof can be made using a straight line instead of a curve of degree 2. While this might be another interesting result, the author has not been able to follow the exposition in [**210**]. (In fact, he knows of nobody who claims to have understood and verified this part of Steinitz's thesis.) The opaqueness of the exposition can best be seen from the last two sentences of Steinitz's introduction to this part of the work (see [**210**, p. 22]):

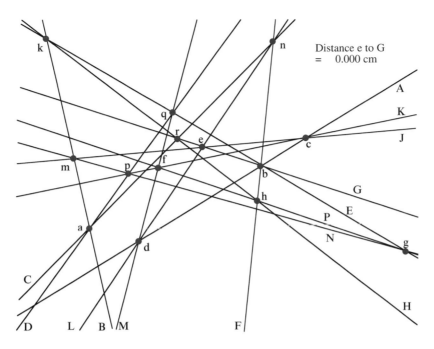

Figure 2.6.5. A *realization* of the configuration (14_3) of Table 2.6.4, in which *all* the incidences are satisfied.

> ... Without any particular assumptions about the configurations, a method will be presented below following which one can reach a linear presentation. However, for each configuration to which we want to apply this method, an additional investigation is necessary since the method becomes *illusory in certain cases* [the author's translation and italics].

In mentioning [**210**] in the survey [**212**, p. 490], Steinitz is equally uninformative. Stating that his method is an extension of Schröter's approach in [**199**], [**201**], he ends the explanation by stating:

> Schröter's method can be generalized so that it is applicable to *most* configurations n_3 [the author's translation and italics].

It seems that the "method of Schröter" is rooted in arguments due to Möbius in the early part of the nineteenth century, in particular in [**172**].

However, even if the proof in [**210**] is valid and if somebody were to make the exposition understandable, this would prove only that every connected configuration has *representations*. It would not be a proof of Conjecture 2.6.1 for *realizations*, as claimed by Steinitz. Indeed, we know from examples such as the one in Figure 2.6.1 that some representable configurations are not realizable; hence Conjecture 2.6.1 cannot be generally valid for realizations.

Exercises and Problems 2.6.

1. Find a geometric construction, analogous to the one given above, for the combinatorial configuration of Table 2.5.1 but with the choice of point q and line D as the exceptional elements.

2. Find the analogous construction for the combinatorial configuration which has as its table the first three rows of the orderly table obtained in the example worked out in Section 2.5 and with point a and line A as exceptional elements. Using suitable software, see whether this configuration has a proper realization.

3. Apply the methods of construction we used here to the configurations $(10_3)_3$ and $(10_3)_4$ of Table 2.2.7.

4. Find a connected combinatorial configuration for which every representation is a #3-subfigure, that is, contains at least three unwanted incidences.

5. Show that there are connected combinatorial configurations (n_3) for which in every representation the number of unwanted incidences is at least $c \cdot n$, for some constant $c > 0$. Open problem: What is the best possible c?

2.7. Astral 3-configurations with cyclic symmetry group

We have seen in Section 1.5 that a 2-astral 3-configuration must have two orbits of points and two orbits of lines. By the convention introduced there we simplify the expressions and call such configurations **astral**, for short.

Lemma 2.7.1. *If an astral 3-configuration has one orbit of points at infinity, it must have reflective symmetries.*

Proof. If such a configuration has no reflective symmetries, then the orbit of points in the finite plane has to coincide with the vertices of a regular polygon; the only alternative would be that they are the vertices of an isogonal polygon—but their equivalence requires reflection. Each of the points at infinity is on three lines, two of which are in the same orbit. Even if these two are related by a rotational (halfturn) symmetry, they must be parallel and of the same length, and by the rotational symmetry the third line parallel to them must pass through the center of the polygon. Thus all lines come in triplets of parallel lines, the middle one serving as a mirror for the other two; these mirror lines are spaced at equal angles, and hence they are mirrors of the configuration. Hence we again are led to reflective symmetries. □

As a consequence of Lemma 2.7.1 we see that astral configurations (n_3) that have a cyclic group of symmetries are necessarily configurations in the Euclidean plane. Astral 3-configurations with a cyclic group of symmetries and no mirrors will be called **chiral**. (Note that this does not mean that all astral configurations contained in the Euclidean plane have a cyclic group of symmetries. We shall consider those with dihedral symmetry in Section 2.8.) The points of a chiral astral configuration are at the vertices of two concentric regular polygons with $m = n/2$ vertices each; the polygons clearly have different sizes. As we shall show next, such 3-configurations depend (up to similarity) on three additional integer parameters. The notation we shall use for these configurations is $m\#(b, c; d)$; a detailed explanation follows, and an illustration is given in Figure 2.7.1.

The lines of one geometric transitivity class are the diagonals of one of the polygons, and those of the other class are diagonals of the other polygon; each line of the configuration contains two points of one polygon and one of the other. The numbers of edges of the polygons bridged (spanned) by the diagonals are the integers b and c of the symbol; usually we shall follow the convention that $m/2 > b \geq c > 0$, but the relative size of b and c is not of intrinsic importance and it is sometimes convenient to disregard the convention. The corresponding points (vertices) are accordingly called the b-points, resp. c-points, and the lines are b-lines and c-lines. Starting from an arbitrary b-point denoted B_0 and proceeding in an arbitrary orientation, we label the other b-points consecutively $B_1, \ldots, B_{m-1}$. Each b-line is then of the form $L_i = \text{aff}(B_i, B_{i+b})$, and it contains a c-point which we label C_i. The c-line that passes through C_0 determines the labeling of the c-lines. In the orientation of the c-points which is induced by the orientation chosen for the b-points, the earlier point of that c-line is C_0, and the latter accordingly is C_c. The remaining c-points are then labeled in the obvious way; the c-lines are labeled by $M_i = \text{aff}(C_i, C_{i+c})$. Here and throughout, all subscripts are to be understood mod m. From a given (n_3) configuration of the kind considered, the values of m, b, and c can be read off instantly. Now we can find a tentative determination of the symbol d in the notation $m\#(b, c; d)$ for the configuration. We consider the b-point that is incident with the c-line $M_0 = \text{aff}(C_0, C_c)$; like all b-points, it already carries a label. We take this label as the value of d in the *preliminary* symbol of the configuration.

The value of d in the final symbol requires a comparison of two possibilities. One is what we have just described, and the other is obtained in the same way but going in the opposite orientation around the b-polygon. As the *final* symbol $m\#(b, c; d)$ for the configuration we shall generally choose that one of the alternatives which has the smaller value of d. As is easily verified, the two values of d add up to $b + c$; hence we may assume that $d \leq (a + b)/2$, which means that in fact only one of the determinations has to be carried out. If it yields such a value of d, we take it; otherwise we subtract it from $b+c$ to get the correct value of d. This is illustrated by the examples in Figure 2.7.2.

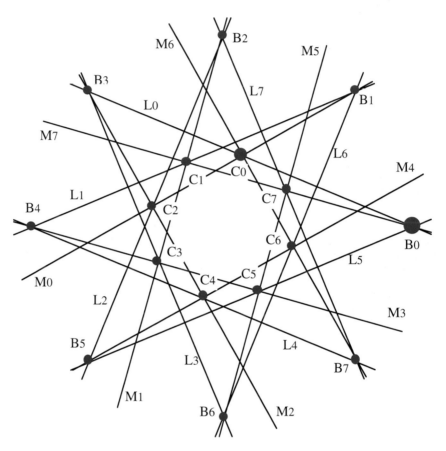

Figure 2.7.1. An example of the labeling and designation of a chiral astral configuration. Following the explanations in the text, this is configuration $8\#(3,2;1)$.

While our conventions assign a unique symbol to each astral (n_3) configuration, the converse is not valid. In general, two configurations are represented by the same symbol $m\#(bc;d)$. They differ by the ratio of the radius of the circle of c-points to that of the b-points; the one with smaller ratio is denoted by a single tag $'$, the other one by double tags $''$. This is illustrated in Figure 2.7.3. Another way of distinguishing the two configurations is by specifying the ratio in which the point C_0 divides the segment B_0B_b; this information is very useful for drawing the configuration, as well as for determining which symbols are possible.

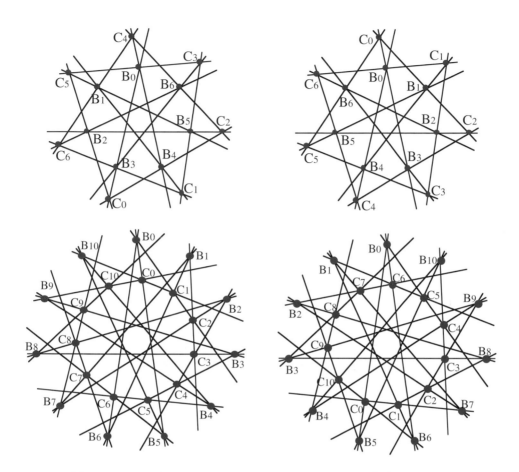

Figure 2.7.2. Additional examples of labeling astral (n_3) configurations. The configuration in the upper row has symbols $7\#(3,2;4)$ and $7\#(3,2;1)$, so the latter is the one conventionally accepted. The configuration in the bottom row has symbols $11\#(5,1;10) = 11\#(5,1;-1)$ since all subscripts can be taken $(\bmod\, n)$ and $11\#(5,1;7)$. Hence the conventional symbol is $11\#(5,1;-1)$.

However, in cases in which either $b = c$ or $2d = b + c$ the symbol $m\#(b,c;d)$ represents only a single configuration. Examples of these situations are shown in Figure 2.7.4, for the symbols $6\#(2,2;1)$ and $11\#(5,1;3)$.

If the highest common factor of m, b, c, d is $f > 1$, then the configuration $m\#(b,c;d)$ is not connected but consists of f copies of the configuration $m/f\#(b/f, c/f; d/f)$. However, exceptions to all the above happen when there are additional "accidental" incidences. For example, an attempt to draw the configuration $12\#(5,1;3)$ leads to the superfiguration shown in Figure 2.7.5(a); it has additional incidences and is, in fact, a configuration (24_4). In Figure 2.7.5(b) we show how sensitive the situation is with respect to correctly drawing the configurations—a seemingly legitimate configuration does not really exist. On the other hand, Figure 2.7.5(a) can be interpreted as a *representation* of the configuration $12\#(5, 1; 3)$, as well as a representation of configurations $12\#(5, 1, -1)$, $12\#(4, 4; 1)$, and $12\#(4, 4; 2)$. Figure 2.7.5(b) serves to illustrate a *topological* realization of the configuration $12\#(5, 1; -1)$.

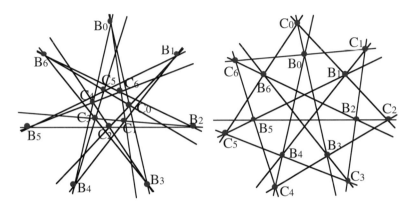

Figure 2.7.3. The two astral configurations with common symbol $7\#(3, 2; 1)$. The one on the left is specified by $7\#(3, 2; 1)'$, the other one by $7\#(3, 2; 1)''$.

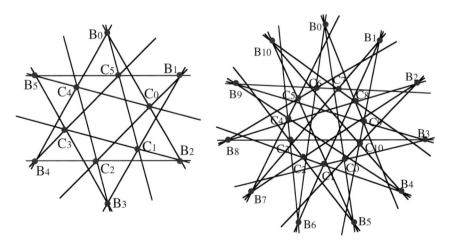

Figure 2.7.4. Astral configurations which are examples of the case in which only a single configuration corresponds to its symbol, here $6\#(2, 2; 1)$ and $11\#(5, 1; 3)$.

A different type of unintended incidences is illustrated by the example in Figure 2.7.6. Here the result is a collection of points and lines which is not a configuration under the definitions we adopted at the beginning, since

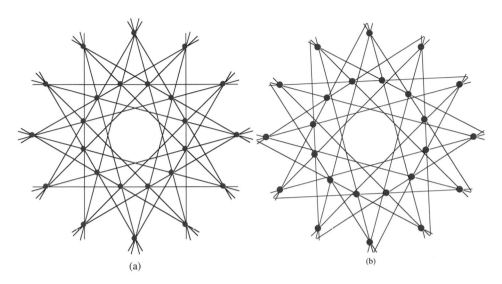

(a) (b)

Figure 2.7.5. The diagram in (a) is supposed to show the configuration $12\#(5,1;3)$; however, additional incidences turn it into an astral (24_4) configuration which, by the conventions we shall specify in Section 3.5, has the symbol $12\#(5,4;1,4)$. Note that the same (24_4) configuration results when drawing any of $12\#(5,1,-1)$, $12\#(4,4;1)$, and $12\#(4,4;2)$. The first of these is illustrated by the pseudoline configuration in (b). Note that these are actually straight lines but that their incidences are faked (ever so slightly). For a different presentation of these cases see Figure 5.8.1 and the explanations given there.

some lines (but not all) are incident with four points, and some points with four lines.

Disregarding the possible presence of unintended incidences, how does one get from the symbol to a drawing, and how does one decide whether a symbol corresponds to any configuration? For the answer to both parts of the question, we can proceed either algebraically or geometrically.

In the algebraic approach, given a symbol $m\#(b,c;d)$, we start with the vertices of a regular m-gon and draw all diagonals of span b (or their extensions, if needed). Points of the other orbit will be the vertices of another regular m-gon, situated on the diagonals of the first one. Their location is determined by the ratio which, in the notation of Figure 2.7.1, is given by the still undetermined ratio of lengths $\lambda = B_0 C_0 / B_0 B_b$. The position of C_c is determined by the same ratio, since $\lambda = B_c C_c / B_c B_{b+c}$. Now, the line $C_0 C_c$ contains the point B_d of the first orbit. Hence, writing the collinearity condition in terms of a determinant, involving the variable λ and the known coordinates of the B points, yields a quadratic equation for λ. Depending on whether there are two, one, or no solutions in real numbers, we obtain

the pair of isomorphic configurations, a single configuration, or no configuration at all. Thus the complete characterization of possible symbols is,
in principle, determinable by the non-negativity of the discriminant of that
quadratic equation. In any particular case, the software used (various versions of Mathematica® on different Macintosh computers) had no problem
finding the value(s) of λ and then drawing the configuration(s). However, no
amount of effort, on the computer or manually, was successful in explicitly
describing the necessary and sufficient conditions on the integer parameters
m, b, c, d for the existence of the configurations. The best we could do is to
deduce several necessary conditions from many specific cases and from an
argument to be described below. In any case, the known conditions for a
symbol $m\#(b, c; d)$ are as follows (this includes the notational conventions
introduced earlier):

$$0 < c \le b < m/2,$$

$$2[(b + c - m)/2] \le c - b + 1 \le 2d \le b + c,$$

$$0 \ne d \ne c,$$

$$2\cos(b\pi/m)\cos(c\pi/m) \le 1 + \cos((b + c - 2d)\pi/m).$$

While the use of calculational and graphic capabilities of appropriate
software (Mathematica®, Matlab®, Maple™, and others) enables one to
find out whether a symbol leads to a configuration, it is of some interest to
note that geometric means can yield the same result. In fact, if the vertices
of a regular m-gon are given, the configurations $m\#(b, c; d)$ can be drawn
with just the classical Euclidean tools. (Naturally, the construction of the
regular m-gon may or may not be possible with Euclidean tools, depending
on the value of m.) Here is how the construction proceeds, illustrated for
$m\#(b, c; d) = 11\#(5, 1; 2)$ by the steps in Figure 2.7.7.

(a) Draw the lines determined by the diagonals of span $b = 5$; this
yields a regular polygon P of type $\{m/b\}$.

(b) Construct the isosceles triangle T determined by two vertices V_1
and V_2 of P, which are separated by span $c = 1$, and the center O
of P.

(c) Construct the circumcircle C of the triangle T described in (b).

(d) Label the sides of the polygon P. We label "0" the two lines of P
that touch T at V_1 and V_2 but do not go through the interior of T.
The other lines of P are numbered by their sequence at the central
convex m-gon determined by these lines. The sides closer to the
center of C are labeled "1", "2", ... in order; the ones farther from
the center of C are labeled " $- 1$", " $- 2$", ...

(e) Find the intersection points of the lines of P with the circle C.

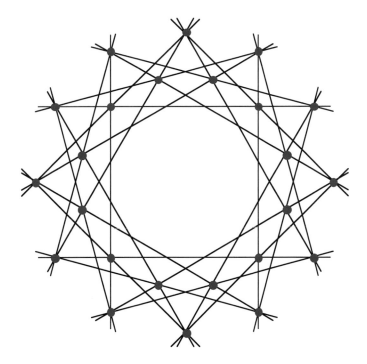

Figure 2.7.6. A drawing of the astral configuration $12\#(3,3;1)$ shows unintended incidences. The resulting family of points and lines is not a *configuration* according to our definitions; it is a *superfiguration*. In fact, by ignoring some incidences, it could be interpreted as a *representation* of the astral configuration $12\#(3,3;1)$.

(f) Label these intersection points by the labels of the lines.

(g) Select one of the points labeled $d = 2$, and draw the line connecting it with one of the points V_1, V_2.

(h) Rotate through all the multiples of $2\pi/m = 2\pi/11$ the point chosen in (g) and the line constructed there. A configuration $m\#(b, c; d) = 11\#(5, 1; 2)$ is obtained.

(i) The same construction as in (g) and (h) but with the other point labeled $d = 2$ yields the other configuration $11\#(5, 1; 2)$. The remaining possibilities of pairing a point labeled "2" with the other V_j yield configurations congruent to the ones in (h) and (i).

(j) An analogous construction but with a point labeled "3" yields the configuration $11\#(5, 1; 3)$. As we shall see in Section 2.10 this configuration is selfpolar.

Naturally, these constructions need justification, which we shall provide below. However, it is appropriate to recall that establishing results by using graphical means can be rigorously justified; see, for example, [**170**].

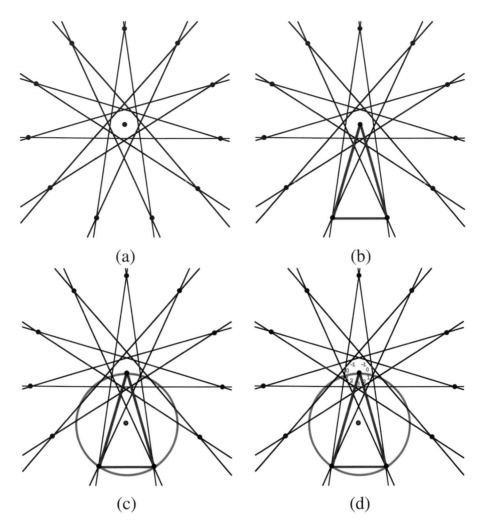

(a)   (b)

(c)   (d)

Figure 2.7.7. (first part). The geometric construction of the configuration 11#(5, 1; 2).

The reasoning follows the above method of construction and is illustrated in Figure 2.7.8 by the example of the configuration 11#(4, 3; 2).

The triangle O $V2$ $V1$ is isosceles. The angle O $V(2)$ $V1$ equals the angle O $V2$ $V1$ since both are peripheral angles over the same arc O $V1$. Let X be the point on the ray $V(2)$ $V1$ such that the angle $V(2)$ O X equals the angle $V2$ O $V1$. Then the triangle O X $V(2)$ has correspondingly equal angles with the triangle O $V1$ $V2$ and hence is similar to it. Therefore it is also isosceles, so OX has the same length as $OV(2)$ and is thus on the circle centered at O and with radius $OV(2)$. As the angle $V(2)$ O X is the same as the angle $V2$ O $V1$, which spans a diagonal of span $c = 3$ of the

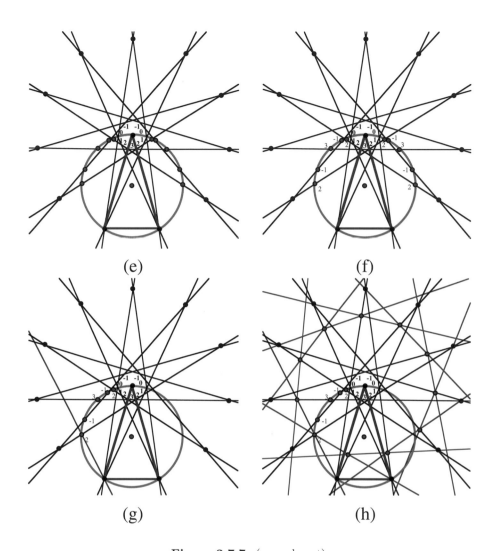

Figure 2.7.7. (second part).

m-gon ($m = 11$), it follows that $V(2)$ X spans the same diagonal on the m-gon determined by the rotates of $V(2)$. The existence of the configuration $m\#(b, c; d) = 11\#(4, 3; 2)$ is established.

Using the description of the determination of the symbol $m\#(b, c; d)$ of an astral configuration (n_3), it is immediate that the reduced Levi graph is as shown in Figure 2.7.9. The simplicity of the reduced Levi graph of such a configuration can be interpreted as the source of the usefulness of such graphs, but it also serves to indicate that the encoding of such an astral configuration by our symbol is natural and not arbitrary.

Astral 3-configurations were first defined in [**110**], but isolated examples occur in earlier publications. The first seem to be by Zacharias [**234**];

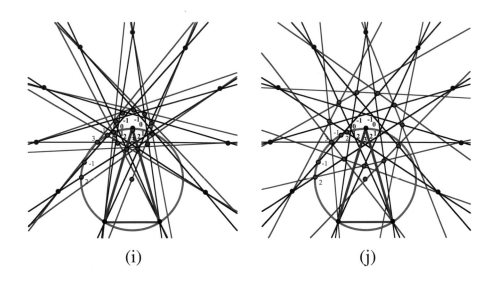

(i)   (j)

Figure 2.7.7. (third part).

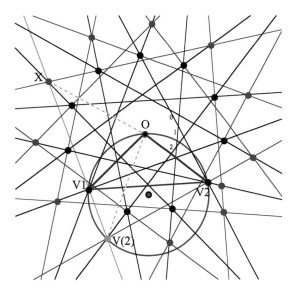

Figure 2.7.8. Starting with the $\{11/4\}$ polygon (black points and lines), the configurations $11\#(4, 3; 2)$ is constructed by the method described above.

he shows examples of astral (10_3), (12_3), and (14_3) configurations and comments on their star-like appearance, but reaches no general conclusions or constructions. Similarly, van de Craats [**219**] shows the astral (10_3) configuration and notes various interesting properties associated with it; he also shows an astral (14_3) configuration and mentions that analogous astral (n_3)

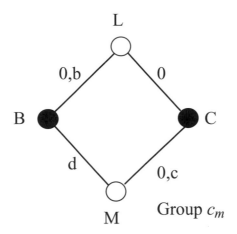

L

B

C

M

Group c_m

Figure 2.7.9. The reduced Levi graph of an astral configuration $m\#(b, c; d)$; the notation is analogous to the one used in Figure 2.7.1.

configurations can be found for all $n = 2m + 2$, where $m \geq 2$. Several other examples can be found in [**22**], as well as in [**117**].

Exercises and Problems 2.7.

1. Derive explicitly the quadratic equation for λ mentioned in the text in the case of $9\#(4, 2; 3)$, and use this to draw this configuration using suitable software.

2. Derive explicitly the quadratic equation for λ in the general case $m\#(b, c; d)$, and try to find criteria on these parameters that will imply that the solutions of the equation are real.

3. Use the geometric construction to draw the configuration $9\#(4, 2; 3)$.

4. Show that the configurations $12\#(5, 1; 2)$ and $12\#(5, 1; 4)$ are congruent. Explain this, and generalize.

5. The configuration $5\#(2, 2; 1)$ has a cyclic automorphism group that acts transitively on its points and lines. Describe this group, and determine whether it acts transitively on the flags of the configuration.

6. The automorphisms group of the astral chiral configurations $5\#(2, 2; 1)$ is transitive on its points. Find other astral chiral configurations with this property. Can you characterize all such configurations?

2.8. Astral 3-configurations with dihedral symmetry group

In contrast to the chiral astral configurations that—in a certain sense—are all formed alike, the dihedral configurations come in several very different varieties.

The first variety consists of configurations that are astral in the extended Euclidean plane but are not contained in the Euclidean plane itself; we shall refer to them as EE configurations. It is clear that such configurations must have one orbit of points at infinity; hence the other orbit of points needs to consist of the vertices of an isogonal polygon. In fact, the polygon must be regular. Indeed, consider any point at infinity and the three lines incident with it, hence mutually parallel. Since there are only two transitivity classes of lines, two of these lines must be in the same orbit; this implies that the third line is situated between these two and is in fact a mirror interchanging the two lines. Therefore the sides of the polygon contained in these lines are congruent; that is, the pairs of points are at equal distance apart. But since each vertex must be on a third line (besides the two determined by the sides of the isogonal polygon), that line must be a mirror as well, and therefore the adjacent sides of the polygon are of equal length. Hence the polygon is regular, and the configuration can be described as follows:

Theorem 2.8.1. *If C is an (n_3) configuration of type EE, hence with dihedral symmetry group, that is, astral in the extended Euclidean plane but not contained in the Euclidean plane itself, then $n = 3m$ for some $m \geq 3$. The points of C are of the vertices of a regular $(2m)$-gon M and the m points at infinity in the directions of the m longest diagonals of M. The lines of C are the ones determined by the m longest diagonals of M, together with the $2m$ lines determined by pairs of points of M at span $m - j$ for some $0 < j < m/2$ with $j \equiv m \pmod 2$. The symmetry group of C is d_{2m}.*

The EE configurations can therefore be characterized by a pair of integers m and j and denoted by $EE(3m; m, j)$, with $0 < 2j < m \geq 3$ and with $j \equiv m(\mathrm{mod}\, 2)$. Several examples of EE configurations are shown in Figure 2.8.1.

Although the symmetry group of an EE configuration is dihedral, the configuration is astral under the cyclic subgroup as well. This is illustrated in Figure 2.8.2, and the corresponding reduced Levi graphs are shown in Figure 2.8.3.

The second variety of dihedral astral 3-configurations may be thought of as **d**ihedrally **d**oubled-up chiral astral configurations, and we shall call them DD configurations. The typical notation is $m\#(b, c; d; \mu)$; it will be explained soon. The DD configurations resemble chiral astral configurations in many respects, but there is one large difference.

First, we give the difference. The construction of chiral astral configurations starts with a set of points at the vertices of a **regular** polygon. In the dihedral case, the $2m$ vertices of any **isogonal** polygon can serve as starting points of a DD configuration $((4m)_3)$. Such vertices fall into two subsets of

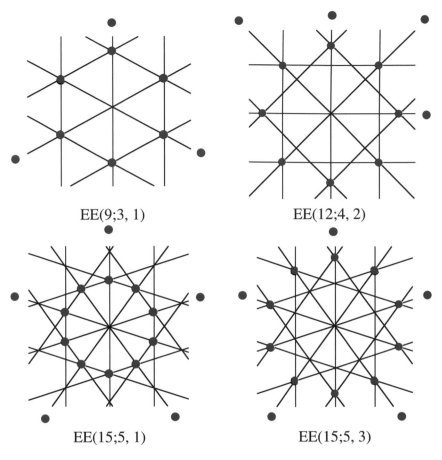

EE(9;3, 1)

EE(12;4, 2)

EE(15;5, 1)

EE(15;5, 3)

Figure 2.8.1. Examples of configurations of type EE. In each case, points at infinity in the directions of the longest diagonals are indicated by the detached dots.

equal size, the m points in each subset being related by rotational symmetries of the whole set. The two subsets of points are images of each other under reflective symmetries of the whole set. The last entry μ in the symbol $m\#(b, c; d; \mu)$ of a dihedral astral (n_3) configuration of type DD refers to the ratio (not exceeding 1) of the angles subtended by the sides of the isogonal $(2m)$-gon used in the construction.

Next, we give the similarities. There are again—naturally, in view of the definition of astrality—two orbits of points and two orbits of lines. Due to the presence of reflections, each orbit of elements has two suborbits, each suborbit consisting of m elements that are equivalent under rotations, without the need for reflections. In the example shown in Figure 2.8.4 and in general, the points in the two suborbits of the first class are denoted by B_j^+ and B_j^- and those in the second class by C_j^+ and C_j^-. If $b \neq c$,

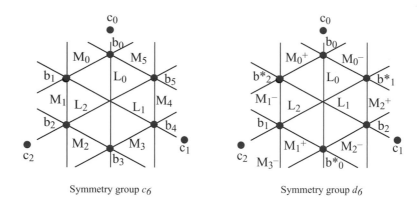

Symmetry group c_6 Symmetry group d_6

Figure 2.8.2. The configuration $EE(9; 3, 1)$ labeled with the symmetry group c_6 (at left) and d_6 (at right). In both cases, the c points and the L lines are mapped onto themselves by halfturns.

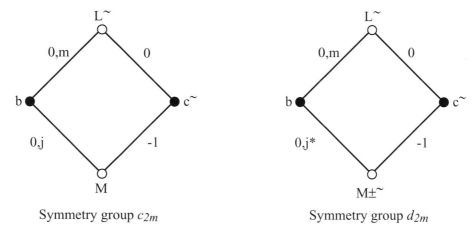

Symmetry group c_{2m} Symmetry group d_{2m}

Figure 2.8.3. The reduced Levi graphs of configurations $EE(3m; m, j)$. All labels are understood mod $2m$.

we shall usually assume $b > c$. The points B_j^+ and B_j^- are on endpoints of diagonals of span b of each of the two m-gons determined by points of the suborbit, while C_j^+ and C_j^- are on diagonals of span c of the other two m-gons. One of the mirrors is the bisector of $B_0^+ B_0^-$; it is indicated in Figure 2.8.4 by the dashed vertical line. The B_j^+ and B_j^- points are obtained by rotation in counterclockwise orientation. The construction of the configuration $m\#(b, c; d; \mu)$ proceeds as follows; it is illustrated in Figure 2.8.4, where $m = 5$, $b = 2$, $c = 1$, $d = 1$, and $\mu = 0.6$.

The points B_0^+ and B_b^+ determine the line L_0^+ and the point C_0^+, which divides the segment $B_0^+ B_b^+$ in a ratio λ; this ratio is fixed throughout the construction but still undetermined. More generally, B_j^+ and B_{j+b}^+ determine

C_j, clearly with the same ratio λ. Then the line M_0^+ in determined by C_0^+ and C_c^+; it passes through B_d^-, and, more generally, C_j^+ and C_{j+c}^+ determine the line M_j^+ that is incident with B_{j+d}^-. This requirement determines the value of λ through a quadratic equation. In turn, the line L_d^- through B_d^- and B_{d-b}^- passes through C_d, and finally C_d^- and C_{d-c}^- are collinear with B_0 on the line M_d^-. As always, the subscripts are understood to be modulo m. These requirements can all be met simultaneously, due to the symmetry of the sets of points involved.

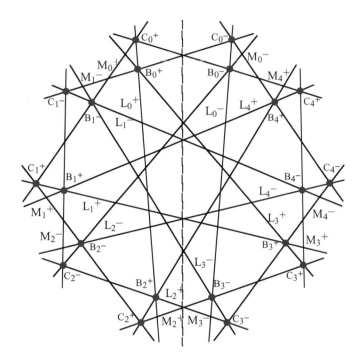

Figure 2.8.4. A dihedral astral configuration (20_3), with symmetry group d_5. The labeling illustrates the description given in the text. The configuration has symbol $5\#(2, 1; 1; 0.6)$.

As in the case of chiral astral configurations, the construction leads to a quadratic equation for λ. Again, the various possibilities and properties encountered with the chiral astral configurations (n_3) are largely present. In particular, depending on the values of the parameters m, b, c, d, μ of the configuration $m\#(b, c; d; \mu)$, there can be two, one, or no real solutions. Moreover, for suitable values of these parameters the resulting construction leads to superfigurations. However, there has been very little done on a systematic investigation of the DD configurations. Several additional examples of such configurations and a case of superfiguration are shown in Figures 2.8.5 and 2.8.6.

There is no information available concerning the range of values of d for given m, b, c and μ or concerning the possible values of λ for given m, b, c, d, μ. Equally missing is any knowledge concerning duality, polarity, selfduality, and selfpolarity of DD configurations.

As with chiral astral configurations, the reduced Levi diagrams for dihedral astral 3-configurations are very simple and straightforward. This is illustrated in Figure 2.8.7, which demonstrates the mutual reinforcing of the notation introduced above, and the graphs.

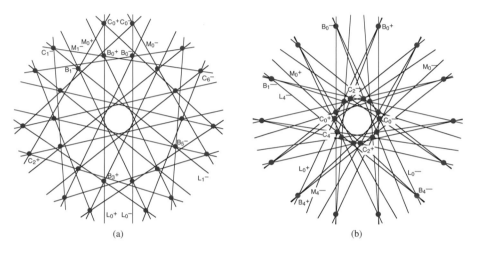

(a)   (b)

Figure 2.8.5. Two dihedral astral configurations (28_3). To reduce clutter, only the labels needed for the determination of the symbol are shown: (a) $7\#(3,2;1;1.0)$ and (b) $7\#(3,2;4;1.0)$.

First examples of the third variety of dihedral astral 3-configurations were discovered only last year and appear in the paper [14] by L. W. Berman and J. Bokowski.[1] We shall designate all configurations of this variety by BB with appropriate parameters attached. Any BB configuration (n_3) has $n = 3m$ for an integer $m \geq 5$. The configuration depends on two other parameters which we call s and t. The meaning of these parameters will be explained as we describe the construction of the configuration $BB(m; s, t)$. We shall illustrate the construction in the case of $BB(5; 2, 2)$, see Figure 2.8.8, but use general terms in the explanation of the steps.

The first step (Figure 2.8.8(a)) is the construction of a regular m-gon P and the selection of the midpoints of its sides; these midpoints are m of the points of the configuration, and the lines L_j determined by the sides of the m-gon are m of the lines. (The vertices of the m-gon play no added role in the construction and are not marked in Figure 2.8.8.)

---

[1] The author had the privilege of receiving a preprint of this paper from the authors.

Figure 2.8.6. For a value of μ close to 0.5, the construction of the configuration $7\#(4, 3; 1; \mu)$ leads to a superfiguration: there are unintended incidences, yielding points on four lines and lines through four points.

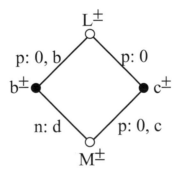

Symmetry group d_{2m}

Figure 2.8.7. The reduced Levi graph of the dihedral astral configuration $m\#(b, c; d; \mu)$. The construction of the graph follows the method given in Section 1.6, based on the labeling of these configurations described above and illustrated in Figures 2.8.4 and 2.8.5. If the inclusion of the parameter in the graph is desirable, it can be attached to the 0 on the edge going from $L^{\pm}$ to $c^{\pm}$.

The second step (Figure 2.8.8(b)) is the selection of a chord of P of span s and the construction of the circumcircle C of the triangle determined by the endpoints of the chord and the center of P. The parameter s needs to be in the range $2 \leq s < m/2$.

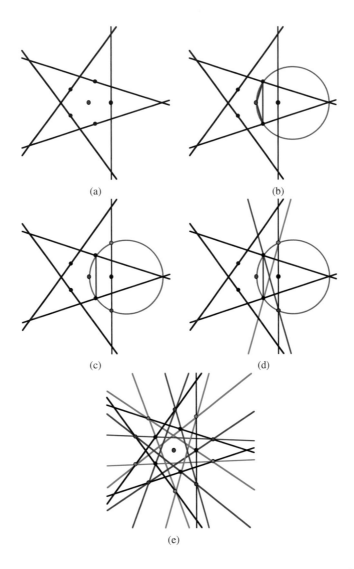

(a)

(b)

(c)

(d)

(e)

Figure 2.8.8. The steps in the construction of the configuration BB(5; 2, 2).

The third step (Figure 2.8.8(c)) consists in determining the intersections of the circle C with the lines L_j constructed in the first step. These intersection points always come in symmetric pairs. In the case $m = 5$ (hence $s = 2$) there is only one such pair; the examples in Figure 2.8.9 show other possibilities. There are always at least $s - 1$ pairs and no more than $2s - 3$. The precise number depends on m and s in a manner that has not been explicitly determined.

In the fourth step (Figure 2.8.8(d)) a selected pair of these intersection points is connected by lines with the endpoints of the chord of span s with which we started in the second step. (To avoid clutter, in Figure 2.8.8(d)

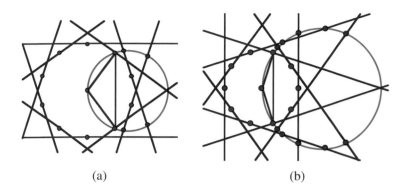

(a)   (b)

Figure 2.8.9. Illustration of the possibilities in the third step of constructing configurations $BB(10; s, t)$. In part (a), $s = 3$ and t is either 2 or 3. In part (b), $s = 4$ and $2 \leq t \leq 6$.

each point of the pair is connected with only one endpoint of the chord.) The parameter t is the label that can be given to the pairs, counting from the endpoints of the chord.

The fifth and final step (Figure 2.8.8(e)) consists in creating the images of the chosen pair of points and the lines generated in the previous step, by rotations about the center of the polygon P through all the multiples of $2\pi/m$.

We give some remarks about the BB configurations. First, just as in the case of the DD configurations (and the chiral astral configurations), in some instances the construction does not yield the expected configuration; instead a superfiguration is obtained. This is illustrated in Figure 2.8.10. Also, the precise relations between the parameters of a BB configuration have not been determined so far. This is illustrated in Table 2.8.1, which shows the (experimentally determined) maximal value of t for given m and s.

The BB configurations considered in [**14**] are presented in a way that is somewhat different from the one followed here. The Berman-Bokowski construction corresponds to the cases of even s only and uses only that pair of intersection points which arises from the intersection of the circumcircle with the line parallel to the chord used to construct the circle. This pair is in general the "middle" pair in the third step of our construction.

Another phenomenon—again shared by other classes of configurations— is the possibility of the configuration being disconnected. This happens, for example, with the configuration $BB(16; 6, 6)$ shown in Figure 2.8.11.

While the construction procedure seems to be working in the examples given above, there is an obvious need for justification in the general case. It is, in fact, quite simple; we explain it for the configuration $BB(m; s, t)$ by

Table 2.8.1. The maximal values of t for given m and s in configurations $BB(m; s, t)$.

m \ s	2	3	4	5	6	7	8	9	10	11	12	13	14	15	16	17	18
5	2																
6	2																
7	2	3															
8	2	3															
9	2	3	6														
10	2	3	6														
11	2	3	6	7													
12	2	3	4*	7													
13	2	3	4	7	10												
14	2	3	4	7	8												
15	2	3	4	7	8	11											
16	2	3	4	7	8	11											
17	2	3	4	7	8	11	12										
18	2	3	4	7	8	9	12										
19	2	3	4	7	8	9	12	15									
20	2	3	4	7	8	9	12	13									
21	2	3	4	7	8	9	12	13	16								
22	2	3	4	7	8	9	12	13	16								
23	2	3	4	7	8	9	12	13	16	19							
24	2	3	4	7	8	9	10*	13	14	17							
25	2	3	4	7	8	9	10	13	14	17	20						
26	2	3	4	7	8	9	10	13	14	17	20						
27	2	3	4	7	8	9	10	13	14	17	18	21					
28	2	3	4	7	8	9	10	13	14	17	18	21					
29	2	3	4	7	8	9	10	13	14	15	18	21	22				
30	2	3	4	7	8	9	10	13	14	15	18	19	22				
31	2	3	4	7	8	9	10	13	14	15	18	19	22	25			
32	2	3	4	7	8	9	10	13	14	15	18	19	22	25			
33	2	3	4	7	8	9	10	13	14	15	18	19	22	23	28		
34	2	3	4	7	8	9	10	13	14	15	18	19	22	23	26		
35	2	3	4	7	8	9	10	13	14	15	18	19	22	23	26	29	
36	2	3	4	7	8	9	10	13	14	15	16*	19	20	23	26	29	
37	2	3	4	7	8	9	10	13	14	15	16	19	20	23	24	27	32

(*) One of the lines is tangent to the circle at its intersection with another line.

using the notation in the illustrative example shown in Figure 2.8.12, where $m = 9, s = 3$, and $t = 3$.

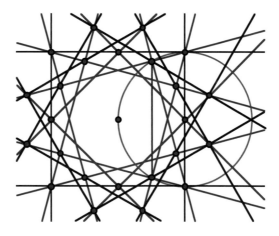

Figure 2.8.10. An example of a superfiguration arising in the construction of a BB type configuration with $m = 12$ and $s = 4$.

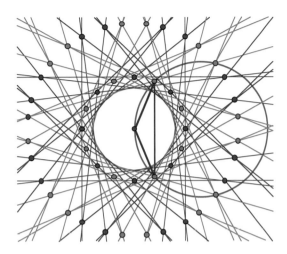

Figure 2.8.11. In the case of BB(16; 6, 6) the construction leads to a disconnected configuration. The two connected components are shown in different colors; each is BB(8; 3, 3).

The chord of span s (used to generate the circumcircle K) spans an angle of $2\pi s/m$ at the center O of K. The line CB^*, the legs of the isosceles triangle generated by the chord and O, and the segment OC are all well determined. Rotating this complex and the circle K about O through an angle of $2\pi s/m$ brings K to K^*, CB^* to C^*B, and OC to OC^*. The five angles denoted by γ are all equal to each other because they are either basis angles of isosceles triangle or they are spanned by congruent arcs of K. Hence the basis CC^* of the isosceles triangle COC^* encloses with the segment OC^* the same angle γ as the line through C^*B; hence that line passes through C, which justifies the construction.

The BB configurations can be described very succinctly by their reduced Levi graphs. This is illustrated in Figure 2.8.13.

It should be noted that the procedure used to justify the construction dealt exclusively with the green lines. This leaves open the possibility of using a different value of t for the red lines. Naturally, the resulting configuration will not be astral.

Another point that needs to be made is the following. For each s, the set of values of possible t is a (non-strictly) decreasing function of m. The experimental results in Table 2.8.1 are a consequence of reasonably complicated trigonometric relations. The main problem in this context is to determine the maximal value of t possible in a BB($m; s, t$) configuration. From numerical evidence (see Table 2.8.2) it seems that this $t_{\max}$ grows approximately as $7s/5$ for sufficiently large m, although this appears a strange dependence.

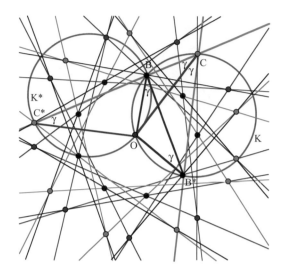

Figure 2.8.12. The validation of the construction of the BB configurations.

Table 2.8.2. The largest value $t_{\max}$ of t possible in configurations $BB(m; s, t)$ for a given s and for all sufficiently large m.

s	$t_{\max}$	for $m \geq$
2	2	5
3	3	7
4	4	12
5	7	11
6	8	14
7	9	18
8	10	24
9	11	42
10	14	24
11	15	29
12	16	36
13	17	50
14	18	92
15	21	41
16	22	48
17	23	61
18	24	84
19	25	78
20	28	60

s	$t_{\max}$	for $m \geq$
21	29	72
22	30	90
23	31	127
24	32	372
25	35	84
26	36	99
27	37	125
28	38	183
29	41	95
30	42	110
31	43	131
32	44	167
33	45	256
34	48	121
35	49	139
36	50	167
37	51	217
38	52	335
39	55	149
40	56	172

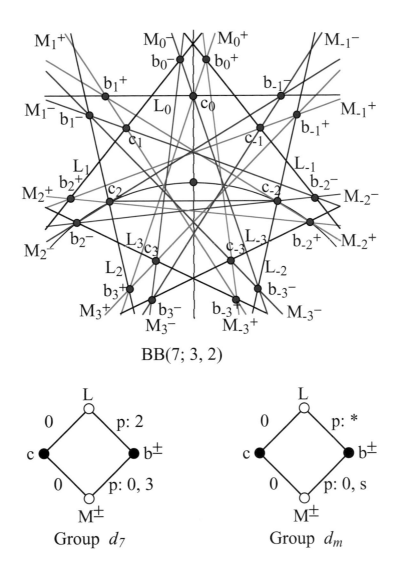

Figure 2.8.13. The labeling of BB configurations and the resulting reduced Levi graphs. The graph at right corresponds to the general configuration $BB(m; s, t)$; the asterisk indicates that no definite relation to the parameters has been found so far.

Exercises and Problems 2.8.

1. Determine what symbol could result for the configuration in Figure 2.8.4 if the roles of B_0^+ and B_0^- were reversed, while still assuming conterclockwise orientation.

2. Determine what symbol could result for the configuration in Figure 2.8.4 if the roles of the B-points and the C-points were reversed, while still assuming conterclockwise orientation.

3. Verify the assignment of symbols to the configurations in Figure 2.8.5.

4. Formulate a general criterion for the configuration $BB(m; s, t)$ to be disconnected.

5. Draw all the different configurations $BB(11; 5, t)$.

6. How many different configurations $4\#(b, c; d; 0.3)$ are there?

7. Find some restrictions on the parameters of the DD and BB kinds of astral configurations.

8. Find disconnected configurations $m\#(b, c; d; \mu)$.

9. Find a geometric construction for configurations $m\#(b, c; d; \mu)$.

10. Find the symbols for the configurations in Figure 2.8.14.

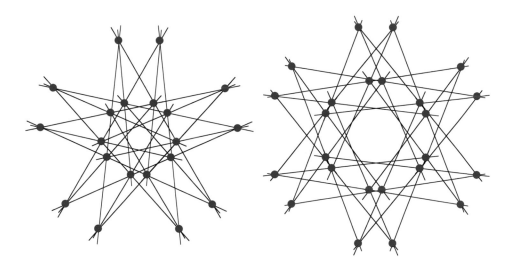

Figure 2.8.14. Two dihedral astral 3-configurations.

2.9. Multiastral 3-configurations

A geometric configuration is said to be of **symmetry type** $[h_1, h_2]$ provided its points form h_1 orbits and its lines form h_2 orbits under the group of its isometric symmetries. We shall also say that such a configuration is $[h_1, h_2]$-**astral**, or, if the precise values of h_1 and h_2 are not important in the discussion, that it is **multiastral**. Clearly, if a configuration of type $[q, k]$ is $[h_1, h_2]$-astral, then $h_1 \geq (k+1)/2$ and $h_2 \geq (q+1)/2$. If h_1 and h_2 have these minimal values, we shall simplify the language and say that the configuration is **astral**. In cases where $h_1 = h_2 = h$, we shall say that the configuration is h-**astral**.

The study of these configurations is much less advanced and promises to be more challenging than the investigation of the 2-astral 3-configurations.

There are two sources of the variety possible for h-astral 3-configurations. On the one hand, similarly to the situation with dihedral astral configurations, in many cases there is at least one parameter that can assume a continuum of different real values. On the other hand, if $h \geq 3$, a line of the configuration can contain points from either two or three different orbits. The case $h = 2$ is radically different from those with $h \geq 3$.

The h-astral 3-configurations come in three varieties:

☐ **projectively h-astral**, that is, configurations that are h-astral in the *extended Euclidean* (that is, *projective*) plane E^{2+} but not in the Euclidean plane E^2 itself,

☐ **h-chiral**, that is, configurations in the Euclidean plane E^2, with a cyclic symmetry group,

☐ **h-dihedral**, that is, configurations in the Euclidean plane E^2, with a dihedral symmetry group.

Throughout, the use of a numerical prefix h- means that there are *at most h* orbits of points and *at most h* orbits of lines, with *equality in at least one case*.

Examples of **projectively astral** configurations are shown in Figures 2.9.1 and 2.9.2. The configuration in Figure 2.9.1 is a realization of the Pappus configuration. Two 3-astral realization of the Desargues configuration (10_3) in E^{2+} are shown in Figure 2.9.2; they are among the illustrations given by Coxeter [**48**]. Two examples of projectively 3-astral configurations (15_3) are shown in Figure 2.9.3. It is clear that similar examples of projectively h-astral configurations could be found for all $h \geq 4$. At least for small h, the complete characterization of projectively astral configurations may be feasible but has not been worked out.

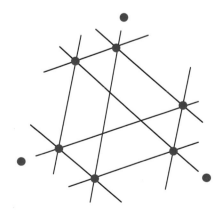

Figure 2.9.1. A 3-astral version of the Pappus configuration (9_3) in the extended Euclidean plane.

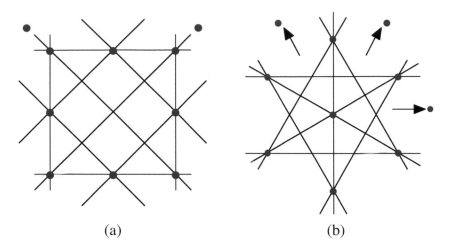

(a)   (b)

Figure 2.9.2. Two projectively 3-astral realizations of the Desargues configuration (10_3) (after [**48**]). In (b) the line at infinity is included.

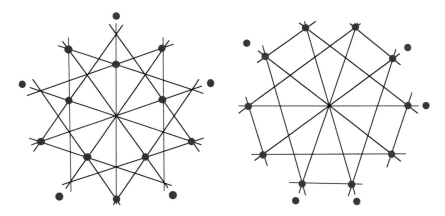

Figure 2.9.3. Two examples of projectively 3-astral configurations (15_3). Notice that the one on the left is $[3, 2]$-astral and the one on the right is $[2, 3]$-astral.

The **h-chiral** astral configurations (n_3) are much more interesting. We have discussed the 2-astral chiral configurations in Section 2.7. For $h \geq 3$ there is more than one possibility. We begin by explaining the notation for h-chiral configurations in which each line is incident with points of two orbits only; the remaining case—some lines incident with three orbits of points—will be described later. The notation used in Section 2.7 will be expanded here; the general form for h-chiral configurations (n_3) of this kind is $m\#(b_1, b_2, \ldots, b_h; b_0; \lambda_1, \lambda_2, \ldots, \lambda_{h-2})$—or $m\#(b_1, b_2, \ldots, b_h; b_0)$ for short. Here $n = hm$, and we have $h-2$ real parameters λ_j besides $h+1$ discrete parameters b_j. Together these parameters lead to a quadratic equation for an additional parameter λ. This equation can have 2, 1, or 0 real solutions—in the last case there are no corresponding real configurations. Our explanation is illustrated in Figure 2.9.4, using a 3-chiral configuration (27_3) as an example.

The detailed study of h-chiral configurations was initiated by Boben and Pisanski [24] under the name "polycyclic configurations" and with slightly different notation. As pointed out in [24], the dual of a configuration $m\#(b_1, b_2, \ldots, b_h; b_0)$ is the configuration $m\#(b_h, b_{h-1}, \ldots, b_1; b_1 + b_2 + \cdots + b_h - b_0)$. For $h = 2$ this reduces to the facts we shall discuss at length in Section 2.10.

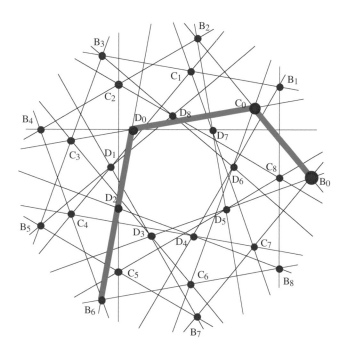

Figure 2.9.4. The characteristic path in a 3-chiral configuration (27_3).

As mentioned earlier, the symbol for an h-chiral configuration (n_3), where $n = hm$, is of the form $m\#(b_1, b_2, \ldots, b_h; b_0; \lambda_1, \lambda_2, \ldots, \lambda_{h-2})$; the parameters are again determined along a **characteristic path**. The entries $b_1, \ldots, b_h$ are the spans of the diagonals in the different regular m-gons that are determined by the path; all the diagonals are oriented in the same way— all clockwise or all counterclockwise—and the real numbers $\lambda_1, \lambda_2, \ldots, \lambda_{h-2}$ denote the ratios in which each diagonal determined by a segment of the path is divided by the endpoint of the segment. The path returns to the starting polygon, but not necessarily to the starting point of the path. The parameter b_0 indicates the vertex of the starting polygon at which the characteristic path ends. These data lead to a quadratic equation for the ratio

λ on the next to last segment; the ratio applicable to the last segment is then completely determined. Thus there are either two or one or no real geometric configurations corresponding to a given symbol. There are also possibilities of unintended incidences similar to the ones we encountered earlier; hence we are in general talking about **representations** of the symbols, rather than **realizations**. In case the parameters $\lambda_1, \lambda_2, \ldots, \lambda_{h-2}$ in a symbol $m\#(b_1, b_2, \ldots, b_h; b_0; \lambda_1, \lambda_2, \ldots, \lambda_{h-2})$ are not relevant or not known, we abbreviate the symbol as $m\#(b_1, b_2, \ldots, b_h; b_0)$.

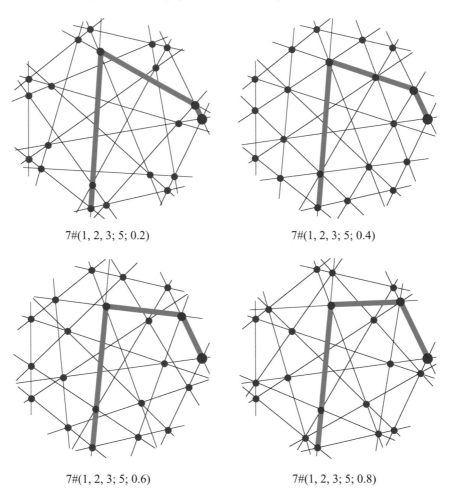

7#(1, 2, 3; 5; 0.2)   7#(1, 2, 3; 5; 0.4)

7#(1, 2, 3; 5; 0.6)   7#(1, 2, 3; 5; 0.8)

Figure 2.9.5. An illustration of the dependence of a 3-chiral configuration (21_3) on the parameter λ_1.

The example in Figure 2.9.4 presents a 3-chiral configuration with symbol $9\#(2, 3, 2; 6; 0.5)$. The points of the three orbits are denoted by $B_j, C_j,$ D_j. The determination of the symbol is highlighted by the three-step characteristic path. Note that the ratio λ_1 can be chosen freely, and in the illustration it was taken as $\lambda_1 = 0.5 = C_0B_0/B_2B_0$. Once the first $h - 2$ ratios λ_j are chosen, the last ratio λ_{h-1} (determining the position of the point of the last orbit on the penultimate diagonal) is determined by a quadratic equation. (For details see [**24**].) In the illustration we have $h = 3$; hence $\lambda_{h-1} = \lambda_2$ (which is about 2/3). Naturally, the symbol is not unique since it depends, besides the λ_j's for $h \geq 3$, on the orbit of the starting point and on the orientation chosen. The influence of the parameter λ_{h-2} is illustrated in Figure 2.9.5.

Using symbols like $u, v, w, \ldots$ for elements of the different orbits of points, we can say that the h-chiral configurations considered so far have lines of type $\{u, u, v\}, \{v, v, w\}, \ldots$. But other possibilities exist in which the incidences of lines with orbits of the points are different. For example, in case $h = 3$, it is possible to have three orbits of lines, all three of the type $\{u, v, w\}$ or else one of the type $\{u, v, w\}$ and the other two of types $\{u, v, v\}$ and $\{u, w, w\}$. Three example of the former variety are shown in Figure 2.9.6, while examples of the second kind are illustrated in Figure 2.9.7; the diagrams in Figure 2.2.1 show the $(9_3)_2$ and $(9_3)_3$ configurations, which are of these two kinds. Notation for the configurations in Figure 2.9.6 is explained in the caption. An apparently convenient notation is proposed for the kind of configurations shown in Figure 2.9.7. It assigns the first symbol to the line and the point that are incident with three orbits of the other kind, and it handles the other symbols in an obvious manner. For the notation one conveniently chooses the one that involves the smallest maximal parameter. No additional details about either of these kinds of configurations are available as of this writing.

Naturally, for $h \geq 4$ it is possible to imagine an increasingly large number of types of h-chiral configurations. However, so far nothing has been done in this direction.

The **h-dihedral** configurations are unexplored as well. A few examples are shown in Figures 2.9.8, 2.9.9, and 2.9.10. The examples in Figure 2.9.8 are obviously typical of an infinite class of analogous constructions. In Figure 2.9.9 only two orbits of points are shown; the points of the third orbit can be chosen at several distinct locations. This kind of configuration can obviously be generalized in a variety of ways. Figure 2.9.10 illustrates the degree of complication possible with h-astral configurations for larger h.

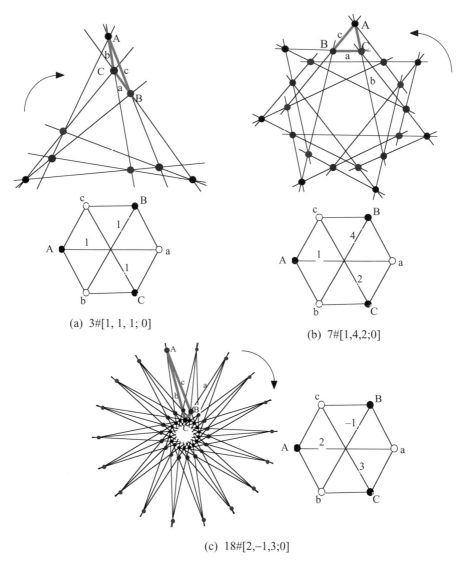

(a) 3#[1, 1, 1; 0]

(b) 7#[1,4,2;0]

(c) 18#[2,−1,3;0]

Figure 2.9.6. Three examples of 3-chiral configurations in which every line meets all three orbits of points and every point meets lines of the three orbits. The characteristic path (which does not have to be closed) leads to a symbol for the configuration. The configuration in (a) is another realization of the Pappus configuration. With each we show a reduced Levi diagram.

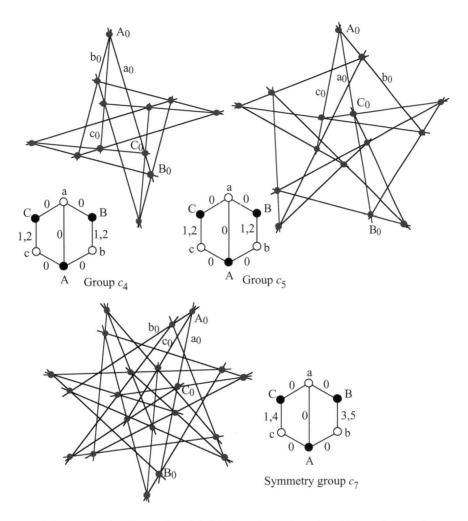

Figure 2.9.7. Examples of 3-chiral configurations in which each line of one orbit is incident with points of each of three orbits, while the other lines are incident with two points from one orbit and one point from another orbit. Each is accompanied by a reduced Levi diagram. As symbols for these configurations we can use $4\#[1,2;1,2]$, $5\#[1,2;1,2]$ and $7\#[1,4;3,5]$. The configuration $5\#[1,2;1,2]$, appears in van de Craats [**219**].

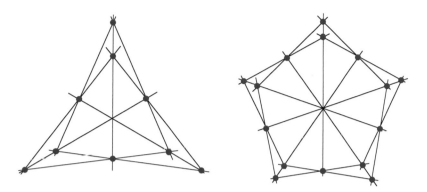

Figure 2.9.8. Two examples of 3-dihedral 3-configurations. The one at left is another realization of the Pappus configuration.

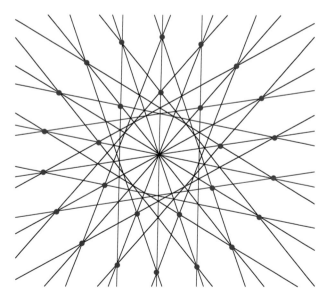

Figure 2.9.9. Adding an orbit of points at suitable intersections leads to several different 3-dihedral configurations. Notice that such configurations are, in fact, [3, 2]-dihedral.

Figure 2.9.10. A $[4, 5]$-dihedral configuration (40_3) found by L. Berman.

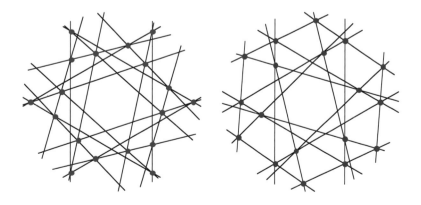

Figure 2.9.11. Two (18_3) 3-chiral configurations.

Exercises and Problems 2.9.

1. Decide whether the two configurations in Figure 2.9.3 are dual— or even polar—to each other. Is either of them isomorphic to the (15_3) configuration in Figure 1.1.1?

2. Determine the number of distinct ways in which it is possible to replace the points in Figure 2.9.9 in such a way that the result is a $[3, 2]$-dihedral configuration.

3. By moving the outer vertices in Figure 2.9.8 along the mirror, a continuum of (projectively) distinct configurations can be obtained; all these are isomorphic. Are there any analogous configurations (15_3) that are not isomorphic to the one in Figure 2.9.8?

4. For $h_1 = 2$ and $h_1 = 3$, determine the possible values of h_2 for which there exist $[h_1, h_2]$-astral configurations of the various kinds. Provide examples for all existing types.

5. Construct examples of 4-chiral configurations (n_3) with the smallest n. Justify your answer. Generalize.

6. Determine the symbols of the two configurations is Figure 2.9.11.

7. Verify that the Cremona-Richmond configuration (15_3) shown in Figure 1.1.1 is of the type represented by the examples in Figure 2.9.7. Find its symbol.

8. Find the criteria for the property of the first two configurations of Figure 2.9.7 (but not the third) such that there are only two orbits of points (and two orbits of lines) under automorphisms.

2.10. Duality of astral 3-configurations

In this section we shall investigate the duality and polarity properties of the chiral astral configurations (n_3). It should be kept in mind that the presentation is based on the assumption that we know all such configurations, although, in fact, we are certain only to the extent that the topic has been explored by numerical calculations. As we have seen in Section 2.7, there corresponds to a symbol $m\#(b, c; d)$ either two or one or no chiral astral configurations (n_3), where $n = 2m$. In the case of two configurations, by their very construction they are isomorphic. But more is true:

Theorem 2.10.1. *Every chiral astral configuration $m\#(b, c; d)$ is selfdual.*

Proof. From the definition given above of the labels of points and lines of such configurations, illustrated in Figure 2.10.1 (which is a copy of Figure 2.7.1), we see that the line L_j contains the points B_j, C_j, B_{j+b} and the line M_j contains the points B_{j+d}, C_j, C_{j+c}. The resulting incidences can then be described by the following criteria:

$B_j \in L_k \Leftrightarrow j - k \equiv 0$ or $b \pmod{m}$,

$B_j \in M_k \Leftrightarrow j - k \equiv d \pmod{m}$,

$C_j \in L_k \Leftrightarrow j - k \equiv 0 \pmod{m}$,

$C_j \in M_k \Leftrightarrow j - k \equiv 0$ or $c \pmod{m}$.

From these relations it follows at once that for every configuration $m\#(b, c; d)$ the mapping δ determined by $\delta(B_j) = L_{-j}$, $\delta(C_j) = M_{-j-d}$, $\delta(L_j) = B_{-j}$, and $\delta(M_j) = C_{-j-d}$ is a selfduality. $\qquad\square$

Another consequence is the following:

If two distinct configurations have the same symbol $m\#(b, c; d)$, then they are dual to each other.

This follows from the fact that they are isomorphic. But even more is true:

Theorem 2.10.2. *If two distinct configurations have the same symbol* $m\#(b, c; d)$, *then they are polars of each other.*

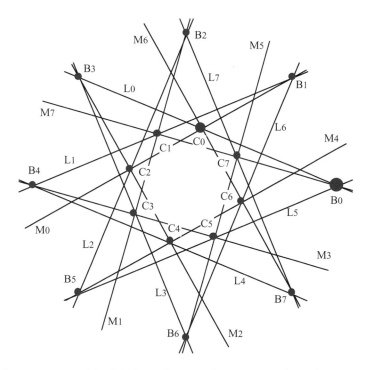

Figure 2.10.1. The labeling of the configuration $8\#(3, 2; 1)$ explained in the text.

Proof. Indeed, polars are combinatorially dual to each other, and the only combinatorially dual astral configuration of an astral configuration

$m\#(b, c; d)$ is either the configuration itself or the other one with the same symbol. Since there are two configurations $m\#(b, c; d)$, neither is polar to itself, but each is polar to the other.   □

This fact is illustrated in Figure 2.10.2.

It is almost selfevident that in general there are other duality maps from a configuration to its dual. For example, Figure 2.10.3 presents the same pair of configuration as Figure 2.10.2(a), with a labeling that shows that the map ε from the rcd configuration to the black one is a duality different from the duality δ described above.

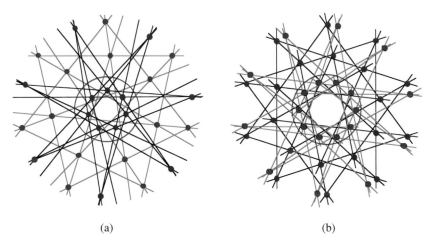

(a)   (b)

Figure 2.10.2. (a) The configuration $7\#(3, 2; 1)'$ (red points and green lines) and its polar $7\#(3, 2, 1)''$ (blue points and black lines). Polarity is with respect to the purple circle. (b) The same for $10\#(4, 3; 2)'$ and $10\#(4, 3; 2)''$.

In the case where only a single configuration $m\#(b, c; d)$ exists (that is, if $b + c = 2d$ or if $b = c$), the configuration is not only selfdual but also selfpolar. The map δ is applicable to all selfdual configurations and is concordant with selfpolarity. The polars (in an appropriate circle) are *congruent* to each other, but only after a reflection in a suitable mirror.

For configurations of this type, the map δ and its rotates are the *only* maps compatible with the polarity. We say that these configurations are **oppositely selfpolar**. This happens for the selfpolar configurations with symbol $m\#(b, b; d)$. Examples are shown in Figure 2.10.4.

Other configurations, called **directly selfpolar configurations**, have symbols of type $m\#(b, c; d)$ with $2d = b + c$. Here the polar pairs are congruent without reflection. There are two subtypes: In the first, both b and c are even, and in the second they are both odd. In the former case the polars actually coincide with each other, while in the latter they are related

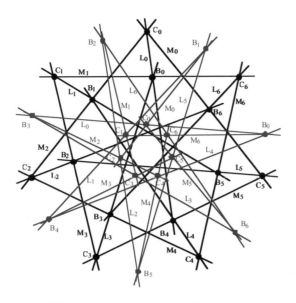

Figure 2.10.3. The dual configurations of Figure 2.10.2(a) illustrate a duality map ε.

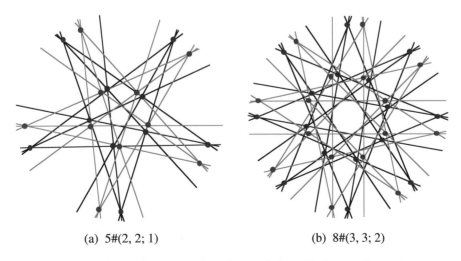

(a) 5#(2, 2; 1)   (b) 8#(3, 3; 2)

Figure 2.10.4. Two examples of oppositely selfpolar configurations, characterized by symbols of the type $m\#(b, b; d)$.

by reflection in the common center (that is, rotation through $180°$). The two subtypes are illustrated in Figures 2.10.5 and 2.10.6.

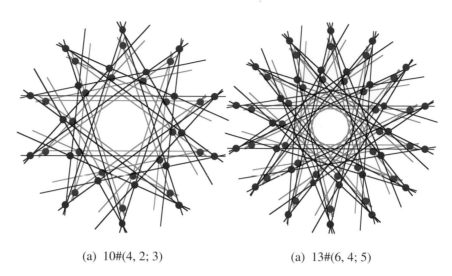

(a) 10#(4, 2; 3)   (a) 13#(6, 4; 5)

Figure 2.10.5. Two examples of directly selfpolar configurations $m\#(b, c; d)$ with b and c even. In this subtype the polars may coincide (for an appropriate circle). In the illustration the circle was chosen to yield different sizes, in order to improve intelligibility.

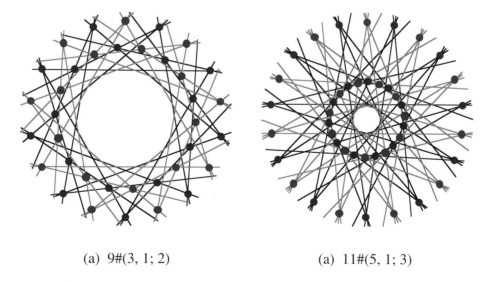

(a) 9#(3, 1; 2)   (a) 11#(5, 1; 3)

Figure 2.10.6. Two examples of directly selfpolar configurations $m\#(b, c; d)$ with b and c odd. In this subtype the polars are congruent but coincide only after reflection in the common center (that is, a rotation of $180°$). We also say that these configurations are **selfpolar***.

Exercises and Problems 2.10.

1. Verify that the correspondence δ is a duality. Determine whether this correspondence establishes a selfduality.

2. Describe the duality introduced by the polarity for the polar configuration in Figure 2.10.2(b); use the labels on the two configurations that are given by their isomorphism.

3. Label the selfpolar configurations in Figures 2.10.4 and 2.10.5 to show that they are selfdual.

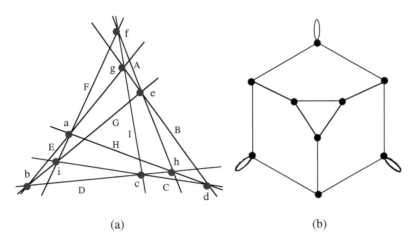

(a)   (b)

Figure 2.10.7. (a) A version of the Pappus configuration (9_3), with a selfduality indicated by uppercase and lowercase letters. (b) An RLG (see Exercise 6) corresponding to the selfduality in (a).

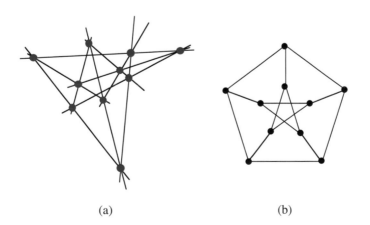

(a)   (b)

Figure 2.10.8. (a) A version of the Desargues configuration (10_3). (b) An RLG of (a).

4. Verify that the Cremona-Richmond configuration (15_3), shown in Figure 1.1.1 and mentioned in Exercise 7 of Section 2.9 is selfdual. Is it selfpolar, and if it is, what is its type?

5. Find criteria for dual pairs of configurations of the various kinds discussed in Sections 2.8 and 2.9.

6. R. Artzy [2] considers selfdual configurations and for a given selfduality δ describes an RLG ("reduced Levi graph"—this is not the same concept we are using throughout the book!) by identifying each element B with its image $\delta(B)$. This clearly depends on the selfduality chosen, but in each case the original Levi graph can be retrieved in a unique way. As observed by Artzy, the RLG may contain loops; this occurs in the case where B and $\delta(B)$ are incident. Artzy illustrates the use of RLGs by investigating special cases of the Desargues configuration. (On this topic see also Killgrove et al. [**138**].) Assign labels to the RLG in Figure 2.10.7(b) to show that it corresponds to the Pappus configuration in Figure 2.10.7(a), with the selfduality δ indicated by the uppercase and lowercase letters.

7. Find a selfduality δ of the Desargues configuration in Figure 2.10.8(a) that leads to the RLG in Figure 2.10.8(b).

8. Is there a meaningful extension to all polar pairs of astral 3-configurations of the distinction between directly and oppositely selfpolar configurations?

9. Describe the polars of the configurations $\mathrm{BB}(m; s, t)$, and determine whether there are any selfpolar configurations among them.

10. (Refresh your memories of elementary geometry.) Given a pair of astral configuration for which it is claimed that they are polar to each other with respect to a circle, how do you find the circle that justifies the assertion? Practice your solution on the selfpolar configurations in Figures 2.10.4, 2.10.5, and 2.10.6.

2.11. Open problems (and a few exercises)

Many unsolved problems and open questions have been mentioned in the preceding sections. While some of these may be challenging and others may hold interest for some people, there are a few problems concerning 3-configurations that seem to be of a fundamental nature; these problems exhibit the paucity of our understanding of what makes geometric configurations work. Some of the problems are related to Steinitz's geometric theorem of Section 2.6.

1. The first problem concerns geometric realizations of connected combinatorial configurations. By Theorem 2.6.1 we know that a (geometric) prefiguration representation is always possible if one incidence is disregarded.

As shown by the examples of the (7_3) and (8_3) configurations, even allowing pseudolines, it is not possible to achieve the last incidence. However, it is well possible that all connected (n_3) configurations with $n \geq 9$ admit realizations as topological *con*figurations or even (for $n \geq 11$) realizations as geometric *pre*figurations. On the other hand, it may well be that for $n = 13$ some counterexamples can already be found for either version of the question. A subsidiary question is to determine the maximal number $t(n)$ of "lines" in a topological configuration (n_3) that may need to be non-straight pseudolines in each *realization* of the configuration in question. It seems that $t(n) \geq cn$ for some $c > 0$.

2. The second problem deals with obstructions to geometric realization of 2-connected 3-configurations with $n \geq 11$ lines. All known examples that include unwanted incidences (superfigurations) contain either a Pappus or a Desargues subfiguration (one incidence of the configuration is missed), or several such subfigurations. Are there any other obstructions to the geometric realizability, or is the presence of at least one of these two a characterization of 3-configurations with unwanted incidences?

3. The third problem, simply stated, is this: Is the combinatorial configuration $(10_3)_4$ using the notation in Section 2.2, the **only** 3-connected configuration (n_3) with $n \geq 9$ that does not have a geometric realization? A negative answer may appear at any time—if somebody hits upon an appropriate example—possibly even with $n = 13$. On the other hand, a positive solution would seem to require several breakthroughs in directions for which we are not even dimly aware of how to start. These would have to include the elimination of *super*figurations (unwanted incidences) as well as *sub*figurations (missing incidences, as in Steinitz's theorem). As a possible example of a negative solution consider the abstract configuration (14_3) derived from the geometric configuration in Figure 2.11.1 on replacing the existing incidences of points A and B with the lines a and b and insisting instead that A be incident with a, and B with b.

4. Is it true that if a 3-configuration admits a geometric realization in the Euclidean plane, then it admits a realization in the rational plane? Or is it (at least) true that every geometrically realizable 3-configuration can also be realized in a plane over a quadratic extension of the rational field? In contrast, it is easy to verify that the #2-superfiguration shown in Figure 1.3.4 is realizable in the Euclidean plane but not in the rational plane.

5. For the various classes of very symmetric 3-configurations (such as astral, 3-chiral, k-dihedral, BB, ...) determine the precise range of the parameters for such configurations.

6. For connected astral configurations $m\#(b, c; d)$, is $m = 12$ the only case in which various superfigurations occur?

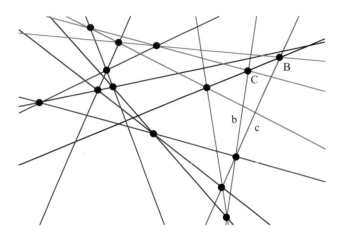

Figure 2.11.1. Is there a geometric realization of the combinatorial configuration (14_3) obtained from the above by keeping all indicated incidences except that C is to be incident with c (and not with b) and B is to be incident with b?

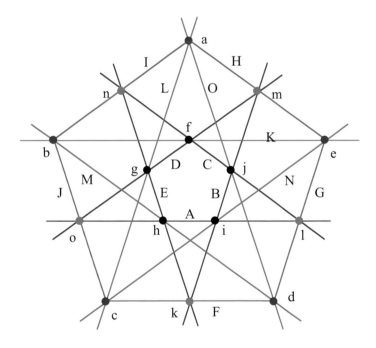

Figure 2.11.2. An intriguing selfdual collection of 15 points and 15 lines.

7. Is there any relation between the automorphism group of a configuration and the symmetries of its possible realizations? In particular, if the automorphisms act transitively on the points (or lines or flags), does there have to exist a realization with non-trivial symmetry?

8. The object in Figure 2.11.2 is not a configuration, but the labeling clearly indicates that it is selfdual; the same can be said for the superfiguration in Figure 1.3.4. These seem to be interesting objects, analogous to configurations in the sense used in this book—but without any systematic framework to support their investigation. A formal proposal to consider such "generalized configurations" was made in [**240**] by K. Zindler as long ago as 1889. An example described by Zindler (as well as in the review [**204**] of [**240**] by H. Schubert) is shown in Figure 2.11.3. However, it seems that Zindler's general challenge has never been met.

9. Decide whether the selfduality of the superfiguration in Figure 2.11.2 is a selfpolarity.

10. Prove that the incidences claimed in Figure 2.11.3 are valid.

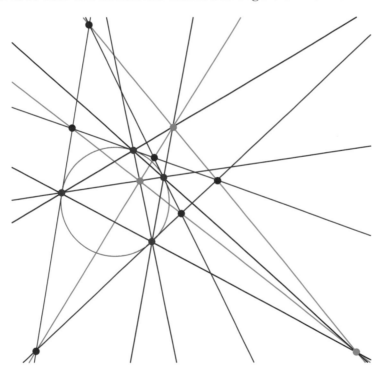

Figure 2.11.3. A "generalized configuration" of 13 points and 13 lines from Zindler [**240**]. It consists of four concyclic points (red) that determine a complete quadrangle (six blue lines) and its three "diagonal points" (green). The four tangents (red) to the circle at the four concyclic points are a complete quadrilateral that determines the six blue points and the three "diagonal lines" (green). The selfpolar "configuration" has six points incident with three lines each and seven points incident with four lines each and dually for lines.

Chapter 3

4-Configurations

3.0. Overview

As Section 3.1 explains, the first publications dealing with 4-configurations appeared before the end of the nineteenth century, but not much development happened till relatively recently. In Section 3.2 we present the general results concerning the existence of topological and geometric 4-configurations. The difference between the present case and that of 3-configurations is quite striking—there still are gaps in the numbers n for which we know that an (n_4) geometric configuration exists.

The various methods of construction of reasonably sized (n_4) geometric configurations—all less than twenty years old—are detailed in Section 3.3. These constructions are then applied in Section 3.4 to determine the values of n for which it is known that a geometric 4-configuration (n_4) exists. Although the development of new methods has made the construction of visually understandable 4-configurations possible in many cases, for some of the small numbers there still are only unattractive diagrams or no known configurations at all.

Section 3.5 sets up the framework for the study of the k-astral 4-configurations; these are the configurations with a very high degree of geometric symmetry.

Based on that, in Section 3.6 we present one of the few complete results about 4-configurations—the complete enumeration of the 2-astral 4-configurations. These are configurations in which there are only two orbits of points and two orbits of lines under Euclidean symmetries of the configuration. This topic is related to (and depends upon) the investigation of the intersection points of diagonals in regular polygons, in itself a subject with a classical flavor but with surprising twists in its unfolding.

Section 3.7 is devoted to 3-astral configurations. The presence of three orbits of points and three orbits of lines results in a family of configurations with properties very different from the ones considered in Section 3.6.

Section 3.8 is concerned with the k-astral 4-configurations with $k \geq 4$. There is again a sea change in properties compared to 2-astral and 3-astral configurations, as well as in our knowledge of the possibilities.

A few problems not mentioned in the earlier sections are presented in Section 3.9.

3.1. Combinatorial 4-configurations

The history of configurations (n_4) is much shorter than that of configurations (n_3) and more easily told.

The first explicit mention of such configurations seems to be in a paper [**139**, p. 440] by Felix Klein in 1879, which deals with quartic curves in the complex plane. He noted that there is a family of 21 points and 21 lines with incidences that make it into a (21_4) configuration, in our terminology—albeit in the complex plane. Although this particular configuration continued to interest mathematicians (various references can be found in Coxeter [**50**], and Burnside [**36**] discovered it independently), it did not have any noticeable direct influence on the study of (n_4) configurations in general. However, later it did play a significant role in the theory of geometric configurations, which we shall discuss in Section 3.2.

The first slightly more general treatment of such configurations was by Georges Brunel (1856–1900) in [**35**], a paper that seems to have escaped the attention of all writers on the topic of configurations (n_4) prior to [**117**][1]. In an earlier paper [**34**] Brunel followed an idea quite popular at that time: a polygon inscribed and circumscribed to itself (with sides understood as lines). Clearly, these are a special class of (combinatorial or geometric) 3-configurations, which we will discuss in Section 5.2. Aware of the need to distinguish between combinatorial and geometric configurations, in [**35**] Brunel pursued this idea further, by considering a "polygon *doubly* inscribed and circumscribed" to itself. In the current terminology we call such polygons "Hamiltonian circuits (or multilaterals)" of the configuration, and we will consider them in more detail in Sections 5.2 to 5.4. Each line of such a doubly selfinscribed and selfcircumscribed "polygon" is incident, besides the two points (vertices of the polygon) that define it as a side of the polygon,

---

[1]Biographical data on Brunel and comments on his work may be found in [**5**] and, in great detail, in [**64**]; see also [**100**].

with precisely two additional vertices of the polygon. Brunel determines that any combinatorial configuration (n_4) must satisfy $n \geq 13$ and gives two constructions.

In the first construction, Brunel presents a configuration table (which is actually an orderly configuration table, in the terminology of Section 2.5) and states that while the verification that this indeed determines a combinatorial configuration (35_4) is easy, the graphical representation requires some effort. (Unfortunately, the remarks in [**64**, p. LXVIII] concerning the geometric realization of this configurations are, at best, misleading.) From Brunel's statement (especially in view of his later comments concerning the other construction) one may conclude that he had found a geometric realization of this configuration. In fact, this configuration turns out to be isomorphic to the geometric configuration (35_4) mentioned in [**122**], communicated to the authors by Ludwig Danzer. (See also [**120**].) Although no reasonable diagram of this configuration seems to be available, the configuration can be described easily enough by a construction of the kind used by Cayley and others in similar contexts a century and a half ago. In the case under discussion, start with seven points in general position in real 4-space; consider the 35 2-planes and 35 3-spaces they generate, and intersect this family by a 2-dimensional plane in general position to obtains the required geometric configuration (35_4). The absence of any reasonable geometric symmetry makes this configuration visually unattractive.

Brunel's second construction yields combinatorial configurations (n_4) on which a cyclic group operates transitively. This includes explicitly specified configurations for $13 \leq n \leq 16$. Unfortunately, the results Brunel presents are marred possibly by typos but also by outright errors. Among the latter, in several cases Brunel lists isomorphic doubly selfinscribed and selfcircumscribed polygons as distinct. For example, in the case $n = 13$ Brunel lists cyclic translates of $\{0, 1, 4, 6\}$ and $\{0, 1, 3, 9\}$ as the two polygons, although the permutation $(0)(1)(2)(3, 4)(5)(6, 9, 8, 10, 12, 7)(11)$ maps the first polygon onto the second. Moreover, it is rather easy to prove that up to isomorphism, there can be only one such combinatorial configuration; this is completely analogous to the proof (in Section 2.2) that the configuration (7_3) is unique. But even allowing for these shortcomings, we see that Brunel anticipated the corresponding results of Merlin [**159**] and even went a bit beyond them. A corrected list would show one cyclic configuration (or polygon) for $n = 13$ and 14, three for $n = 15$, and two for $n = 16$. This coincides with the recent list of cyclic configurations given by Betten and Betten [**16**], to which we shall return soon. Brunel also noted that translates of $\{0, 1, 4, 6\}$ yield a configuration (n_4) for all $n \geq 13$; this anticipated by nearly a century a result of Gropp [**79**].

Merlin mentions in [**159**] that configurations (n_4) have not been investigated systematically, although some isolated ones were discovered by F. Klein [**139**], W. Burnside [**36**], and others. Like Brunel, he constructs a combinatorial configuration (13_4); moreover, he proves its uniqueness and minimality. He also constructs a configuration (14_4) and proves it is unique. Merlin states that there are exactly **three** distinct configurations (15_4), which, however, are not presented. In fact, he is mistaken. As shown by Betten and Betten [**16**], there are **four** different configurations (15_4), three of which are cyclic and coincide with the three doubly selfinscribed and self-circumscribed polygons of Brunel (who did not comment on the possibility of noncyclic configurations (15_4) or (n_4) in general). In the same context, Merlin makes two additional errors:

(i) He claims that his three configurations (15_4) can be distinguished by the number of vertex-disjoint triangles present in them, which he claims to be 5, 1, and 0, respectively. In fact, all four configurations (15_4) have five such triangles, the maximal possible number.

(ii) He states that his configurations (13_4), (14_4), and (15_4) have orderly configuration tables; this is correct—see Section 2.5—and has been proved by Steinitz in [**210**] for all configurations (n_k). However, Merlin then claims that it follows that there is no Hamiltonian circuit for any of them—which is wrong. Steinitz's orderliness result has no such implications, and cyclic 4-configurations such as Brunel's explicit constructions in [**35**] (of which Merlin was unaware) provide counterexamples to Merlin's claim.

By a construction analogous to the one devised by Martinetti (in [**152**]; see Section 2.3) for configurations (n_3), Merlin shows that for every $n \geq 30$ there are combinatorial configurations (n_4). In fact, it is easy to show that there are such configurations for all $n \geq 13$; for example, as noted by Brunel and mentioned above, for all $n \geq 13$ it is enough to consider cyclic translates of the "line" $\{0, 1, 4, 6\}$.

Concerning the number $N(n)$ of distinct combinatorial configurations (n_4), the only known values are those given by Betten and Betten [**16**]: the old $N(13) = N(14) = 1$ and their new results $N(15) = 4$, $N(16) = 19$, $N(17) = 1972$, and $N(18) = 971171$. These new numbers seem not to have been independently verified, except for the value $N(17) = 1972$ (see [**33**]).

The configurations (13_4) and (14_4) can be obtained as cyclic configurations with generating "line" $\{0, 1, 4, 6\}$. The four configurations (15_4) can be characterized as follows: The three cyclic configurations are generated by the "lines" $\{0, 1, 4, 6\}$, $\{0, 1, 5, 7\}$, and $\{0, 1, 3, 7\}$, given already by Brunel. The other three configurations given by Brunel yield isomorphic configurations (two to the first and one to the second). Betten and Betten [**16**] give other generators for the three cyclic configurations: $\{0, 2, 8, 12\}$, $\{0, 1, 9, 11\}$, and

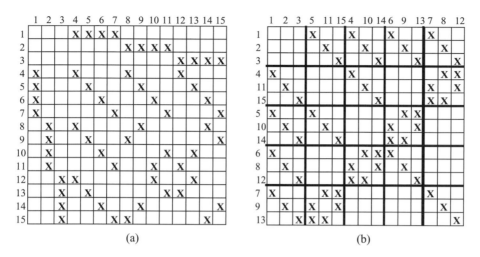

Figure 3.1.1. (a) A Levi incidence matrix of the noncyclic (15_4) configuration constructed by Betten and Betten [**16**]. (b) A selfdual incidence matrix of this configuration.

$\{0, 1, 9, 13\}$, respectively; these are shown in [**16**] by Levi incidence matrices (see Section 1.4)—but matrices that do not exhibit the *cyclic* character of the configurations. Their fourth configuration (n_4) is clearly illustrated in [**16**] by a Levi incidence matrix shown in Figure 3.1.1(a). As it is the only noncyclic configuration (15_4), it is necessarily selfdual. An incidence matrix exhibiting one of the selfdualities is shown in Figure 3.1.1(b); it is obtained by suitable permutations of the rows and columns of the matrix in Figure 3.1.1(a).

Brunel's generating "lines" of the three cyclic configurations (15_4) given in [**35**] have an advantage over the ones given by Betten and Betten [**16**], even though they are isomorphic for the (15_4) configurations: Brunel's can serve as generating lines for combinatorial configurations (n_4) for **all** $n \geq 15$.

Concerning the (16_4) combinatorial configurations, it should be noted that the three generating lines of the cyclic (15_4) configurations listed above do serve to generate cyclic (16_4) configurations—but the three resulting configurations are isomorphic. There is one other configuration (16_4), also cyclic, specified in Betten and Betten [**16**] by its generating line $\{0, 1, 6, 13\}$; Brunel renders the same configuration, but with a typo; when corrected, its generating line is $\{0, 1, 3, 12\}$, or equivalently, $\{0, 1, 3, -4\}$. The generating lines $\{0, 1, 6, 13\}$ or $\{0, 1, 3, 12\}$ do not yield a cyclic configuration for all $n > 16$; however, if the generating line is taken in the form $\{0, 1, 6, -3\}$ or $\{0, 1, 5, -2\}$, which are equivalent for (16_4), then they works for all such n. Obviously, any generating line for a cyclic configuration is also a generating line for all sufficiently large n.

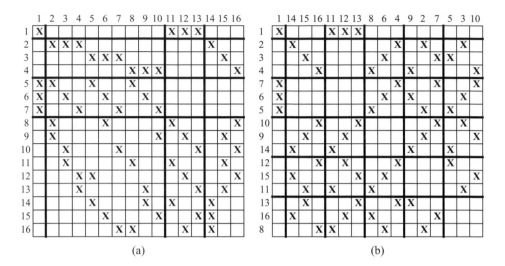

Figure 3.1.2. (a) A (16_4) combinatorial configuration as illustrated in [**16**] by its Levi incidence matrix. (b) A symmetric incidence matrix of the same configuration, illustrating its selfduality.

Besides the two cyclic configurations, Betten and Betten [**16**] describe 17 noncyclic combinatorial configurations (16_4); they state that these 19 are the complete list but give no details of the determination of this claim. There seems to have been no independent confirmation of this list. As with all the listings in [**16**], it seems that no attention was given to finding presentations of the configurations as symmetric as possible; in particular, there is no mention of duality or selfduality. Beyond the cyclic configurations already mentioned and the (15_4) configuration in Figure 3.1.1(a), this is illustrated by one of the seventeen (16_4) configurations illustrated in Figure 8 of [**16**]. This example is shown in Figure 3.1.2(a).

Betten and Betten [**16**] state (or at least imply) that there are only two cyclic configurations (17_4); their generating lines given are equivalent to the ones mentioned above, $\{0, 1, 4, 6\}$ and $\{0, 1, 5, -2\}$.

As we shall see in Section 3.2, except for one of the (17_4) configurations, none of the *combinatorial* configurations (n_4) with $n \leq 17$ is even *topologically* realizable (see Section 3.2). Merlin [**159**] shows that the configurations (13_4), (14_4), and the three cyclic (15_4) configurations are not geometrically realizable. But he also notes that *geometric* configurations (n_4) do exist for infinitely many values of n. His construction uses "stacks" of 3-configurations and vertical lines through their vertices to construct $[4, 3]$-configurations and then stacks of duals of the projections of these into the plane to construct 4-configurations. While this yields geometric configurations (n_4) for infinitely many values of n, there are infinitely many n that are not covered.

Much new information on the question of existence of topological and geometric 4-configurations has become available recently. We discuss it in the following sections.

Exercises and Problems 3.1.

1. Decide whether the (35_4) configuration of Brunel is cyclic or not.

2. Prove that the three cyclic configurations (15_4) generated by the "lines" $\{0, 1, 4, 6\}$, $\{0, 1, 5, 7\}$, and $\{0, 1, 3, 7\}$, given by Brunel, are distinct (non-isomorphic).

3. Prove that the three cyclic configurations (15_4) generated by the "lines" $\{0, 1, 4, 6\}$, $\{0, 1, 5, 7\}$, and $\{0, 1, 3, 7\}$ are isomorphic to the three generated by $\{0, 2, 8, 12\}$, $\{0, 1, 9, 11\}$, and $\{0, 1, 9, 13\}$, respectively.

4. Investigate the duality properties of the three cyclic configurations (15_4).

5. Validate the claim that the three generating lines in Exercise 2 yield isomorphic configurations (16_4).

6. Show that the cyclic (16_4) configurations with starting lines $\{0, 1, 4, 6\}$ and $\{0, 1, 6, 13\}$ are not isomorphic.

3.2. Existence of topological and geometric 4-configurations

As mentioned in Section 3.1, both Brunel [**35**] in 1897 and Merlin [**159**] in 1913 discussed *geometric* 4-configurations in the real Euclidean plane and were clear about the distinction between combinatorial and geometric configurations. However, neither did actually show a drawing of any geometric configuration.

The first published diagram of a geometric configuration (n_4) appeared only in [**122**], published in 1990. It is reproduced here as Figure 3.2.1. As it happens, it is a realization of Klein's configuration (21_4), introduced in [**139**] and mentioned in Section 3.1. The paper [**122**] marked the beginning of the research of *geometric* configurations (n_4); the results of these investigations form the topic of the remaining part of Chapter 3. The results are intimately connected to the study of topological configurations (n_4), and we shall first describe the known facts concerning these configurations.

The arguments given by Merlin [**159**] to establish the non-existence of geometric configurations (n_4) for $n \leq 15$ do not carry over to topological configurations. However, we have

Theorem 3.2.1 (Bokowski and Schewe [**27**]). *For $n \leq 16$ there are no topological configurations (n_4).*

This is the best possible, since we also have

Theorem 3.2.2 (Bokowski, Grünbaum, and Schewe [**26**]). *Topological configurations* (n_4) *exist for every* $n \geq 17$.

In contrast to this situation, we have

Theorem 3.2.3 (Bokowski and Schewe [**28**]). *For* $n \leq 17$ *there are no geometric configurations* (n_4).

Figure 3.2.1. A geometric configuration (21_4).

Theorem 3.2.4 (Bokowski and Schewe [**28**]). *There exist geometric configurations* (n_4) *for all* $n \geq 18$ *except possibly for* $n = 19, 22, 23, 26, 37, 43$.

Theorems 3.2.3 and 3.2.4 demonstrate how the understanding of the (n_4) configurations has developed during the past twenty years. In [**122**] it was conjectured that there are no geometric configurations (n_4) with $n \leq 21$ other than the configuration in Figure 3.2.1. Similar conjectures were repeated in various other publications, such as [**112**], [**113**], [**114**]. However, the recent discovery (see [**118**]) of a (20_4) configuration led to a modified conjecture that geometric configurations (n_4) exist only for $n \geq 20$. But this was also short-lived and was resolved in the negative by the discovery of a geometric (18_4) configuration by J. Bokowski and L. Schewe [**28**]. Thus Theorems 3.2.3 and 3.2.4 settle the 20-year quest for the smallest geometric configuration (n_4).

The history of Theorem 3.2.4 illustrates the rapid improvement in the understanding of configurations (n_4). The first version, in [**118**], established that connected (n_4) configurations exist for every $n \geq 21$ except possibly if $n = 32$ or $n = p$ or $n = 2p$ or $n = p^2$ or $n = 2p^2$ or $n = p_1 p_2$, where p, p_1, p_2 are odd primes and $p_1 < p_2 < 2p_1$. The number of exceptional cases was soon reduced (in [**112**]) to a finite number: There are (n_4) configurations for all $n \geq 21$ except possibly if n has one of the following thirty two values: 22, 23, 25, 26, 29, 31, 32, 34, 37, 38, 41, 43, 46, 47, 49, 53, 58, 59, 61, 62, 67, 71, 77, 79, 89, 97, 98, 103, 113, 131, 178, 179. Newly found construction methods [**114**] reduced the list of possible exceptions to the following ten values: $22, 23, 26, 29, 31, 32, 34, 37, 38, 43$. All this was while the general belief was that 21 is the smallest number of points in an (n_4) configuration. After Theorem 3.2.3 was established and additional constructions found, the result became that connected (n_4) configurations exist if and only if $n \geq 18$, except possibly if n has one of the eight values $18, 19, 22, 23, 26, 34, 37, 43$. Finally, the discovery of an (18_4) configuration and a construction of a (34_4) configuration led to the result stated above [**28**].

The proofs of Theorems 3.2.3 and 3.2.4 will be given in the next two sections; here we shall give outlines and some details of the proofs of Theorems 3.2.1 and 3.2.2.

The proof of Theorem 3.2.1 given in [**27**] is easy for $n \leq 15$. The case $n = 16$ is much more complicated and forms the bulk (six pages) of that paper. It follows a large number of a priori possible topological subconfigurations and in each case leads to a contradiction. We have to refer the reader to the original paper. In contrast, the case $n \leq 15$ is easily explained and for fixed k is applicable to all combinatorial configurations (n_k) with n sufficiently small. We present the proof from [**27**] with only minor adaptations.

Assume that a combinatorial (n_k) configuration is realized by pseudolines in the projective plane. Due to the possibility of locally perturbing pseudolines at points that are not vertices of the configuration, we may assume that in the *arrangement* generated by the perturbed pseudolines each vertex of the arrangement is incident with either k or 2 pseudolines. Since each of the former accounts for $k(k-1)/2$ pairwise intersections of pseudolines, the total number of vertices of the modified arrangement is $f_0 = n + n(n-1)/2 - nk(k-1)/2 = n(n - k^2 + k + 1)/2$. Similarly, the number of edges of the modified arrangement is $f_1 = n(n - k^2 + 2k - 1)$. From Euler's theorem for the projective plane it follows that the number of cells (faces) of the arrangement is $f_2 = f_1 - f_0 + 1 = n(n - k^2 + 3k - 5)/2$. On the other hand, arrangements of pseudolines have no digons; hence counting incidences of edges and cells yields $3f_2 \leq 2f_1$. Therefore we have

Table 3.2.1. A configuration table of the only (17_4) configuration that admits a topological realization.

1	1	1	1	2	2	2	3	3	3	4	4	4	8	9	10	10
2	5	8	11	5	6	7	5	6	7	5	6	7	13	13	11	12
3	6	9	12	8	9	11	12	8	9	11	10	12	15	14	14	16
4	7	10	13	14	15	16	15	16	17	17	13	14	17	16	15	17

$f(n) = -n^2 + nk^2 + nk - 5n + 6 \leq 0$ as a necessary condition for the existence of a topological realization of a combinatorial (n_k). For fixed k, this function $f(n)$ of n has its only maximum for $n = (k^2 + k - 5)/2$ and is decreasing for all larger n. Simple checking shows that for $k = 4$ we have $(4^2 + 4 - 5)/2 < 8$ and $f(15) = 6 > 0$; hence (n_4) is not topologically realizable for $n \leq 15$. Since $f(16) = -10 < 0$, this criterion is not applicable for $n = 16$. On the other hand, this result shows that there are no topologically realizable configurations (n_5) for $n \leq 24$, nor are there any topological (n_6) for $n \leq 36$.

Turning now to Theorem 3.2.2, the first thing to observe is that geometric configurations are, obviously, examples of topological configurations. Hence, assuming that Theorem 3.2.4 can be proved without reliance on Theorem 3.2.2 (as is in fact the case), we need only provide examples of topological configurations for those values of $n \geq 17$ for which there are no known geometric configurations. These values are $n = 17, 19, 22, 23, 26, 37, 43$. We shall now show such examples, together with a few others that we find appropriate for various reasons. Most of these examples are modified from [26].

In Figure 3.2.2 we show a topological configuration (17_4) that is a realization of the configuration given by Table 3.2.1. This is taken from [26], where a proof is outlined according to which this combinatorial configuration (17_4) is the only one admitting a topological realization. It should be noted that this realization has 4-fold rotational symmetry in the extended Euclidean plane. It is not known whether there are realizations with any symmetry in the Euclidean plane proper or whether there are additional combinatorial automorphisms. Since the configuration is the only topologically realizable (17_4) configuration, it is necessarily self-dual. (Although it seems not to be well known, topological configurations in the projective plane do have dual configurations. This can be inferred from results in [76].)

A topological configuration (18_4) is shown in Figure 3.2.3. This configuration is not isomorphic to the geometric configuration (18_4) we shall see in the next section, and it is not known whether it can be realized geometrically. On the other hand, it has a six-fold rotational symmetry.

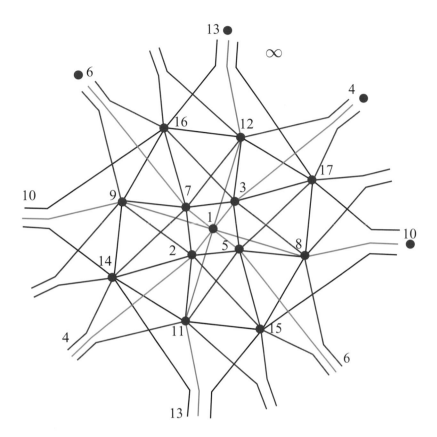

Figure 3.2.2. A topological configuration (17_4). It is a realization of the unique combinatorial configuration (17_4), specified in Table 3.2.1, that has a topological realization.

In contrast, all known topological configurations (19_4) and (23_4) have only trivial symmetry groups. Examples of these configurations are shown in Figure 3.2.4.

The examples of topological configurations presented so far have been ad hoc, obtained essentially through (lots of) trial and error. Their rather ungainly appearance is a reminder of their genesis. In contrast, the examples of configurations (22_4) and (26_4) shown in Figure 3.2.5 are members of a systematic family: They are *topological* examples of *astral* configurations; the *geometric* members of the family will be studied in detail in several sections, starting with Section 3.5. The two examples in Figure 3.2.5 are representatives of configurations (n_4) possible for all even $n \geq 22$. In the terms of astral configurations we shall discuss in Sections 3.5 and 3.6, these configurations have *spans* 4 and 5; other possibilities exist, increasing in number with increasing n. Additional information will be given in the discussion of geometric astral configurations and in Section 5.8.

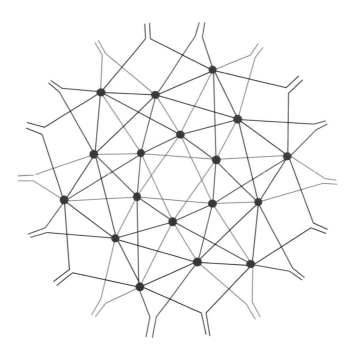

Figure 3.2.3. An example of a topological configuration (18_4) with six-fold rotational symmetry in the Euclidean plane. Adapted from [**26**].

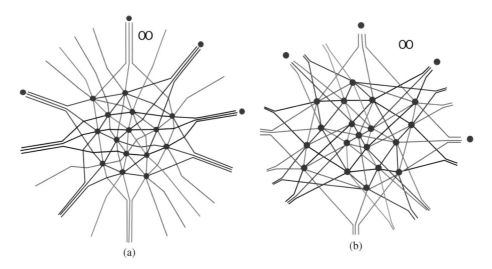

(a)   (b)

Figure 3.2.4. Topological configurations (a) (19_4) and (b) (23_4). Adapted from [**26**].

The examples we provide for (37_4) and (43_4) are special cases of a much more general construction that is actually very simple. Assuming we have a (p_k) topological configuration and a (q_k) topological configuration for some $k \geq 2$, we can construct an (n_k) topological configuration, where

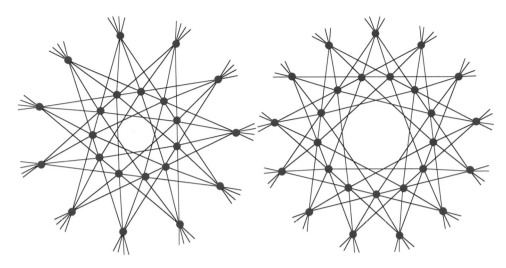

Figure 3.2.5. Topological configurations (22_4) and (26_4), with 11-fold, resp. 13-fold, dihedral symmetry. They are typical of topological astral configurations (n_4) possible for all even $n \geq 22$.

$n = p + q - 1$, in the following way: Delete one pseudoline from the former configuration and a point from the latter, and make the k pseudolines that are now incident with only $k - 1$ points pass through the k points that are incident with only $k - 1$ pseudolines. The case of (24_4) and (20_4) geometric configurations is shown in Figure 3.2.6, leading to a (43_4) topological configuration; the significant points and lines are shown in red. Another (43_4) configuration could be obtained by pairing in the same way an (18_4) configuration with a (26_4) configuration. The same kind of construction with (20_4) and (18_4) configurations (either geometric or topological) yields the last of the required topological configurations, (37_4); alternatively, the topological (17_4) configuration could be paired with the geometric (21_4) configuration. A different topological configuration (37_4) is shown in [**26**]. We shall revisit the same idea for the construction of geometric configurations in the next section.

Since the combinatorial (n_4) configurations for $n = 13, 14, 15, 16$ as well as a large majority of such configurations for $n \geq 17$ cannot be realized by topological configurations, two kinds of questions arise naturally.

First, what relaxation of incidence requirements would be sufficient to enable the construction of topological near-configurations realizing these combinatorial configurations?

Second, what are the obstructions preventing topological realizations of some of the combinatorial configurations? For the smallest combinatorial configurations (such as $(7_3), (13_4), (21_5), (31_6), (49_8), \dots$) the existence of *ordinary* points in any family of pseudolines not all of which pass through

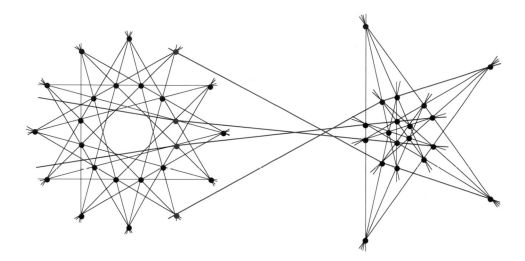

Figure 3.2.6. A topological configuration (43_4). The red points are collinear on the deleted line, and the red (pseudo)lines were concurrent at a deleted point of the (20_4) configuration.

the same point (see Lemma 2.1.1) can be interpreted as such an obstruction. Indeed, it implies that these configurations cannot have topological realizations since all intersections of pairs of pseudolines would have to be "used up" in points incident with multiple pseudolines, leading to an absence of ordinary points.

The inequality $-n^2 + nk^2 + nk - 5n + 6 \leq 0$ mentioned above as a necessary condition for the existence of an (n_k) topological configuration is another kind of obstruction. It shows that combinatorial configurations (n_k) with $n \leq k^2 + k - 5$ cannot be topologically realized. Since $n \geq k^2 - k + 1$ in all cases, that shows that for each k certain values of n lead to topologically non-realizable configurations (n_k). However, it must be noted that for quite a few of the relevant pairs n, k there exist no combinatorial configurations either—and there is no necessary and sufficient criterion for their existence.

Exercises and Problems 3.2.

1. Construct the configuration table dual to Table 3.2.1, and show that it is realized by the configuration in Figure 3.2.2.

2. Prove that the topological (18_4) configuration in Figure 3.2.3 is not isomorphic to the geometric (18_4) configuration shown in Figures 3.3.4 and 3.3.5.

3. Find a topological (18_4) configuration that is dual to the one in Figure 3.2.3.

4. Construct a topological (37_4) configuration.

5. There seems to be no a priori reason that would preclude the existence of topological (19_4) or (23_4) configurations with halfturn symmetry. Do any exist?

6. Determine how many topological configurations (26_4) with dihedral symmetry d_{13} exist.

7. Which multiastral combinatorial configurations (n_4) have topological realizations?

3.3. Constructions of geometric 4-configurations

The fact that the first graphic realization of *any* (n_4) configuration (see Figure 3.2.1) is less than twenty years old attests to the difficulties that have to be overcome in realizations of such configurations in any intelligible manner. One reason for this situation is that an (n_4) geometric configuration implies the (non-trivial) satisfaction of $2n$ collinearity conditions, while on the other hand, any finite set of n points (not all collinear) has an affine image that depends on $2n - 6$ parameters. Hence there must be some dependences— obvious or hidden—between the collinearity conditions in every geometric configuration (n_4). For a relevant discussion of this topic see Michelucci and Schreck [**169**].

In contrast to the situation concerning (n_3) configurations we have presented in Section 2.4, there is no reasonable method or algorithm to go from a combinatorial configuration (n_4) to a topological or geometric configuration— even if any of these do exist—nor are any criteria known to distinguish topological configurations which admit geometric realizations from those that do not. Hence, if we wish to find geometric 4-configurations, we are, by necessity, forced to resort to more or less ad hoc arguments. This does not preclude constructing by the same method large (even infinite) families of examples; however, *finding* such methods or isolated examples is more of an art than a deductive science.

In this section we shall describe several kinds of such constructions. The various families or constructions will be designated in the form (sm), where s is a suitable integer (or another short symbol); the reason for such a name is that for appropriate values of m, the construction leads to a configuration (n_4) with $n = sm$ (or some other value that depends on m).

Following this preamble, let us turn to some concrete cases. In most instances, the construction starts from some given configuration and yields a 4-configuration.

The first construction, which we call $(\mathbf{5m})$, starts with an arbitrary (m_3) configuration C; in the example in Figure 3.3.1 this is the (9_3) configuration shown with blue points and lines. We select in the plane a line L (black line

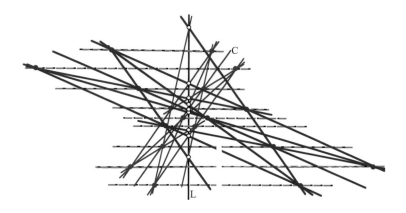

Figure 3.3.1. An illustration of the $(5m)$ construction.

in Figure 3.3.1) which misses all the points of C and is neither parallel nor perpendicular to any line determined by any two points of C. We construct three additional copies of C by stretching C through three different ratios in the direction perpendicular to L; only one such copy is shown (red points and lines) in Figure 3.3.1 in order to avoid crowding. The resulting configuration C^* consists of the four replicas of C, together with the m intersection points of C with L (shown as hollow dots, which are also intersection points with L of the copies of C), and of the m lines perpendicular to L (shown dashed green) which pass through the points of C (and the other copies). Hence this construction yields a configuration C^* of type (n_4), with $n = 5m$. Since—by Theorem 2.1.3—(m_3) configurations are well known to exist if and only if $m \geq 9$, this establishes the existence of configurations (n_4) for all $n \geq 45$ which are divisible by 5. Very important for the sequel is the observation that, as follows from the construction, each such configuration C^* contains a set of m parallel lines. Moreover, this construction yields "movable" configurations in the sense explained in Section 5.7.

It should be noted that this construction—as well as the ones discussed below—leads in some cases to unwanted incidences, that is, to *prefigurations*. However, this can in all cases be avoided by selecting appropriate parameters for the construction.

Our second construction is called **$(5/2m)$**. It starts with a $(2m_3)$ configuration C that has a line L of mirror symmetry with the following properties: No point of C lies on the mirror L, no point on L belongs to more than two lines of C, and no line of C is perpendicular to L. It follows from the mirror property of L that there are m points of L at which pairs of lines of C meet. From C another copy is obtained by shrinking C towards L by a certain factor f (say $f = 1/2$) and then adding the m intersection points of the lines of

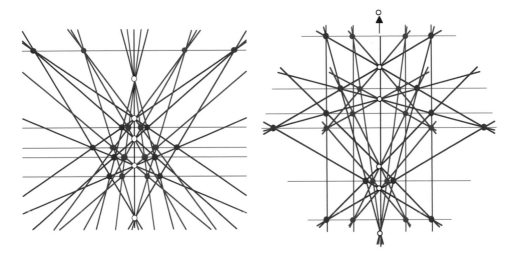

Figure 3.3.2. A (25_4) configuration with five parallel lines and a (30_4) configuration with six parallel lines. The one on the left starts with a (10_3) configuration, the other one with a dihedral astral (12_3) configuration (blue points and lines); copies of these are obtained by shrinking in ratio $f = 1/2$ towards the vertical line of symmetry (black line). Adding the five or six intersection points on the line of symmetry (hollow points, in the right one at infinity) and five or six horizontal lines (green) completes these typical $(5/2m)$ constructions.

the two copies with L and the m lines perpendicular to L that pass through the points of the two configurations. This is illustrated for (10_4) and (12_4) in Figure 3.3.2, yielding configurations (25_4) and (30_4), respectively. We note that this construction also yields configurations $(5m_4)$ with m parallel lines. Moreover, this construction is **movable**; that is, non-trivial parts of it can be changed in a continuous manner without changing other non-trivial parts. (As already mentioned, we shall discuss movable configurations in Section 5.7.) This implies that the cross-ratio of the four points on each of the new (horizontal) lines (which is the same for all m of these lines) can be made equal to any predetermined value by an appropriate choice of f. An example of a (14_3) configuration to which the $(5/2m)$ construction is applicable is shown in Figure 3.3.3.

A construction of the only known (18_4) configuration was discovered very recently by J. Bokowski and L. Schewe; it is illustrated in Figure 3.3.4, and two different realizations of the same configuration are shown in Figure 3.3.5. This configuration can be considered the smallest member of an infinite family; we shall call this the **(6m) construction** or **family**. The idea of looking for such a family came from noticing that the original rendering of the configuration (in Figure 3.3.4) contains a well-known subconfiguration

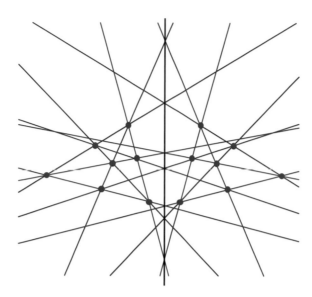

Figure 3.3.3. A (14_3) configuration that can be used to construct a configuration (35_4) by the method in Figure 3.3.2; this (35_4) configuration will have seven parallel lines.

(9_3), which we encountered in Figure 1.1.6; see Figure 3.3.6. This observation led to the construction of a whole family of analogous configurations. The $(6m)$ construction is explained by the example of the typical case illustrated in Figure 3.3.7. The precise membership in the $(6m)$ family has not been determined so far, but the family includes members (n_4) for every $n = 6m$ with odd $m \geq 3$. An additional example is shown in Figure 3.3.8.

The next case to consider is (20_4), first described in [**118**], shown in Figure 3.3.9. It too was discovered as a single configuration, and the family to which it belongs was found only later; for obvious reasons we call this the **(4m) family** or **construction**. At the time of its discovery the construction seemed quite strange; particularly surprising is the use of two chiral configurations of the same handedness in order to obtain a mirror symmetric configuration. By now we have a much better understanding of the process, although a general proof of the validity of the construction is still not available.

Extensive experimental evidence led to the general understanding explained below. It leads to the conclusion that geometric configurations (n_4) exist for all $n = 4m$, with $m \geq 5$.

The construction can be described as follows; the explanation is illustrated in Figures 3.3.10 and 3.3.11. We start (see parts (a) in these illustrations) with an astral configuration $m\#(b, c; d)$, which we denote C, where $b \geq c > d > 0$ in the notation detailed in Section 2.6. We call this the

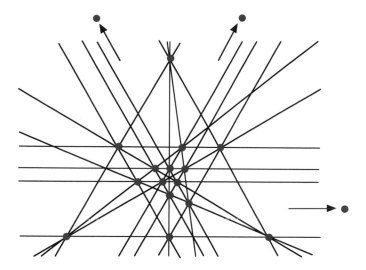

Figure 3.3.4. The only known geometric configuration (18_4) (after [**28**]).

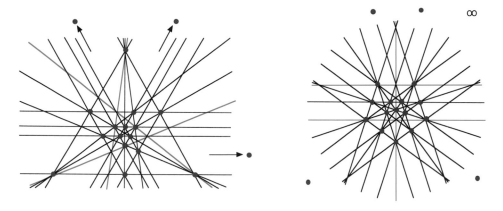

Figure 3.3.5. Two versions of the configuration (18_4) (red points and black lines) from Figure 3.3.4. In each version, adding the three green lines yields a simplicial arrangement of 21 lines (denoted $A(21, 2)$ in the catalog [**119**]).

"outer part" of the construction, and we note that the outermost points of the configuration C determine diagonals of span c. The other m points of C determine diagonals of span b; through each of the outermost points of C there passes one of these diagonals. The lines of symmetry of the two diagonals of span c at each outermost point of C (one of these is shown by the green line in (a)) can be used as mirrors to reflect the m inner points of C as well as the diagonals of span b (see parts (b) and (c)). The m new points become the outermost points of the "inner part" of the configuration we are constructing. To find the last m (inner) lines, we connect each of the

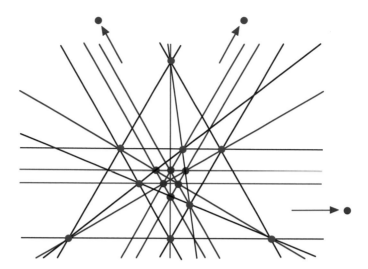

Figure 3.3.6. The configuration (18_4) from Figure 3.3.4 arises from a copy of the configuration $(9_3)_2$ taken from Figure 1.1.6, shown here in red points and black lines, by the addition of nine additional points and lines (shown in blue).

new "outermost" points with one of the original inner points—specifically, we connect it to the $(b+1)^{\text{st}}$ of these points, counting in the same orientation as used in calculating the symbol $m\#(b,c;d)$. This is indicated by the purple segments in parts (c). The new lines (see parts (d)) pass through previous intersections of two lines, creating the last m points of the $((4m)_4)$ configuration.

It is worth stressing that if the starting outer configuration is the self-polar $m\#(b,b;d)$ as in Figures 3.3.9 and 3.3.10, then the inner configuration is another copy (similar to the outer one) of $m\#(b,b;d)$. On the other hand, if $b > c$ as in the illustration in Figure 3.3.11, then the outer and inner parts are the two isomorphic and mutually polar configurations with symbol $m\#(b,c;d)$.

It is also worth mentioning that if $d > c$, then this construction (or any analogous one the author could think of) does not seem to work. This includes the case of selfpolar configurations $m\#(b,c;d)$ with $d = (b+c)/2$.

Another infinite family, which we designate as the **(5/6m) construction**, is constructed as follows, starting from a 3-astral configuration (n_4) with $n = 6m$, where $m \geq 5$. Let us assume this configuration satisfies the following conditions:

(i) It has $2m$-gonal dihedral symmetry.

(ii) The configuration is encoded by the symbol $(2m)\#(s_1,t_1;s_2,t_2;s_3,t_3)$, where s_i is the span of the ith family of diagonals of the ith level polygon P_i

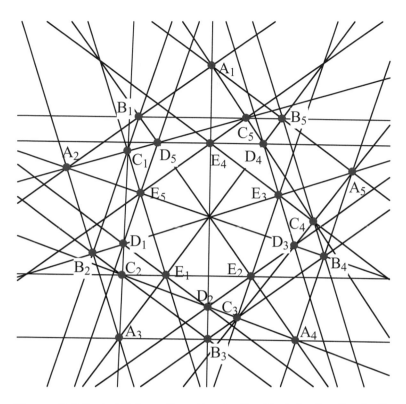

Figure 3.3.7. A (30_4) configuration in the $(6m)$ family, the family that includes the configuration (18_4) in Figure 3.3.4. More generally, the construction of a $((6m)_4)$ configuration starts with a regular m-gon $A_1, \ldots, A_m$, where $m \geq 3$ is odd. The point B_i is the midpoint of A_i and A_{i+1}, and C_i is selected on B_i, B_{i+1} so that the line C_i, C_{i+1} passes through A_{i+2}. Then D_i is determined on C_i, C_{i+1} so that $D_i C_i / C_{i+1} C_i = C_i B_{i+1} / B_i B_{i+1}$, and E_i is the midpoint of D_i and D_{i+1}. Finally, m points at infinity (not shown) are added, in the directions $A_i A_{i+1}$. Lines are $A_i A_{i+1}, B_i B_{i+1}, C_i C_{i+1}, D_i D_{i+1}, E_i E_{i+1}$, and $A_i B_{i+2}$. All subscripts are understood mod m.

and t_i is the order of the intersection point, counting from the midpoint of the diagonal s_i and considering only diagonals of span s_i of the polygon P_i. For more details see Section 3.6.

   (iii) s_1 and t_3 are distinct, and both are even; this implies $m \geq 5$.

   (iv) t_1 and s_3 are odd.

   (v) s_2 and t_2 have the same parity.

   Condition (iii) implies that both kinds of diagonals ending at points of P_1 have even lengths. Therefore, omitting every other point of P_1 and all the lines incident with these points leads to a loss of m points and $2m$ lines. (Note that, as shown in Figures 3.3.13 and 3.3.14, "level 1" does not mean that P_1 is the "outermost level".) The claim is that the above conditions

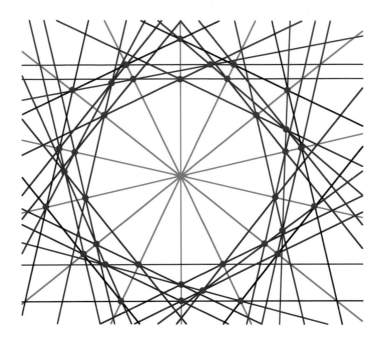

Figure 3.3.8. Shown here is the $(6m)$ construction in the case of the regular 7-gon (black lines), leading to a (42_4) configuration. The seven points at infinity are again not shown; they are in the directions of the quadruplets of parallel black and blue lines.

imply that one can add to the remaining lines and points m suitable lines through the center to obtain a $((5m)_4)$ configuration. The examples in Figures 3.3.13 and 3.3.14 illustrate the construction.

The reason why the construction works is the following. All points of a 3-astral configuration with $(2m)$-gonal dihedral symmetry (that is, based on a regular $(2m)$-gon G) are on lines through the center that are mirrors for the symmetries of the configuration. The points of G are on mirrors that enclose angles that are multiples of π/m. More specifically, the two types of points on an s_i diagonal of P_1 are spaced an even multiple of π/m if s_i and t_i have the same parity and an odd multiple of π/m if these parities are different.

In view of the above, conditions (iii), (iv), and (v) imply that viewed from the center, the points of level 1 are not aligned with the points of the other two levels. Hence these latter points are aligned and provide the m lines required for the formation of a $((5m)_4)$ configuration.

Since configurations $(2m)\#(2, 1, 4, 2, 1, 4)$ and $(2m)\#(2, 3, 4, 2, 3, 4)$ exist for all $m \geq 5$, our existence claim is justified. In fact, for every $m \geq 5$ there exist additional possibilities. This is illustrated in Figure 3.3.13, for $m = 5$. This case was the starting point of this construction. Some of the

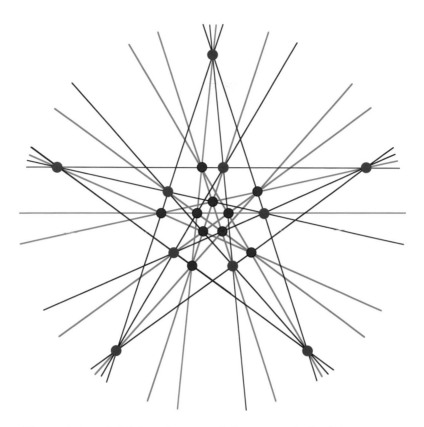

Figure 3.3.9. A (20_4) configuration belonging to the $(4m)$ family. Here $m = 5$. The construction uses two astral (10_3) configurations; one is shown with red points and black lines, the other with blue points and green lines.

configurations in Figure 3.3.13 were first constructed, independently and by ad hoc methods, by T. Pisanski and J. Bokowski.

A few other configurations in the $(5/6m)$ family are illustrated in Figure 3.3.14.

The constructions we have seen so far started from given configurations that had to satisfy certain conditions. The resulting (n_4) configurations always had as n a composite number—more specifically, a multiple of 4 or 5 or 6. Now we shall describe constructions that are applicable quite generally but are apt to give (n_4) configurations with other values of n.

The general construction, which we call the $(3m+)$ **construction**, has the interesting feature that it is more easily visualized and explained in 3-space; the resulting configuration is then readily projected into the plane. We start with an (m_4) configuration C in the plane. We assume that this is the (x, y)-plane in a Cartesian (x, y, z)-system of coordinates and that C has $p \geq 1$ lines parallel to the x-axis and $q \geq 1$ lines parallel to the y-axis, such

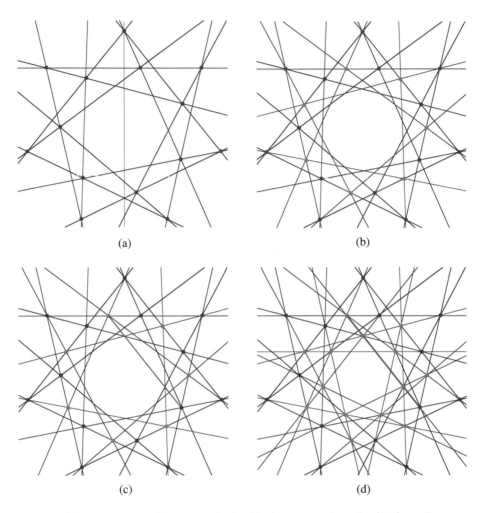

(a)   (b)

(c)   (d)

Figure 3.3.10. The steps in the $(4m)$ construction of a (28_4) configuration from the (14_3) configuration $7\#(2,2;1)$, as explained in the text.

that no two of them have a point of the configuration in common. (Note
that by an affine transformation—which does not change incidences—any
two sets of parallel lines can be made orthogonal. The orthogonality is
assumed only in order to simplify the description.) We select a real number
$h > 1$ and keep it constant throughout the discussion; it is convenient (but
not necessary) to think of $h = 10$. We construct two copies of C. One
is C', obtained from C by stretching C in the ratio $(h-1)/h$ (that is, in
fact, shrinking it) towards the y-axis, stretching it in the ratio $(h+1)h$
towards the x-axis, and then translating it to the level $z = 1$. A schematic
representation of a section parallel to the x-axis is shown in Figure 3.3.15.
The other is C'', obtained similarly but by using the ratio $(h+1)/h$ for
stretching towards the y-axis, $(h-1)h$ for the ratio towards the x-axis, and

Figure 3.3.11. Another illustration of the construction. We start with a (16_3) configuration $8\#(3,2;1)$ and obtain a (32_4) configuration. Note that the outer and inner parts are not similar but are polar to each other.

translation to the plane $z = -1$. Thus, C' is obtained from C by the map

$$f(x, y, 0) = (x(h-1)/h, y(h+1)/h, 1),$$

and C'' is obtained by

$$g(x, y, 0) = (x(h+1)/h, y(h-1)/h, -1).$$

It is easy to check that for each point $A = (x, y, 0)$ the points $A, f(A)$, and $g(A)$ are collinear and that the points $h(A) = (0, 2y, h)$ and $h^*(A) = (2x, 0, -h)$ are collinear with them. Now, for any four points A_j ($j = 1, 2, 3, 4$) of C that are on a line L parallel to the x-axis—that is, have

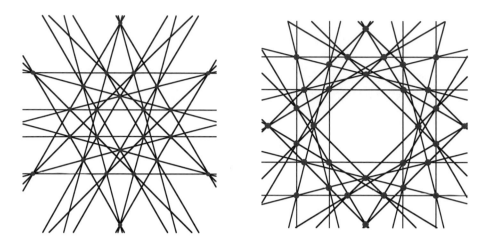

Figure 3.3.12. Configurations (24_4) and (32_4) from the (4_m) family; the latter is different from the one in Figure 3.3.11.

the same y-coordinate—the point $h(A_j)$ will be the same since it does not depend on the x-coordinate. Therefore we can conclude that by deleting the line L from the configuration C and its parallels in C' and C'', while adding the lines from A_j to $h(A_j)$, the points A_j and the corresponding points in C' and C'' will remain incident with four lines, and the new point $h(A_j)$ will also be incident with four lines. We deleted three lines and added four and also added one point. Thus, from the starting (m_4) configuration we obtained a configuration (n_4) where $n = 3m + 1$. Analogously, any four points of C collinear on a line parallel with the y-axis may lead to an additional increase in the number of points and lines; the assumed disjointedness of the two families of parallels is needed here to assure that no 5-point line arises. Proceeding similarly with some or all lines parallel to either the x-axis or the y-axis, we see that from (m_4) we can obtain configurations (n_4) for each n such that $3m + 1 \le n \le 3m + p + q$.

Next, we have the **deleted unions constructions** (DU-1) and (DU-2). Consider any configurations $C_1 = ((n')_4)$ and $C_2 = ((n'')_4)$, such that the cross-ratio of points of C_1 on a certain line coincides with the cross-ratio of lines through a certain point of C_2. Then omitting the line and the point in question and adjusting the positions and sizes of the deleted configurations appropriately, we obtain a configuration with $n' + n'' - 1$ points. In every case one can use for C_2 a polar of C_1 to go from (n_4) to $((2n-1)_4)$. This is construction (DU-1); illustrations are provided in Figures 3.3.16 and 3.3.17. For (DU-2) we need to delete two disjoint lines and two unconnected points, respectively. An illustration is given in Figure 3.3.18. Again, the only requirement is that the cross-ratios of the appropriate quadruplets of points and of lines be equal.

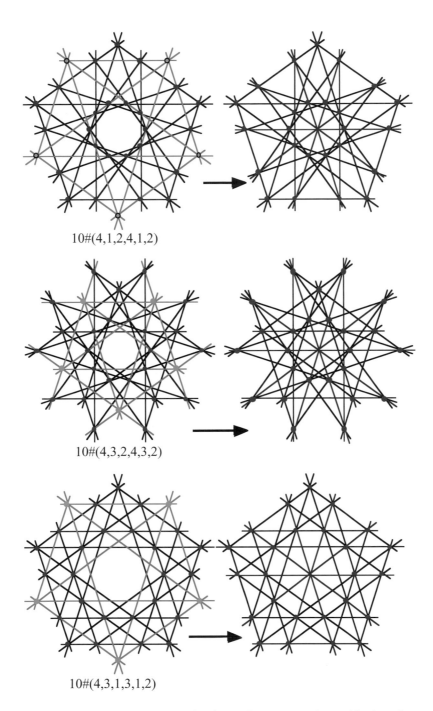

Figure 3.3.13. The three (25_4) configurations obtainable by the $(5/6m)$ construction.

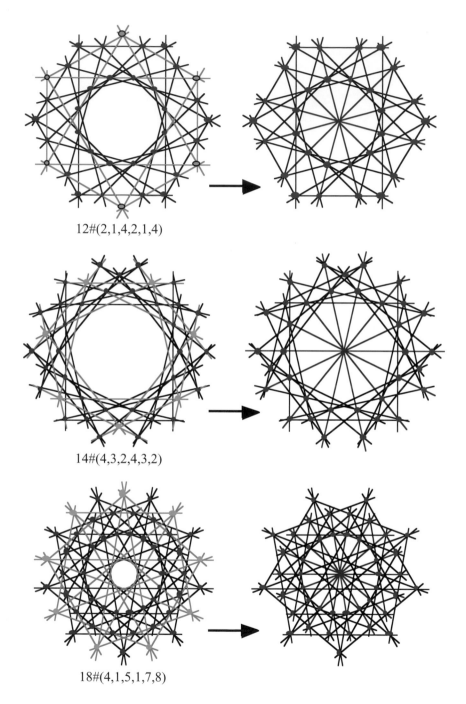

12#(2,1,4,2,1,4)

14#(4,3,2,4,3,2)

18#(4,1,5,1,7,8)

Figure 3.3.14. Configurations (30_4), (35_4), and (45_4) belonging to the $(5/6m)$ family.

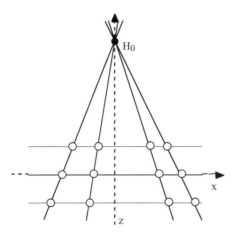

Figure 3.3.15. A schematic illustration of the $(3m+)$ construction.

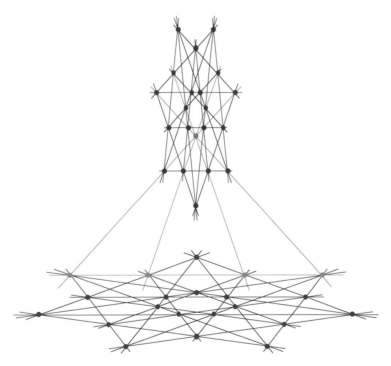

Figure 3.3.16. Configuration (41_4) from two copies of configuration (21_4) using (DU-1).

With this we have completed the description of the various constructions that will enable us to find geometric configurations (n_4) for almost all values of $n \geq 18$. The proof of this assertion, which we have already formulated as Theorem 3.2.4, will be given in the next section. In it we shall utilize various configurations with very high symmetry—astral, multiastral, and other.

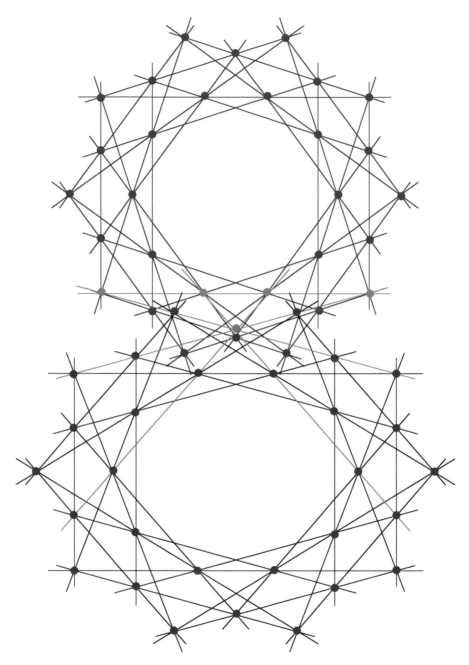

Figure 3.3.17. Configuration (59_4) from two copies of configuration (30_4) using (DU-1). Here $p + q = 8$.

Since their construction and properties are both interesting and complicated, we are not describing them here; instead, we shall devote several sections to them later.

Figure 3.3.18. Configuration (46_4) from two copies of configuration (24_4), using (DU-2).

Exercises and Problems 3.3.

1. Carry out the construction of the (35_4) configuration described in Figure 3.3.3.

2. Determine whether any of the three configurations (25_4) in Figure 3.3.13 are isomorphic.

3. Devise a general proof for the validity of the $(4m)$ construction, as detailed in the text.

4. Formulate the analog of the $(3m+)$ construction that leads from 3-configurations to 3-configurations. Illustrate by a simple example.

5. The $(6m)$ construction is applicable to regular star-polygons as well. Explore the case of a pentagram and of one of the regular star-heptagons.

6. Explain why the (DU-1) construction cannot be applied to get a (43_4) configuration from (20_4) and (24_4) configurations.

3.4. Existence of geometric 4-configurations

We start with a quick summary description of the construction methods detailed in Section 3.3.

The $(5m)$ construction is illustrated in Figure 3.3.1. It starts with an arbitrary (m_3) configuration and yields a $((5m)_4)$ configuration.

The $(5/2m)$ construction is illustrated in Figure 3.3.2. It starts with appropriate configurations $((2m)_3)$ and yields a $((5m)_4)$ configuration; the criteria for usable (m_3) configurations are given in Section 3.3

The $(4m)$ construction starts with an astral configuration $((2m)_3)$ and yields a 4-orbit dihedral configuration $((4m)_4)$. As explained in Section 3.3, it works for most (but not all) such configurations with $m \geq 5$.

The $(6m)$ construction starts with a 3-orbit configuration $((3m)_3)$ and yields a 6-orbit configuration $((6_m)_4)$. It assumes that $m \geq 3$ is odd. Some details are given in Section 3.3.

$(3m+)$ denotes the construction described in detail in the caption to Figure 3.3.7. It starts with an (m_4) configuration and yields a $((3m+p+q)_4)$ configuration.

We discussed deleted union constructions (DU-1) and (DU-2). Using (DU-1), from suitable configurations $C_1 = ((n_1)_4)$ and $C_2 = ((n_2)_4)$ we obtain a configuration with $n_1 + n_2 - 1$ points and as many lines. In particular, we can go from any (n_4) to $((2n - 1)_4)$ configuration. For (DU-2) we delete two disjoint lines and two unconnected points and obtain $((2n - 2)_4)$ from (n_4).

In addition to these, we use the notation $(t\text{-}A.m)$ for the multiastral configuration with t orbits and with symmetry group d_m. This implies that each orbit has m points. Details of these configurations and the notation used for them appear in Section 3.5. Also, $(2\text{-}A.m)$ denotes astral configurations. If no other indication is given, the references are to the "trivial" choices of parameters such as $(1, 2, 3, 1, 2, 3)$ or $(1, 2, 3, 4, 2, 1, 4, 3)$.

It is relatively simple to show that (n_4) configurations exist for all $n \geq 210$. Indeed, by the $(5m)$ construction there is for each $m \geq 9$ a $((5m)_4)$ configuration with $p = m$ parallel lines. It follows by the $(3m+)$ construction that for all $m \geq 9$ and $1 \leq p \leq m$ there exists a $((15m + p)_4)$ configuration. Since $15m = 5(3m)$, by the $(5m)$ construction we can add $p = 0$ to the range of p. Thus (n_4) configurations exist all values of n such that $15m \leq n \leq 16m$; for $m \geq 14$ these ranges are contiguous or overlapping, and so the claim is established.

For smaller values of n we have to rely on the various constructions described above and in Section 3.3. We found it simplest to arrange the

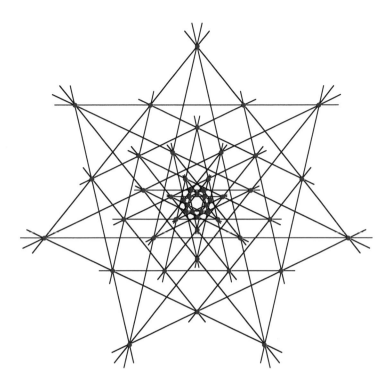

Figure 3.4.1. Depiction of a $(7\text{-}A.m)$ configuration (49_4), with symbol $7\#(2, 1, 2, 1, 3, 2, 3, 2, 1, 2, 1, 3, 2, 3)$.

necessary data in Table 3.4.1 in which we list examples of configurations (n_4) for each n. In most cases there are other configurations we could have listed—the present choice is largely accidental.

The arguments presented above, together with the data in Table 3.4.1, constitute a proof of Theorem 3.2.4.

The known constructions explained above for the configurations (n_4) with small n (such as $18, 20, 21, 24, 25$) all rely on r-fold rotational symmetry with $r \geq 3$. As a consequence, none of these constructions can be carried out in the rational projective plane. While there is no proof available showing that some or all these configurations are not realizable in the rational projective plane, it is a challenging problem to decide for which n such a realization is possible. An easy argument shows that if we start with rational configurations, then $(5m)$ constructions can be performed so as to yield rational configurations, and similarly for $(5/2m)$ and $(3m+)$ constructions.

Table 3.4.1. Descriptions of the construction of (n_4) configurations for $n \leq 304$.

n	Reference or explanation
18	$(6m)$ for $m = 3$; Figure 3.3.4
19	Not known
20	$(4m)$ for $m = 5$; Figure 3.3.9;
21	$(3\text{-}A.m)$, $7\#(3, 2, 1, 3, 2, 1)$; Figure 3.2.1
22	Not known
23	Not known
24	$(2\text{-}A.m)$, $12\#(5, 4, 1, 4)$, Figure 3.6.2; $(3\text{-}A.m)$
25	$(5/2m)$ for $m = 10$; starting with $(10_3)_{10}$. Figure 3.3.2.
26	Not known
27	$(3\text{-}A.m)$
28	$(4m)$ for $m = 7$; Figure 3.3.10
29	Bokowski (unpublished)
30	$(3\text{-}A.m)$
31	Bokowski (unpublished)
32	$(4m)$ for $m = 8$. Figures 3.3.11 and 3.3.12
33	$(3\text{-}A.m)$
34	$(DU\text{-}2)$ from two (18_4) configurations
35	$(5/2.m)$ for $m = 7$; starting from a (14_3) configuration shown in Figure 3.3.3
36	$(2\text{-}A.m)$ several possibilities with $m = 18$; $(3\text{-}A.m)$, $(4\text{-}A.m)$
37	Not known
38	$(DU\text{-}2)$ from two (20_4) configurations
39	$(3\text{-}A.m)$
40	$(4\text{-}A.m)$
41	$(DU\text{-}1)$ from a (21_4) configuration shown in Figure 3.3.16
42	$(3\text{-}A.m)$
43	Not known
44	$(4\text{-}A.m)$
45	$(5\text{-}A.m)$ for $m = 9$; e.g. $9\#(1, 2, 3, 4, 2, 1, 2, 3, 4, 2)$; $(3\text{-}A.m)$
46	$(DU\text{-}2)$ from a (24_4) configuration shown in Figure 3.3.18
47	$(DU\text{-}1)$ from a (24_4) configuration
48	$(2\text{-}A.m)$; $(3\text{-}A.m)$; $(4\text{-}A.m)$; $(6\text{-}A.m)$
49	$(7\text{-}A.m)$; Figure 3.4.1
50	$(5\text{-}A.m)$, $10\#(1, 4, 3, 2, 3, 1, 4, 3, 2, 3)$
51	$(3\text{-}A.m)$
52	$(4\text{-}A.m)$
53	$(DU\text{-}1)$ from a (27_4) configuration
54	$(3\text{-}A.m)$

Table 3.4.1. (Continued)

n	Reference or explanation
55	$(5\text{-}A.m)$
56	$(4\text{-}A.m)$
57	$(3\text{-}A.m)$
58	(DU-2) from a (30_4) configuration
59	(DU-1) from a (30_4) configuration
60	$(2\text{-}A.m)$; $(3\text{-}A.m)$; $(4\text{-}A.m)$; $(5\text{-}A.m)$;
	$(6\text{-}A.m)$, $10\#(1,3,2,4,2,3,4,1,3,2,3,2)$
61	(DU-1) applied to configurations (21_4) and (41_4)
62	(DU-1) applied to configurations (21_4) and (42_4)
63	$(3\text{-}A.m)$
61–63	$(3m+)$ from a (20_4) configuration
64–66	$(3m+)$ from a (21_4) configuration
67	(DU-1) from (33_4) and (35_4) configurations,
	obtained by $(5/2m)$ for $m = 14$
68	$(4\text{-}A.m)$
69	$(3\text{-}A.m)$
70	$(5\text{-}A.m)$
71	(DU-1) from a (36_4) configuration
72	$(2\text{-}A.m)$; $(3\text{-}A.m)$; $(4.\text{-}A.m)$; $(6\text{-}A.m)$
73–76	$(3m+)$ from a (24_4) configuration, $p + q = 4$
75	$(5/2m)$, $m = 30$; $(5\text{-}A.m)$
76–80	$(3m+)$ from $(25_4) = (5/2m)$, $m = 10$, $p + q = 5$
81	$(3\text{-}A.m)$, $m = 27$; $(9\text{-}A.m)$, $m = 9$,
	$9\#\{3,4,2,1,4,1,4,3,2,3,4,2,1,4,1,4,3,2\}$
82–87	$(3m+)$ from a (27_4) configuration, $p + q = 6$
88–95	$(3m+)$ from a (29_4) configuration, $p + q = 6$
91–98	$(3m+)$ from a (30_4) configuration, $30\#(4,6,9,4,6,9)$, $p + q = 8$
99	$(3\text{-}A.m)$, $m = 33$
100–105	$(3m+)$ from a (33_4) configuration, $p + q = 6$
106–112	$(3m+)$ from $(35_4) = (5/2m)$, $m = 14$, $p + q = 7$
109–114	$(3m+)$ from a (36_4) configuration, $12\#(1,2,3,1,2,3)$, $p + q = 6$
115	$(5/2.m)$, $m = 46$; $(5\text{-}A.m)$
116	$(4\text{-}A.m)$, $m = 29$
117	$(3\text{-}A.m)$, $m = 39$
118–123	$(3m+)$ from a (39_4) configuration, $13\#(1,5,3,1,5,3)$, $p + q = 6$
121–128	$(3m+)$ from a $(40_4) = (5/2m)$, $m = 16$, $p + q = 8$
127–132	$(3m+)$ from a (42_4) configuration, $14\#(1,3,5,1,3,5)$, $p + q = 6$
133–139	$(3m+)$ from a (44_4) configuration,
	$11\#(1,2,5,4,2,1,4,5)$, $p + q = 7$

Table 3.4.1. (Continued)

n	Reference or explanation
136–144	$(3m+)$ from $(5/2m) = (45_4)$, $m = 18$, $p + q = 9$
145–152	$(3m+)$ from a (48_4) configuration,
	$12\#(1, 2, 5, 4, 2, 1, 4, 5)$, $p + q = 8$
151–160	$(3m+)$ from $(50_4) = (5/2m)$, $m = 20$, $p + q = 10$
157–164	$(3m+)$ from a (52_4) configuration,
	$13\#(1, 2, 5, 4, 2, 1, 4, 5)$, $p + q = 8$
165	$(5m)$ (33_3)
166–173	$(3m+)$ from a (55_4) configuration,
	$11\#(1, 2, 3, 4, 5, 1, 2, 3, 4, 5)$, $p + q = 8$
172–177	$(3m+)$ from a (57_4) configuration, $19\#(1, 4, 7, 1, 4, 7)$, $p + q = 6$
178–185	$(3m+)$ from a (59_4) configuration, $p + q = 8$, Figure 3.3.17
181–192	$(3m+)$ from $(60_4) = (5/2m)$, $m = 24$, $p + q = 12$
193–200	$(3m+)$ from a (64_4) configuration,
	$16\#(1, 3, 7, 5, 3, 1, 5, 7)$, $p + q = 8$
199–207	$(3m+)$ from a (66_4) configuration,
	$11\#(1, 2, 3, 4, 5, 4, 2, 1, 4, 3, 4, 5)$, $p + q = 9$
208–213	$(3m+)$ from a (69_4) configuration, $23\#(1, 3, 5, 1, 3, 5)$, $p + q = 6$
211–224	$(3m+)$ from $(5m)$, $m = 14 = p + q$
225	$(5m)$ from a (45_3) configuration
226–240	$(3m+)$ from $(5m)$, $m = 15 = p + q$
241–256	$(3m+)$ from $(5m)$, $m = 16 = p + q$
256–272	$(3m+)$ from $(5m)$, $m = 17 = p + q$
271–288	$(3m+)$ from $(5m)$, $m = 18 = p + q$
286–304	$(3m+)$ from $(5m)$, $m = 19 = p + q$

Exercises and Problems 3.4.

1. Decide whether a suitable affine (or projective) image of the (18_4) configuration shown in Figure 3.3.4 can be put in the rational plane.

2. Determine for which n one can find a configuration (n_4) in the plane over a quadratic extension of the rationals.

3.5. Astral 4-configurations

In this section we start our investigation of a special class of 4-configurations which we call k-astral for some $k \geq 2$. They are of interest for several reasons. To begin with, such configurations were the first 4-configurations for which geometric realizations were found (see, for example, [**122**], [**111**], [**24**], and the other publications that will be mentioned later). Next, they have a clear-cut definition that leads to a natural notation, as well as to the construction of the configuration given its symbol. Finally, they exhibit a variety of phenomena that add interest to their study, such as the relation of a configuration to its dual (actually, its polar) configuration and questions of realizability versus representation.

k-astral configurations have appeared under several different names and with several different definitions—not all of which coincide in all cases. In several publications configurations we call k-astral have been termed *celestial*. The intention in the present account of these configurations is to have an easily implementable decision algorithm for checking the membership of either a given configuration to the class, or of a symbol for correspondence to a geometric configuration.

Definition 3.5.1. An (n_4) configuration C is k-**astral** provided all the following conditions are satisfied:

(**A1**) $k \geq 2$ and $n = k \cdot m$, for some $m \geq 7$.

(**A2**) The points of C are at the vertices of k regular convex m-gons, with common centers and such that all angles subtended from this center by the various points of C are multiples of π/m.

(**A3**) C has symmetry group d_m; the vertices of each k-gon form an orbit.

(**A4**) Each line of C contains two points from each of two m-gons (point orbits); each point is incident with two lines each from two line orbits.

We have already encountered various configurations that are k-astral, for example, the ones in Figures 1.1.2 and 1.5.1(a). Two additional examples are shown in Figure 3.5.1.

Some comments deserve to be made regarding k-astral configurations.

(i) In most cases the k regular m-gons have different sizes; however, in some cases with $k \geq 3$ there may be pairs of polygons with the same size. We shall give examples later.

(ii) The conditions in Definition 3.5.1 could be weakened at the expense of complicating the verification.

(iii) It will turn out to be convenient to consider the case of connected k-astral configurations separately from the case of disconnected ones.

Theorem 3.5.1. *Each k-astral configuration C can be assigned a symbol $m\#(s_1, t_1; s_2, t_2; \ldots; s_k, t_k)$ in such a way that C is the only configuration from which that symbol arises. At most $2k$ distinct symbols correspond to each configuration; such symbols are said to be equivalent. The family of equivalent symbols can be obtained from any one of them by cyclic permutations of an even number of steps of the $2k$ entries in parentheses or by reversal of these.*

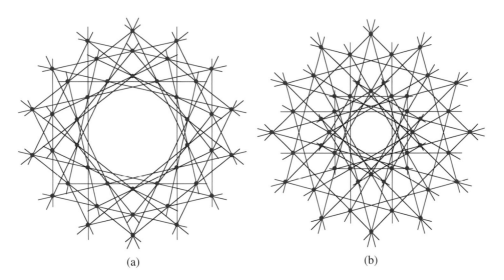

(a)   (b)

Figure 3.5.1. Depictions of (a) a 3-astral configuration (42_4) with symbol $14\#(5,3;2,4;1,3)$ and (b) a 4-astral configuration (48_4) with symbol $12\#(5,4;3,2;4,5;2,3)$.

Proof. The proof consists of a description of the steps leading from the configuration to one of the symbols and observing the stages at which distinct symbols may arise. The main tools in the derivation are a notation for the intersection points of the diagonals determined by each of the regular m-gons and the "characteristic paths" along lines of the configuration.

For a regular convex m-gon M, the **span** s of a diagonal S is the number of edges of M between the endpoints of S, taken as the smaller of the two possible numbers; hence $s \leq m/2$. Despite talking about "endpoints", by "diagonal" we understand both the elementary-geometric meaning of the term as a segment, as well as the whole line determined by this segment. In Figure 3.5.1, the outer polygon has diagonals of spans 3 and 5 for both configurations (a) and (b).

The intersections of a diagonal S of span s with the other diagonals of span s of the same polygon M are denoted by the symbol $(s//t)$, where t is the position of the intersection points on S, counting from the midpoint of S. (Instead of $(s//t)$, the notation $[[s,t]]$ has also been used.) Thus, for example, each endpoint of S has symbol $(s//s)$. The intersection points are not limited to the diagonal considered as a segment but continue "outside" and exist for all $t < m/2$. If m is even, one may consider the point-at-infinity on S as having $t = m/2$. An illustration of the notation $(s//t)$ is given in Figure 3.5.2.

The use of polar coordinates is particularly convenient for the intersection points $(s//t)$, since it is easily seen that in the setting of Figure 3.5.2, such a point has coordinates $(\cos(s\pi/m)/\cos(t\pi/m),\ t\pi/m)$. If the endpoints of the diagonal are not on the unit circle but at distance r, then the first polar coordinate needs to be multiplied by r.

A **characteristic path** P of a (connected) k-astral configuration C consists of k *segments* of lines of the configuration, determined as follows. The procedure we describe here is illustrated by the example in Figure 3.5.3.

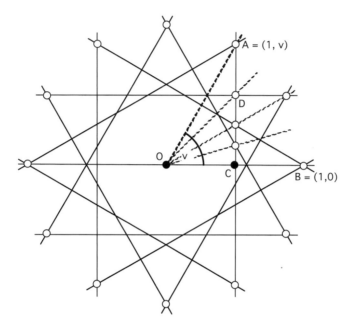

Figure 3.5.2. The determination of the symbol $(s//t)$ of an intersection point of diagonals of a regular m-gon. Here $m = 12$, the diagonals are of span $s = 4$, and the vertices of the m-gon with unit radius have polar coordinates $(1, \nu)$, where ν is a multiple of π/m. In the diagram $\nu = 4\pi/m$, angle DOB is $3\pi/m$, so that $s = 4$, $t = 3$. This gives for D the symbol $(s//t) = (4//3)$. The right triangles OCA and OCD imply that $OD = (\cos(s\pi/m)/\cos(t\pi/m),\ t\pi/m)$; hence D has polar coordinates $(\cos(s\pi/m)/\cos(t\pi/m),\ t\pi/m)$.

As the first step we orient all lines of C in the same sense, generally taken to be counterclockwise as seen from the center. Next, we choose an arbitrary point P_0 of C and through it an arbitrary line L_1 for which P_0 is the earlier of the two points in the same orbit; this involves the choice of one line from the two orbits of points through P_0. On L_1 we take the first point (in the orientation we adopted) of the *other* orbit of points incident with L_1 and denote it P_1. We choose as line L_2 a line through P_1 that is in the orbit different from L_1 and for which P_1 is the earlier point in its orbit. On L_2 we choose the earlier point in the orbit different from the one of P_1 and denote it P_2. Continuing in the same way, we select the line L_{j+1} through the already selected point P_j that is in the orbit different from L_j and for which P_j is the earlier point among the points on L_{j+1} belonging to the orbit different from the orbit of the earlier point P_j. This continues until we reach the line L_k and the point P_k. (In Figure 3.5.3 we have $k = 3$.) This point P_k necessarily belongs to the same orbit as the starting point P_0; in the illustration P_k coincides with P_0, but this is not necessarily the case. Figure 3.5.4 illustrates the possibility of P_k being different from P_0. By using the notation $(s_j//t_j)$ for the point P_j, the characteristic path $P_0, L_1, P_1, L_2, P_2, \ldots, L_k, P_k$ determines a symbol $m\#(s_1, t_1; s_2, t_2; \ldots; s_k, t_k)$ for the configuration.

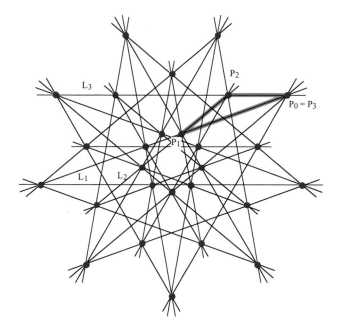

Figure 3.5.3. An illustration of the construction of a characteristic path (shown in green) and the corresponding symbol of the configuration, according to the description given in the text. Since $P_1 = (4//2)$, $P_2 = (3//4)$, $P_3 = (2//3)$, and $k = 3$ while $m = 9$, the resulting symbol of this (27_4) configuration is $9\#(4, 2; 3, 4; 2, 3)$.

What are the possible alternative symbols for a configuration? We arbitrarily chose the orientation of the lines, the starting point of the characteristic path, and the starting line through that point. The choice of orientation does not lead to any new symbols since a k-astral configuration has *dihedral* symmetry group d_k. However, the other two choices obviously matter and in general lead to $2k$ distinct symbols—k choices of the orbit of the starting point of the path and two choices for the starting line through that point. As an illustration we show in Figure 3.5.5 the four characteristic paths and the resulting four equivalent symbols for the 2-astral configuration (48_4).

Figure 3.5.4. Another illustration of the construction of a characteristic path and the symbol of the configuration. Since $P_1 = (4//3)$,
$P_2 = (2//3)$, $P_3 = (1//3)$, and $k = 3$ while $m = 9$, the resulting symbol
of this (27_4) configuration is $9\#(4, 3; 2, 3; 1, 3)$, and $P_3 \neq P_0$.

The various equivalent symbols for a given k-astral configuration arise
in one of the following two ways. For a given characteristic path, selecting
on this path a different point as the starting point clearly permutes the
symbols $(s_j//t_j)$ cyclically, that is, by an even number of steps in the symbol
$m\#(s_1, t_1; s_2, t_2; \dots ; s_k, t_k)$ of the configuration. This yields up to k distinct
symbols. On the other hand, if we consider a diagonal of span s, the symbol
$(s//t)$ for the tth intersection point (counting from the midpoint) can be
interpreted as saying that on the orbit of all points $(s//t)$ the same diagonal
line has span t and the original endpoints (that gave span s to the diagonal)
now have symbol $(t//s)$. This means the following: Starting with a given
characteristic path but traversing it in the opposite direction will reverse
the roles of s_j and t_j in all the diagonals, as well as the order of the points.
Hence this leads to the reversal of all the entries in the original symbol, thus
accounting for (up to) an additional k symbols.   □

The construction of the symbols for a k-astral configuration (n_4) leads
to several notable properties of the symbols and the configurations. For ease
of reference we list them as a continuation of the entries in Theorem 3.5.1.

(A5) Since the symbol $(s//s)$ denotes the endpoints of a diagonal of
span s (hence would not constitute a step in the characteristic path), any two

adjacent entries in the symbol $m\#(s_1, t_1; s_2, t_2; \dots ; s_k, t_k)$ must be different; this includes the requirement that s_1 and t_k are distinct.

Next, as obvious from the reasoning concerning the symbol $(s//t)$ and visible in Figure 3.5.2, the polar angle of the point (s/t) differs from 0 by a multiple of π/m. The parity of that multiple is the same as the parity of $s + t$. Since the endpoint of a characteristic path leads to a point in the orbit of the starting point and since the polar angles of any two such points differ by a multiple of $2\pi/m$, it follows that the sum of all entries in the parentheses of a symbol $m\#(s_1, t_1; s_2, t_2; \dots ; s_k, t_k)$ must be even, or equivalently, that

$$(\textbf{A6}) \qquad\qquad \delta = \frac{1}{2} \sum_{1 \le j \le k} (s_j - t_j) \text{ is an integer.}$$

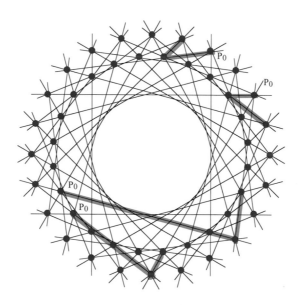

Figure 3.5.5. Four characteristic paths (green) for a 2-astral (48_4) configuration; all proceed counterclockwise. In order to avoid excessive clutter, in each path only the starting point is labeled. The path on top starts in the outer ring of points; it leads to the symbol $24\#(5, 2; 7, 8)$, since the first point of the inner ring that is encountered by the path has symbol $(5//2)$ and the first point met after that in the outer ring has symbol $(7//8)$. The other characteristic paths lead to the symbols $24\#(7, 8; 5, 2)$, $24\#(2, 5; 8, 7)$, and $24\#(8, 7; 2, 5)$, respectively, in counterclockwise order of the starting points.

If condition (A6) is not satisfied in a symbol that fulfils all other requirements, then the last point of the characteristic path ends midway between points of the orbit of the starting point—and consequently has only two lines incident with it, just as the starting point is incident with only two lines. We shall discuss this in more detail later, but we can already supply in Figure 3.5.6 an example of such a situation.

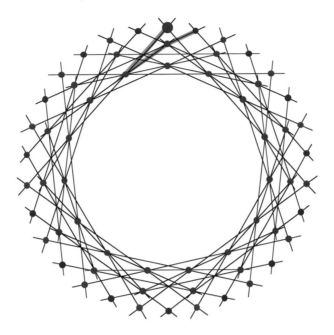

Figure 3.5.6. The symbol $15\#(4,2;1,3;2,3)$ satisfies all conditions for a valid symbol of a 3-astral 4-configuration (45_4), except (A6). The characteristic path (green) that starts at the top point (large red) leads to a point (blue) at an in-between position. Both the starting point and the endpoint of the path are incident with just two lines each—hence the symbol does not correspond to any 4-configuration.

One additional—and very important and useful—property of all k-astral 4-configurations follows from the comments we made after the introduction of the $(s//t)$ notation. Since the radius of a point $(s//t)$ of a regular convex m-gon with circumradius r is $r \cdot \cos(s \cdot \pi/m)/\cos(t \cdot \pi/m)$, the distance of the point P_j from the center is (assuming the starting point of the characteristic path is at unit distance from the center)

$$\prod_{1 \leq i \leq j} (\cos(s_i \cdot \pi/m)/\cos(t_i \cdot \pi/m)).$$

Since the endpoint of any characteristic path is in the same orbit as the starting point, this yields

$$(\mathbf{A7}) \qquad \prod_{1 \leq j \leq k} \cos(s_j \cdot \pi/m) = \prod_{1 \leq j \leq k} \cos(t_j \cdot \pi/m).$$

It is easy to verify that in all examples of k-astral configurations presented in this section the condition (A7) is fulfilled.

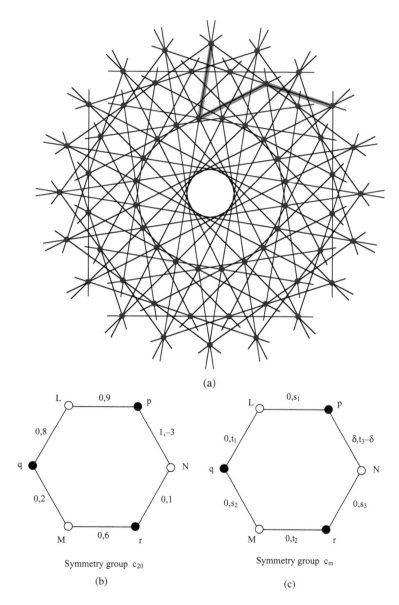

(a)

Symmetry group c_{20}

(b)

Symmetry group c_m

(c)

Figure 3.5.7. (a) The 3-astral configuration $20\#(9, 8; 2, 6; 1, 4)$ and a characteristic path. (b) The corresponding reduced Levi graph of $20\#(9, 8; 2, 6; 1, 4)$. (c) The reduced Levi graph of the 3-astral configuration $m\#(s_1, t_1; s_2, t_2; s_3, t_3)$.

The appropriateness of the characteristic path approach to the notation for k-astral 4-configurations can be seen in the straightforward translation of the characteristic path into the reduced Levi diagram of the configuration. Without entering into lengthy descriptions of the procedure (which is essentially taken from Boben and Pisanski [24]), we show a typical example in Figure 3.5.7. The configuration $20\#(9, 8; 2, 6; 1, 4)$ and a characteristic path leading to this symbol are shown in part (a), while part (b) presents a reduced Levi diagram of this configuration. In part (c) we show the reduced Levi diagram of the general case of a 3-astral configuration $m\#(s_1, t_1; s_2, t_2; s_3, t_3)$. The corresponding situation for a k-astral configuration differs only in the length of the circuit, so that it contains k white and k black points. The value of d is determined by (A6).

Next, we explore what happens if the $2k$ entries between parentheses of a symbol $m\#(s_1, t_1; s_2, t_2; \ldots ; s_k, t_k)$ of a k-astral configuration C are changed by a cyclic permutation that moves them an *odd* number of steps. What—if anything—can we say about a configuration C^* that would correspond to $m\#(t_1, s_2; t_2, \ldots, s_k; t_k, s_1)$?

Considering the well-known relations between points and lines polar to them with respect to a circle of a given radius and center (illustrated in Figure 3.5.8; see also, for example, [52, Chapter 6]), we see that for a configuration corresponding to the symbol $m\#(t_1, s_2; t_2, \ldots, s_k; t_k, s_1)$, the distance of the line L_j of C^* that is polar to the point P_j of C with respect to a circle of unit radius should satisfy

$$\text{distance}(O, L_j) = OP_j^* = \prod_{1 \leq i \leq j} (\cos(t_i \cdot \pi/m)/\cos(s_i \cdot \pi/m)) = 1/OP_j$$

$$= 1 \Big/ \prod_{1 \leq i \leq j} (\cos(s_i \cdot \pi/m)/\cos(t_i \cdot \pi/m))$$

$$= \prod_{1 \leq i \leq j} (\cos(t_i \cdot \pi/m)/\cos(s_i \cdot \pi/m)).$$

Hence distances from the center O of all lines of the putative configuration C^* are correct for them being the polars of the points of C, and since incidences and symmetry are all preserved under polarity, we can conclude:

Theorem 3.5.2. *If the symbol $m\#(s_1, t_1; s_2, t_2; \ldots ; s_k, t_k)$ corresponds to a k-astral 4-configuration, then the symbol $m\#(t_1, s_2; t_2, \ldots, s_k; t_k, s_1)$ corresponds to a k-astral 4-configuration that is polar to the former with respect to the unit circle with center at the common center of both configurations.*

We make here a notational remark. Unless there is a definite reason to do otherwise, we shall always strive to use the lexicographically highest symbol for each k-astral 4-configuration.

Several concepts simplify the listing and classification of possible k-astral configurations. One is based on the observation that if we switch the positions of two entries separated by an odd number of other entries in the symbol of an astral configuration, the modified symbol automatically satisfies all

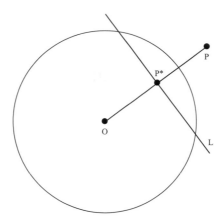

Figure 3.5.8. If the point P and the line L are polars of each other with respect to the circle of radius r and center O, then the distance between O and L is the same as the distance OP^* and the relation between the distances is $OP \cdot OP^* = r^2$.

the conditions listed above, except possibly (A5). By repeated application of this observation while avoiding a violation of (A5), we arrive at the conclusion that it is sensible to introduce the **cohort concept** and **notation**. For a k-astral configuration with symbol $m\#(s_1, t_1; s_2, t_2; \ldots ; s_k, t_k)$ the **cohort symbol** is $m\#\{s, t\} = m\#\{\{s_1, s_2, \ldots, s_k\}, \{t_1, t_2, \ldots, t_k\}\}$; it stands for all the valid assignments of suitable permutations of the s_i's and permutations of the t_i's into a symbol for a k-astral configuration. For example, for the configuration in Figure 3.5.1(a) the symbol is $14\#(5, 3; 2, 4; 1, 3)$, and the cohort symbol is $14\#\{\{5, 2, 1\}, \{4, 3, 3\}\}$. This cohort symbol indicates, and is shared by, the six distinct 3-astral configurations $14\#(5, 4; 2, 3; 1, 3)$, $14\#(5, 3; 2, 4; 1, 3)$, $14\#(5, 3; 2, 3; 1, 4)$, $14\#(5, 4; 1, 3; 2, 3), 14\#(5, 3; 1, 4; 2, 3)$, and $14\#(5, 3; 1, 3; 2, 4)$.

The second comes from the observation that all the conditions, except possibly (A5), are satisfied if in the cohort symbol $m\#\{s, t\}$ the sets s and t are the same. As an example, the configuration we used in Figure 3.5.3 has symbol $9\#(4, 2; 3, 4; 2, 3)$; hence $s = t = \{4, 3, 2\}$. Since condition (A7) is satisfied without the need to make any calculations, we shall say that the cohort $9\#\{\{4, 3, 2\}, \{4, 3, 2\}\}$ is **trivial**. On occasion we shall use "trivial" also for an individual configuration in a trivial cohort. For odd k, a typical representative of a trivial cohort is $m\#(a, b; c, \ldots ; u, v; w, a; b, c; \ldots, u; v, w)$, while for even k we can use $m\#(a, b; c, d; \ldots ; v, w; b, a; d, c; \ldots ; w, v)$. We should mention here that there cannot be any 2-astral configurations of the trivial type.

We shall say that a cohort symbol $m\#\{s, t\}$ of k-astral configurations is **systematic** provided m and the elements of s and t depend on one or more parameters in such a way that the validity of (A7) can be ascertained using only trigonometric identities and without the need to calculate specific values of the parametrized s_i's and t_i's. As we shall illustrate in Section 3.6, the cohort with $m = 6k$, $s = \{3k - j, j\}$, $t = \{3k - 2j, 2k\}$ is a systematic 2-astral cohort.

If a k-astral configuration belongs neither to a trivial cohort nor to a systematic cohort, we shall say that the configuration and its cohort are **sporadic**. For $k = 2$ all sporadic configurations are known, and we list them in Section 3.6. However, for $k = 3$ we already have examples of such configurations (as discussed in Section 3.7) but no complete characterization.

If a cohort symbol $m\#\{s, t\}$ of a k-astral configuration contains the same integer in both s and t, a cohort symbol of a $(k-1)$-astral configuration may result if this integer is deleted from both s and t. As is easily verified, all the conditions for a $(k-1)$-astral symbol are automatically verified, except possibly (A5). We call **clade** of $m\#\{s, t\}$ all the cohorts resulting from one or several applications of this procedure. This will be illustrated in Section 3.7.

$$* \ * \ * \ * \ * \ *$$

We end this section with an unsolved problem of methodology in the study of k-astral configurations. We required in the definition that the symmetry group of every astral configuration be d_k. In fact, the other conditions show that this happens automatically if we require that the cyclic group c_k be a symmetry group of the configuration. The characteristic path, the symbol, and the reduced Levi graph of each k-astral configuration are all based on the cyclic group. The reason is that (as far as the author knows) nobody has come up with a reasonable version of all these based on the dihedral symmetry group. The construction of a reduced Levi graph that is based on the dihedral symmetry is certainly feasible—but does not appear to be useful. Why?

Exercises and Problems 3.5.

1. Devise symbols for the configurations in Figure 3.5.9.

2. What are the symbols for the objects in Figure 3.5.10? Devise a characteristic path in each and find out.

3. Superimpose each object in Figure 3.5.10 with a copy rotated $12°$ about the center. What is the result? Can you find a symbol for it?

4. Find the dual configurations of the ones in Figure 3.5.9.

5. Find a symbol for the 4-configuration in Figure 3.5.11.

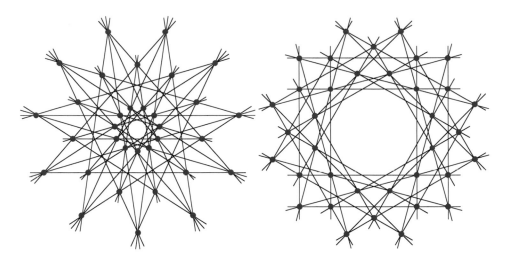

Figure 3.5.9. Two 3-astral 4-configurations.

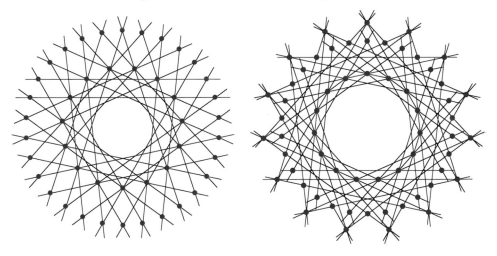

Figure 3.5.10. Not configurations!

Figure 3.5.11. A configuration (60_4).

3.6. 2-Astral 4-configurations

Following the terminology introduced in Section 3.5, a geometric 4-configuration (that is, an (n_4) configuration for some integer n) is called **2-astral** provided there are precisely two orbits of points and two orbits of lines under the symmetry group of the configurations and the other conditions spelled out in Section 3.5 are satisfied. Since $k = 2$ is the smallest value of k possible in a 4-configuration, following the convention proposed in Section 1.5, we shall call such configurations **astral** for short. An astral 4-configuration cannot have points at infinity, since any line through such a point would have to have three points of a single orbit in the finite part of the plane. Hence we need to consider only what happen in the Euclidean plane.

The astral 4-configurations have been completely characterized. To present this characterization, we need an appropriate notation; this was set up in Section 3.5. Here we shall present the list of these astral configurations (Theorem 3.6.1). Before giving the proof that our list is complete, we have to digress into explanations of some of the detailed results about the intersection of diagonals in regular polygons—a topic that has its own interesting and convoluted history. Finally, a proof of completeness of the list will be given; the first such proof is that of L. Berman [**6**], [**7**].

The notation for astral 4-configurations has evolved in several stages since the first publication on the topic in [**110**]. The notation used here, introduced in Section 3.5, is the one that was found most suitable for the present purpose as well as for the generalization to k-astral 4-configurations that we shall consider in Section 3.7. The notation is explained by the example of a (48_4) astral configuration shown in Figure 3.5.5. One of its symbols is $24\#(8, 7; 2, 5)$; the configuration belongs to the cohort $24\#\{\{8, 2\}, \{7, 5\}\}$; this cohort contains only one other configuration, with symbol $24\#(8, 5; 2, 7)$. Both configurations are shown in Figure 3.6.1.

In the next two figures we show the smallest astral 4-configurations. The unique (24_4) configuration is shown in Figure 3.6.2, while the six configurations (36_4) appear in Figure 3.6.3. Additional illustrations appear in several other sections but in particular in Section 5.8.

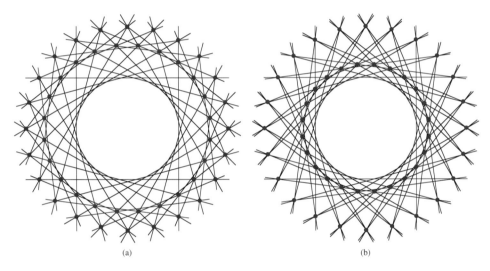

(a)   (b)

Figure 3.6.1. The only two 2-astral configurations (48_4) in the cohort $24\#\{\{8, 2\}, \{7, 5\}\}$. (a) The configuration $24\#(8, 7; 2, 5)$. (b) The configuration $24\#(8, 5; 2, 7)$.

After these preliminaries, here is the detailed result.

Theorem 3.6.1.[2] *Astral 4-configurations $m\#(s_1, t_1; s_2, t_2)$ must satisfy all the conditions from Section 3.5 and in particular the equation (A7):*

$$\cos(s_1\pi/m) \cdot \cos(s_2\pi/m) = \cos(t_1\pi/m) \cdot \cos(t_2\pi/m).$$

The symbols of these configurations are

   (i) *The systematic configurations with symbols $(6k)\#(3k - j, 2k; j, 3k - 2j)$ for $k \geq 2$, $1 \leq j < 3k/2$, with $j \neq k$.*

   (ii) *The systematic configurations with symbols $(6k)\#(2k, j; 3k - 2j, 3k - j)$ for $k \geq 2$, $1 \leq j < 3k/2$, with $j \neq k$. By the general results of Section*

---

[2]The author is indebted to L. Berman and T. Pisanski for a number of comments and corrections. These led to the present formulation, which we hope is more informative and useful than the statements in previous publications.

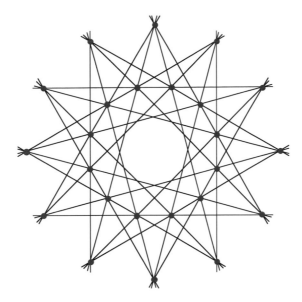

Figure 3.6.2. The smallest astral 4-configuration. It is a sporadic and selfdual (24_4) configuration, with symbol $12\#(5, 4; 1, 4)$.

3.5, *these configurations are polar to the ones in* (i) *with the same values of* k *and* j.

(iii) *The* 27 *symbols of the sporadic configurations listed in Table* 3.6.1 *and their multiples.*

For even k and $j = k/2$, the configurations in (i) and (ii) are selfpolar, hence coincide. If $k = f \cdot g$ and $j = f \cdot h$, with $f \geq 2$, $g \geq 2$, then both $(6k)\#(3k - j, 2k; j, 3k - 2j)$ and $(6k)\#(2k, j; 3k - 2j, 3k - j)$are disconnected. Each consists of f equidistributed copies of $(6g)\#(3g - h, 2g; h, 3g - 2h)$ or $(6g)\#(2g, h; 3g - 2h, 3g - h)$ and is denoted by $(f)(6g)\#(3g - h, 2g; h, 3g - 2h)$ or $(f)(6g)\#(2g, h; 3g - 2h, 3g - h)$, respectively.

For simpler formulation, we can say that the configurations in (i) and (ii) are in the cohorts of $(6k)\#\{\{3k - j\}, \{3k - 2, 2k\}\}$ for $k \geq 2$, $1 \leq j < 3k/2$, with $j \neq k$.

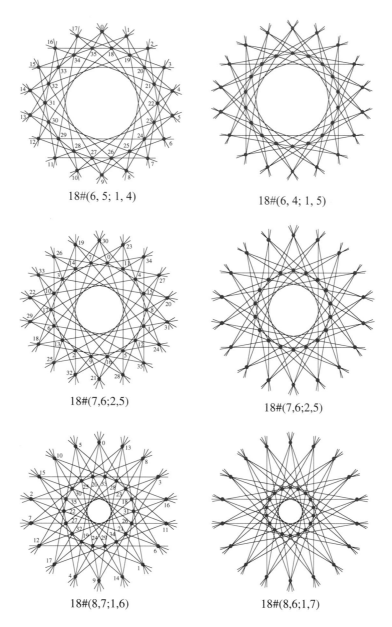

18#(6, 5; 1, 4)   18#(6, 4; 1, 5)

18#(7,6;2,5)   18#(7,6;2,5)

18#(8,7;1,6)   18#(8,6;1,7)

Figure 3.6.3. The six configurations (36_4) belong to three cohorts: $18\#\{\{6,1\},\{5,4\}\}$, $18\#\{\{7,2\},\{6,5\}$, $18\#\{\{8,1\},\{7,6\}\}$. Near each configuration we show the lexicographically highest among its symbols. Although it is not obvious from the symbols or the diagrams, the three configurations on the left are isomorphic. This isomorphism is established by the labels near their vertices. Since these configurations are isomorphic, their polars (shown on the right) are also isomorphic to each other; they are not isomorphic to the other three configurations.

Table 3.6.1. The complete list of connected sporadic astral 4-configurations. The three stand-alone symbols denote selfpolar configurations, and the paired symbols correspond to configurations polar to each other.

$$30\#(7,6;1,4) \qquad 30\#(7,4;1,6)$$
$$30\#(8,6;2,6)$$
$$30\#(11,10;1,6) \qquad 30\#(11,6;1,10)$$
$$30\#(12,10;6,10)$$
$$30\#(12,11;2,7) \qquad 30\#(12,7;2,11)$$
$$30\#(13,12;1,8) \qquad 30\#(13,8;1,12)$$
$$30\#(13,12;7,10) \qquad 30\#(13,10;7,12)$$
$$30\#(14,12;4,12)$$
$$30\#(14,13;6,11) \qquad 30\#(14,11;6,13)$$
$$42\#(13,12;1,6) \qquad 42\#(13,6;1,12)$$
$$42\#(18,17;6,11) \qquad 42\#(18,11;6,17)$$
$$42\#(19,18;5,12) \qquad 42\#(19,12;5,18)$$
$$60\#(22,21;2,9) \qquad 60\#(22,9;2,21)$$
$$60\#(25,24;5,12) \qquad 60\#(25,12;5,24)$$
$$60\#(27,26;3,14) \qquad 60\#(27,14;3,26)$$

Here, too, the cohorts notation allows a more condensed listing:

$$30\#\{\{7,1\},\{6,4\}\}, \qquad 30\#\{\{8,2\},\{6,6\}\}, \qquad 30\#\{\{11,1\},\{10,6\}\},$$
$$30\#\{\{12,6\},\{10,10\}\}, \quad 30\#\{\{12,2\},\{11,7\}\}, \quad 30\#\{\{13,1\},12,8\}\},$$
$$30\#\{\{13,7\},\{12,10\}\}, \quad 30\#\{\{14,4\},\{12,12\}\}, \quad 30\#\{\{14,6\},\{13,11\}\},$$
$$42\#\{\{13,1\},\{12,6\}\}, \quad 42\#\{\{18,6\},\{17,11\}\}, \quad 42\#\{\{19,5\},\{18,12\}\},$$
$$60\#\{\{22,2\},\{21,9\}\}, \quad 60\#\{\{25,5\},\{24,12\}\}, \quad 60\#\{\{27,3\},\{26,14\}\}.$$

The proof of the theorem will be interwoven with an account of the history of its development. In view of all the interest in configurations during the last quarter of the nineteenth century (as well as the sporadic interest later), it is hard to understand that *no graphical representation* of **any** 4-configuration appeared in print prior to [**122**] in 1990. The configuration shown above as Figure 3.6.2 was one of the configurations shown in that paper. Another was the (21_4) configuration that gave the paper its title; we shall encounter it again in Section 3.7.

In the early 1990s the author found several 4-configurations in addition to the ones in [**122**], with two or three orbits of points (and of lines); these were found by drawing with such software as was available to him (mainly MacDraw), until he was initiated to Mathematica® through friendly persuasion by Stan Wagon. (A few other k-astral 4-configurations with various k's were communicated to the author by J. F. Rigby.) With programs in

Mathematica® it was possible to "experimentally" find all possible astral configurations with reasonably small numbers of vertices. This led to the understanding that there are systematic infinite families of such configurations, as well as an apparently finite number of sporadic configurations. The author became convinced that he had a complete description and presented this in seminars and courses during the 1990s; the results were published in 2000 [**111**], together with formal demonstrations of the geometric realizability of these configurations. This covers the existence aspect of Theorem 3.6.1.

The main tool for the proof of completeness was the observation that an astral configuration $m\#(s_1, t_1; s_2, t_2)$ has a realization by straight lines if and only if the same points are reached starting from one of the regular polygons regardless of which of two diagonals we are following. In other words, the points described by $(s_1//t_1)$ must coincide with the points $(t_2//s_2)$. (Note that the designation $(s_2//t_2)$ used in determining the symbol of the configuration refers to the diagonals as looked upon from the other polygon.) This leads to the following necessary condition for the existence of an astral configuration $m\#(s_1, t_1; s_2, t_2)$:

(1) $\cos(s_1\pi/m) \cdot \cos(s_2\pi/m) = \cos(t_1\pi/m) \cdot \cos(t_2\pi/m)$.

Due to the dihedral symmetry of such configurations, this is also a sufficient condition for the existence. Moreover, criterion (1) is easily implemented for computational searches; the results of these calculations led to the classes listed in Theorem 3.6.1.

For the convenience of use of Theorem 3.6.1 we list in Table 3.6.2 the cohort symbols of the systematic astral configurations (n_4) with $n \leq 100$.

Once the characterization of the astral configurations has been guessed, it is easy to see that the symbols listed in Table 3.6.2 correspond to actual geometric configurations and are not results of an approximation error in the computations.

Indeed, for the symbols in part (i) we have to show that

(2) $\cos((3k - j)\pi/(6k)) \cdot \cos(\pi(6k)) = \cos(2k\pi/(6k)) \cdot \cos((3k - 2j)\pi/(6k))$.

In view of the trigonometric identity

(3) $(\cos a) \cdot (\cos b) = \frac{1}{2}(\cos(a + b) + \cos(a - b))$,

validity of (2) is equivalent to

$\frac{1}{2}(\cos 3k\pi/(6k) + \cos((3k - 2j)\pi/(6k))) = (\cos \pi/3) \cdot \cos((3k - 2j)\pi/(6k))$.

Since $\cos \pi/2 = 0$ and $\cos \pi/3 = 1/2$, this is valid for all k and j; hence (2) is correct. The same calculation shows that the symbols in (ii) correspond to astral geometric configurations as well. The fact that the above arguments

Table 3.6.2. The cohort symbols of all systematic astral (n_4) configurations with $n \leq 100$. Disconnected configurations are in italics.

$12\#\{\{5,1\},\{4,4\}$
$18\#\{\{8,1\},\{7,6\}\}, \qquad 18\#\{\{7,2\},\{6,5\}\}, \qquad 18\#\{\{6,1\},\{5,4\}\}$
$24\#\{\{11,1\},\{10,8\}\}, \qquad 24\#\{\{10,2\},\{8,8\}\}, \qquad 24\#\{\{9,3\},\{8,6\}\},$
$24\#\{\{8,2\},\{7,5\}\}$
$30\#\{\{14,1\},\{13,10\}\}, \quad 30\#\{\{13,2\},\{11,10\}\}, \quad 30\#\{\{12,3\},\{10,9\}\},$
$30\#\{\{11,4\},\{10,7\}\}, \quad 30\#\{\{10,3\},\{9,6\}\}, \qquad 30\#\{\{10,1\},\{8,7\}\}$
$36\#\{\{17,1\},\{16,12\}\}, \quad 36\#\{\{16,2\},\{14,12\}\}, \quad 36\#\{\{15,3\},\{12,12\}\},$
$36\#\{\{14,4\},\{12,10\}\}, \quad 36\#\{\{13,5\},\{12,6\}\}, \qquad 36\#\{(12,4\},\{11,7\}\},$
$36\#\{\{12,2\},\{10,8\}\}$
$42\#\{\{20,1\},\{19,14\}\}, \quad 42\#\{\{19,2\},\{20,16\}\}, \quad 42\#\{\{18,3\},\{15,14\}\},$
$42\#\{\{17,4\},\{14,13\}\}, \quad 42\#\{\{16,5\},\{14,11\}\}, \quad 42\#\{\{15,6\},\{14,9\}\},$
$42\#\{\{14,5\},\{13,8\}\}, \qquad 42\#\{\{14,3\},\{12,9\}\}, \qquad 42\#\{\{14,1\},\{11,10\}\}$
$48\#\{\{23,1\},\{22,16\}\}, \quad 48\#\{\{22,2\},\{20,16\}\}, \quad 48\#\{\{21,3\},\{18,16\}\},$
$48\#\{\{20,4\},\{16,16\}\}, \quad 48\#\{\{19,5\},\{16,14\}\}, \quad 48\#\{\{18,6\},\{16,12\}\},$
$48\#\{\{17,7\},\{16,10\}\}, \quad 48\#\{\{16,6\},\{15,9\}\}, \qquad 48\#\{\{16,4\},\{14,10\}\},$
$48\#\{\{16,2\},\{13,11\}\}$

did not rely on particular values of the cosines involved shows that (i) and (ii) are symbols of systematic configurations.

The existence of the sporadic configurations proceeds somewhat analogously but needs to rely on information specific to the angles involved. For example, concerning the configuration $30\#(8,6;2,6)$ we note that (1) becomes

$$\cos 8\pi/30 \cdot \cos 2\pi/30 = (\cos 6\pi/30)^2,$$

which by (3) is equivalent to

$$\tfrac{1}{2}(\cos 10\pi/30 + \cos 6\pi/30) = (\cos 6\pi/30)^2.$$

Since $\cos \pi/3 = 1/2$ and $\cos \pi/5 = \tfrac{1}{4}(1 + \sqrt{5})$, this reduces to

$$(12 + 4\sqrt{5})/16 = (6 + 2\sqrt{5})/8,$$

which is obviously true.

Using other explicit algebraic values for cosines, similar arguments can be made for the other sporadic configurations with symbols that start with 30 or 60. Among the values that can be used are

$$\cos 2\pi/30 = (-1 + \sqrt{5} + \sqrt{6(5 + \sqrt{5})})/8,$$
$$\cos 4\pi/30 = (1 + \sqrt{5} + \sqrt{6(5 - \sqrt{5})})/8,$$
$$\cos 8\pi/30 = (1 - \sqrt{5} + \sqrt{6(5 + \sqrt{5})})/8,$$

and so on.

For the symbols that involve 42, it is convenient to follow a slightly different path. The validity of the first of these symbols, $42\#(13, 12; 1, 6)$, is by (1) and (3) equivalent to

$$\cos \pi/3 + \cos 2\pi/7 = \cos 3\pi/7 + \cos \pi/7,$$

that is,

$$1 + 2 \cos 2\pi/7 + 2 \cos 4\pi/7 + 2 \cos 6\pi/7 = 0.$$

But this is simply an expression of the fact that the centroid of a regular heptagon, centered at the origin and with one vertex at $(1, 0)$, is itself at the origin. A completely analogous reasoning shows the validity of the other symbols involving 42.

What is still missing is a proof that there are no other astral 4-configurations. Since these configurations are determined by intersections of diagonals of regular polygons and since these have been extensively studied and completely determined, in the late 1990s it seemed to the author that it should be very easy to supply the proof of completeness.

In reality this task proved far from simple, and it was first successfully carried out in 2001 in the Ph.D. work of L. Berman [**7**], [**6**]. Berman's rather complicated argumentation relied on the complete description of intersections of diagonals of regular polygons, given by Poonen and Rubinstein [**185**] in 1998. Theirs was a new proof (and a much more convenient presentation) of material that has been contained, to a large extent, in earlier publications of G. Bol [**30**] in 1933 (with some misprints) and J. F. Rigby [**189**] in 1980.[3] For regular n-gons with prime n, or with any odd n, it has been proved by many authors that there are no intersections of more than two diagonals; references to these papers and other related material can be found in [**189**] and especially in [**185**].

However, independently of these developments, an approach that is easier to apply for our purposes was published by G. Myerson [**173**] in 1993; it came to the author's attention only recently. Myerson's result (his Theorem 4) that is relevant to our proof can be formulated as follows.

Theorem 3.6.2 (Myerson [**173**]). *The equation*

$$\sin \pi/6 \cdot \sin t = \sin(t/2) \sin(\pi/2 - t/2)$$

is valid for all t. The only other solutions of the equation

$$(4) \qquad\qquad \sin x_1 \pi \cdot \sin x_2 \pi = \sin x_3 \pi \cdot \sin x_4 \pi$$

---

[3]In contrast to other writers on the topic, Rigby considers the multiple intersections of diagonals outside the n-gon as well. However, his intended [**189**, p. 222] investigation of outside intersections of four or more diagonals seems not to have been published and remains an open problem.

Table 3.6.3. The complete list of sporadic solutions of equation (4) as given by Myerson in [**173**].

Label	x_1	x_2	x_3	x_4
1	1/21	8/21	1/14	3/14
2	1/14	5/14	2/21	5/21
3	4/21	10/21	3/14	5/14
4	1/20	9/20	1/15	4/15
5	2/15	7/15	3/20	7/20
6	1/30	3/10	1/15	2/15
7	1/15	7/15	1/10	7/30
8	1/10	13/30	2/15	4/15
9	4/15	7/15	3/10	11/30
10	1/30	11/30	1/10	1/10
11	7/30	13/30	3/10	3/10
12	1/15	4/15	1/10	1/6
13	2/15	8/15	1/6	3/10
14	1/12	5/12	1/10	3/10
15	1/10	3/10	1/6	1/6

in rational x_1, x_2, x_3, x_4 *with* $0 < x_1 < x_3 \leq x_4 < x_2 \leq \frac{1}{2}$ *are given in Table 3.6.3.*

The result of Theorem 3.6.2 gives an immediate solution to the completeness question of Table 3.6.1. Indeed, we only have to recall that $\sin \alpha = \cos(\pi/2 - \alpha)$ in order to see that the rows of Table 3.6.3 correspond (in an appropriate permutation) to the rows of Table 3.6.1. For example, rows with labels 1, 2, 3 correspond to the entries involving 42 of the earlier table, while those labeled 4, 5, and 14 correspond to the last three rows of Table 3.6.1.

This completes the proof of Theorem 3.6.1.

Exercises and Problems 3.6.

1. Verify the complete correspondence between Myerson's list in Table 3.6.3 and the list of sporadic symbols in Table 3.6.1.

2. Verify the validity of the existence claims made above for all sporadic configurations.

3. Draw the configuration $36\#(15, 12; 3, 12) = (3)12\#(5, 4; 1, 4)$. Is it selfpolar?

4. Prove that the configurations $18\#(6, 5; 1, 4)$ and $18\#(6, 4; 1, 5)$ shown in Figure 3.6.3 are not isomorphic.

5. Determine whether the pairs of polar configurations in Figure 3.6.3 are in appropriate orientation to exhibit the polarity or if one member of the pair has to be rotated.

3.7. 3-Astral 4-configurations

The 3-astral 4-configurations have a lot in common with the 2-astral configurations we studied in Section 3.6, but they also have many properties and peculiarities that are not present in the earlier case. This is the main reason for treating them in a separate section.

It seems to the author that the 3-astral configurations are arguably the most interesting type of configurations. The reason for this assessment is that they are more general in the opportunities for investigation than the 2-astral 4-configurations considered in the preceding section, but they are still experimentally quite accessible. As we shall show, there are many open problems that seem very attractive, as well as tractable with an appropriate investment of effort. Naturally, k-astral configurations with $k \geq 4$ have their own attraction and appeal, but with increasing k, they are harder to investigate and, in any case, much less is known about them.

The first graphic presentation in a published paper of any 4-configuration[4] was that of a 3-astral (21_4) configuration in [122]; here we show it in Figure 3.7.1. As with 2-astral configurations, 3-astral configurations will be illustrated in various section; several are shown in Section 5.8.

The notation we use for the k-astral configurations is the one introduced in Section 3.5, based on characteristic paths. In the case of 3-astral configurations there are, in general, 6 different characteristic paths, leading to distinct symbols. We preferentially choose the symbol that is lexicographically the highest.

For 3-astral configurations our approach is completely analogous to the treatment of 2-astral configurations in Section 3.6. It would be nice if at this stage we could formulate a theorem analogous to Theorem 3.6.1 and give necessary and sufficient conditions for symbols corresponding to geometric 3-astral 4-configurations. However, we have only partial information; the most important criterion is condition (A7) from Section 3.5, the analog of the trigonometric condition in Theorem 3.6.1:

---

[4]Kárteszi [133] in 1986 and Zeitler [238] in 1987 came very close to finding such 4-configurations. In the diagrams they show that one only has to delete some lines and make all copies of one of the shown lines to get a 3-astral 4-configuration, with $m = 10$ in the former and $m = 12$ in the other.

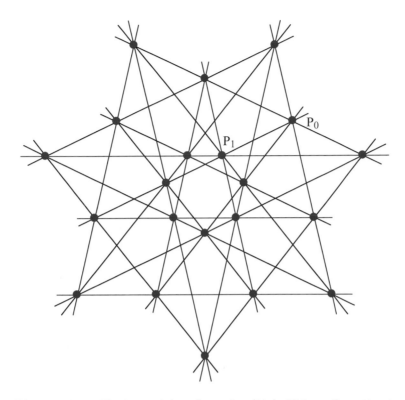

Figure 3.7.1. The 3-astral 4-configuration (21_4). This configuration is 3-astral with symbol $7\#(3,2;1,3;2,1)$ obtained from the characteristic path that starts at P_0 and has as its next point P_1. This configuration belongs to the "trivial" type.

Theorem 3.7.1. *If $m\#(s_1,t_1;s_2,t_2;s_3,t_3)$ is the symbol of a geometric 3-astral 4-configuration, then*

$$(*) \quad \cos s_1\pi/m \cdot \cos s_2\pi/m \cdot \cos s_3\pi/m = \cos t_1\pi/m \cdot \cos t_2\pi/m \cdot \cos t_3\pi/m.$$

This is an expression of the fact that each characteristic path has its endpoint in the same orbit as its starting point. Conversely, if a symbol satisfies (*) and the natural conditions listed in Section 3.5, then *in general* there exists a geometric 3-astral 4-configuration with the symbol in question.

The natural conditions just mentioned are

(**) No entry is equal to either of the two adjacent entries, the first and last considered as adjacent.

(***) Each entry is smaller than $m/2$.

(****) The sum of all s_j and t_j entries in the symbol is even.

There are two deep deficiencies in this theorem. One unsatisfactory aspect of Theorem 3.7.1 (and of the analogous statements one can make for $k \geq 4$) is that we do not have any analogue of Myerson's Theorem

3.6.2; hence we cannot devise a list of all the configurations in question. As is stated in [**173**], Myerson's methods could probably lead to a complete, explicit list of solutions of (*), but this appears to be a momentous task—a task that has not been carried out. This is the first big problem concerning 3-astral 4-configurations.

The other problem is euphemistically covered by the italicized words "in general". We shall discuss later in the section the known and the unknown results in this direction.

For the presentation of *known* solutions of (*) satisfying all the necessary conditions, it is convenient to distinguish three kinds of symbols $m\#(s_1, t_1; s_2, t_2; s_3, t_3)$ or the corresponding cohorts $m\#\{\{s_1, s_2, s_3\}, \{t_1, t_2, t_3\}\}$, which we shall call "trivial", "systematic", and "sporadic". The terminology was introduced in Section 3.5, and here we only briefly remind the reader of the meaning of these terms.

• In accordance with this terminology, **trivial** symbols (and 3-astral configurations) have the form $m\#(b, c; d, b; c, d)$, where b, c, d are different positive integers, each less than $m/2$. Since the terms on the two sides cancel each other without any calculations and the other conditions are automatically satisfied, the label "trivial" seems appropriate—not in any derogatory sense but as describing the mathematically simplest case. In other words, those astral configurations for which the cohort symbol $m\#\{s, t\}$ is of the special form $m\#\{s, s\}$ are trivial. Figure 3.7.1 shows an example of a trivial 3-astral configuration. From the general properties of equivalent symbols and symbols of dual configuration discussed in Section 3.5, it follows that all trivial 3-astral configurations are selfdual; the same applies to all trivial k-astral configurations with odd k. Indeed, $m\#(b, c, d, b, c, d)$ has as dual $m\#(c, d, b, c, d, b)$, which is the same as the original; the argument is analogous for other odd k. Obviously, the polar of a trivial configuration is itself trivial.

• **Systematic** symbols are those that contain infinite families for which the validity can be verified by *formal* trigonometric calculations, without the need to determine *values* of the trigonometric functions that depend on specific parameters. At present, four such families $m\#\{s, t\}$ of 3-astral 4-configurations are known, mostly through unpublished work of L. Berman.

(1) $m = 2q,$ $s = \{q - p, q - 2r, p\},$ $t = \{q - 2p, q - r, r\},$

(2) $m = 3q,$ $s = \{q + p, q - p, p\},$ $t = \{q, q, 3p\},$

(3) $m = 6q,$ $s = \{3q - p, r, p\},$ $t = \{3q - 2p, 2q, r\},$

(4) $m = 10q,$ $s = \{5q - p, 2p, p\},$ $t = \{|5q - 4p|, 4q, 2q\}.$

Here p, q, and r are any positive integers, and the possibilities have to be understood in the sense of cohorts; that is, all permutations within s and within t, as well as interchanging s and t, are allowed provided conditions (**) and (***) are satisfied; condition (****) is fulfilled automatically. If all the entries are distinct, the cohort contains 12 distinct configurations; equality of some entries reduces this number.

For example, in family (4) with $m = 20$, $q = 2$. For $p = 1$ we have $s = \{9, 2, 1\}$ and $t = \{8, 6, 4\}$; this leads to a total of 12 distinct symbols: $20\#(9, 8; 2, 6; 1, 4)$, $20\#(9, 8; 2, 4; 1, 6)$, $20\#(9, 6; 2, 8; 1, 4)$, $20\#(9, 6; 2, 4; 1, 8)$, $20\#(9, 4; 2, 8; 1, 6)$, $20\#(9, 4; 2, 6; 1, 8)$, and six more in which the positions of 1 and 2 are interchanged. Switching the two sets of parameters yields an additional 12 symbols, but no new configurations, since these symbols are equivalent to the earlier dozen. For $p = 3$ we have $s = \{7, 6, 3\}$ and $t = \{8, 4, 2\}$ leading again to 12 different symbols. For $p = 2$ and $p = 4$ we get trivial symbols only, while $p \geq 5$ exceeds the bound in (***). On the other hand, the cohort of $9\#\{\{4, 2, 1\}, \{3, 3, 3\}\}$ consists of just two configurations, shown in Figure 3.7.2. As mentioned above, in some cases the resulting symbols become trivial.

The verification that the above symbols of the four families satisfy condition (*) is quite straightforward. We illustrate this only for family (2), for which condition (*) reduces to the verification of

$$\cos(q+p)\pi/m \cdot \cos(q-p)\pi/m \cdot \cos p\pi/m = (\cos q\pi/m)^2 \cdot \cos 3p\pi/m.$$

Since $q = m/3$, the value of $\cos q\pi/m = \cos\pi/3 = 1/2$, and standard trigonometric identities yield

$$2\cos(q+p)\pi/m \cdot (\cos q\pi/m + \cos(q - 2p)\pi/m) = \cos 3p\pi/m,$$

so

$$\cos(q+p)\pi/m + 2\cos(q+p)\pi/m \cdot \cos(q - 2p)\pi/m = \cos 3p\pi/m,$$
$$\cos(p\pi/m + \pi/3) + \cos(2q - p)\pi/m + \cos 3p\pi/m = \cos 3p\pi/m,$$

and finally

$$\cos(p\pi/m + \pi/3) + \cos(-p\pi/m + 2\pi/3) = 0,$$

which is obviously true.

Similar calculations validate the other families of symbols.

• **Sporadic** symbols and configurations are those that do not belong to any of these two families. For example, $18\#(5, 4; 1, 3; 1, 2)$ shown in Figure 3.7.3 is sporadic—*at least for the time being*. The reason for this qualification is that although the symbol is not trivial nor does it belong to one of the four known systematic families, it may well be part of a still undiscovered infinite (systematic) family.

In Table 3.7.1 we give a list of cohorts of the sporadic 3-astral configurations (n_4) with $n \leq 108$. It was obtained by numerically solving equation

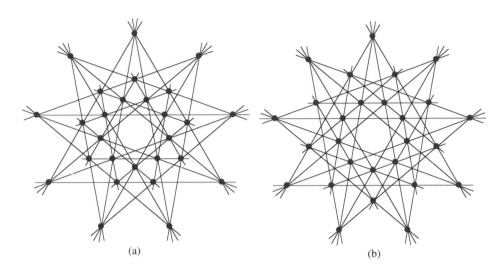

(a)   (b)

Figure 3.7.2. Depiction of the only two distinct members in the cohort $9\#\{\{4, 2, 1\}, \{3, 3, 3\}\}$ of family (2) for $q = 3$, $p = 1$. (a) $9\#(4, 3, 2, 3, 1, 3)$; (b) $9\#(4, 3, 1, 3, 2, 3)$.

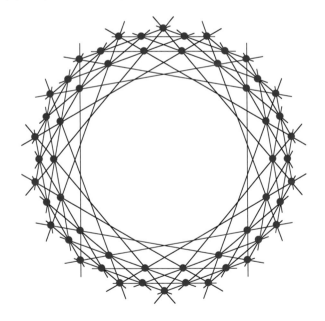

Figure 3.7.3. The sporadic configuration $18\#(5, 4; 1, 3; 1, 2)$.

(*) and eliminating duplicates and symbols that correspond to trivial or systematic 3-astral configurations. An unexpected result of these computations is that such sporadic configurations exist only for n that are multiples of 12.

We make one additional comment concerning Table 3.7.1. Some of the cohort symbols contain the same symbol in both parts. This implies that the crucial relation (*) will be satisfied even if the symbol is deleted from both parts. In such a case the **reduced** cohort symbol belongs to a 2-astral 4-configuration. This brings us to other open problems:

▷ Do there exist any other systematic families besides the ones listed above?

▷ Is the list of connected sporadic 3-astral configurations finite?

It is worth noting that there is no known visible cue in a given 3-astral configuration whether it is trivial, systematic, or sporadic. It takes working out its symbol and looking at the criteria in order to decide where it belongs.

We turn now to the second deficiency in Theorem 3.7.1. It parallels the problems with the Steinitz theorem on 3-configurations encountered in Chapter 2 and did not arise for 2-astral 4-configurations.

Table 3.7.1. A list of all sporadic 3-astral 4-configurations (n_4) with $n \leq 72$. In the cohort notation used, each entry corresponds to a cohort of configurations. Reducible cohorts are indicated by an asterisk following the symbol. Equal uppercase letters indicate that the configurations reduce to the same 2-astral configuration. In the case of the symbols for $m = 30$, some have a common factor 2; however, they are not disconnected, since in each case the symbol corresponding to entries one-half of the ones given would violate condition (****). Equal lowercase letters attached to the symbols indicate that the symbols share one of the two parentheses; while it is not clear what this commonality implies, it is signaled to ease possible investigations.

$m = 18$

$\{5,1,1\}, \{4,3,2\}$	$\{6,2,1\}, \{5,4,2\}^*$
$\{6,3,2\}, \{5,5,1\}$	$\{7,1,1\}, \{6,4,3\}$
$\{7,4,2\}, \{6,6,1\}$	$\{7,6,1\}, \{7,5,4\}^*A$
$\{8,2,1\}, \{8,6,5\}a$	$\{8,3,2\}, \{7,5,5\}$
$\{8,4,3\}, \{7,7,1\}$	$\{8,5,4\}, \{7,6,6\}b$
$\{8,6,1\}, \{8,5,4\}^*bA$	$\{8,6,3\}, \{7,7,5\}$
$\{8,7,2\}, \{8,6,5\}^*a$	

$m = 24$

$\{6,2,1\}, \{5,3,3\}$	$\{8,2,1\}, \{7,5,1\}^*A$
$\{8,3,2\}, \{7,5,3\}^*A$	$\{8,3,3\}, \{7,6,1\}$
$\{8,6,2\}, \{7,6,5\}^*A$	$\{9,2,1\}, \{8,5,3\}$
$\{9,8,2\}, \{9,7,5\}^*A$	$\{10,3,2\}, \{9,7,1\}$
$\{10,5,3\}, \{9,8,1\}$	$\{10,6,5\}, \{9,9,1\}$
$\{10,7,5\}, \{8,8,8\}$	$\{10,8,2\}, \{10,7,5\}^*A$
$\{10,9,3\}, \{10,8,6\}^*B$	$\{11,2,1\}, \{8,8,8\}$
$\{11,3,2\}, \{9,8,7\}$	$\{11,3,3\}, \{10,7,6\}$

Table 3.7.1. (Continued)

$m = 24$

$\{11, 5, 3\}, \{10, 9, 2\}$	$\{11, 6, 2\}, \{9, 9, 7\}$
$\{11, 6, 5\}, \{9, 9, 8\}$	$\{11, 8, 2\}, \{11, 7, 5\}^*A$
$\{11, 8, 3\}, \{10, 9, 7\}$	$\{11, 9, 3\}, \{11, 8, 6\}^*B$
$\{11, 10, 2\}, \{11, 8, 8\}^*$	

$m = 30$

$\{7, 2, 1\}, \{6, 4, 2\}^*A$	$\{7, 3, 1\}, \{6, 4, 3\}^*A$
$\{7, 5, 1\}, \{6, 5, 4\}^*A$	$\{8, 2, 1\}, \{6, 6, 1\}^*B$
$\{8, 3, 2\}, \{6, 6, 3\}^*B$	$\{8, 4, 2\}, \{6, 6, 4\}^*aB$
$\{8, 4, 2\}, \{7, 6, 1\}a$	$\{8, 5, 2\}, \{6, 6, 5\}^*$
$\{8, 7, 1\}, \{8, 6, 4\}^*A$	$\{8, 7, 2\}, \{7, 6, 6\}^*bB$
$\{9, 1, 1\}, \{8, 4, 3\}$	$\{9, 2, 1\}, \{7, 6, 3\}$
$\{9, 4, 2\}, \{7, 7, 3\}$	$\{9, 7, 1\}, \{9, 6, 4\}^*eA$
$\{9, 8, 2\}, \{9, 6, 6\}^*cB$	$\{10, 1, 1\}, \{8, 6, 4\}$
$\{10, 2, 1\}, \{7, 6, 6\}b$	$\{10, 2, 1\}, \{8, 7, 2\}^*bC$
$\{10, 3, 1\}, \{8, 7, 3\}^*dC$	$\{10, 3, 1\}, \{9, 6, 1\}^*dD$
$\{10, 3, 2\}, \{9, 6, 2\}^*D$	$\{10, 4, 1\}, \{8, 7, 4\}^*C$
$\{10, 4, 2\}, \{7, 7, 6\}$	$\{10, 4, 3\}, \{9, 6, 4\}^*eD$
$\{10, 4, 3\}, \{9, 7, 1\}e$	$\{10, 6, 1\}, \{8, 7, 6\}^*C$
$\{10, 6, 3\}, \{9, 8, 2\}c$	$\{10, 6, 4\}, \{8, 7, 7\}$
$\{10, 7, 1\}, \{10, 6, 4\}^*A$	$\{10, 7, 3\}, \{9, 7, 6\}^*D$
$\{10, 8, 2\}, \{10, 6, 6\}^*B$	$\{10, 8, 3\}, \{9, 8, 6\}^*D$
$\{10, 9, 1\}, \{9, 8, 7\}^*C$	$\{11, 1, 1\}, \{8, 7, 6\}$
$\{11, 2, 1\}, \{10, 6, 2\}^*E$	$\{11, 3, 1\}, \{9, 6, 6\}c$
$\{11, 3, 1\}, \{9, 8, 2\}c$	$\{11, 3, 1\}, \{10, 6, 3\}^*cE$
$\{11, 4, 1\}, \{10, 6, 4\}^*E$	$\{11, 5, 1\}, \{10, 6, 5\}^*E$
$\{11, 6, 1\}, \{10, 8, 2\}$	$\{11, 7, 1\}, \{10, 7, 6\}^*fE$
$\{11, 7, 1\}, \{11, 6, 4\}^*fA$	$\{11, 7, 3\}, \{10, 9, 2\}$
$\{11, 7, 6\}, \{10, 10, 2\}j$	$\{11, 8, 1\}, \{10, 8, 6\}^*E$
$\{11, 8, 2\}, \{11, 6, 6\}^*B$	$\{11, 8, 4\}, \{10, 10, 1\}$
$\{11, 8, 7\}, \{10, 10, 6\}$	$\{11, 9, 1\}, \{10, 9, 6\}^*E$
$\{11, 10, 1\}, \{11, 8, 7\}^*gC$	$\{11, 10, 3\}, \{11, 9, 6\}^*D$
$\{12, 1, 1\}, \{10, 8, 4\}$	$\{12, 2, 1\}, \{10, 7, 6\}f$
$\{12, 2, 1\}, \{11, 6, 4\}f$	$\{12, 2, 1\}, \{11, 7, 1\}^*fF$
$\{12, 3, 2\}, \{11, 7, 3\}^*F$	$\{12, 4, 2\}, \{10, 7, 7\}h$
$\{12, 4, 2\}, \{11, 7, 4\}^*hF$	$\{12, 5, 2\}, \{11, 7, 5\}^*F$
$\{12, 6, 1\}, \{10, 8, 7\}i$	$\{12, 6, 1\}, \{10, 10, 1\}^*iG$
$\{12, 6, 2\}, \{10, 10, 2\}^*jG$	$\{12, 6, 2\}, \{11, 7, 6\}^*jF$
$\{12, 6, 3\}, \{10, 10, 3\}^*kG$	$\{12, 6, 3\}, \{11, 9, 1\}k$
$\{12, 6, 4\}, \{10, 10, 4\}^*lG$	$\{12, 6, 6\}, \{11, 8, 7\}g$
$\{12, 6, 6\}, \{11, 10, 1\}g$	$\{12, 7, 1\}, \{10, 10, 4\}l$

Table 3.7.1. (Continued)

$m = 30$

$\{12,7,1\},\{12,6,4\}^*lA$	$\{12,7,6\},\{10,10,7\}^*mG$
$\{12,7,6\},\{11,10,4\}m$	$\{12,8,2\},\{12,6,6\}^*gB$
$\{12,8,2\},\{11,8,7\}^*gF$	$\{12,8,2\},\{11,10,1\}g$
$\{12,8,6\},\{10,10,8\}^*G$	$\{12,9,2\},\{11,9,7\}^*F$
$\{12,9,6\},\{10,10,9\}^*nG$	$\{12,10,1\},\{12,8,7\}^*qC$
$\{12,10,2\},\{11,10,7\}oF$	$\{12,10,2\},\{11,11,4\}o$
$\{12,10,3\},\{12,9,6\}^*nD$	$\{12,11,1\},\{10,10,10\}p$
$\{12,11,1\},\{12,10,6\}^*pE$	$\{12,11,4\},\{12,10,7\}^*H$
$\{12,11,6\},\{11,10,10\}^*G$	$\{13,2,1\},\{10,10,6\}r$
$\{13,2,1\},\{12,6,6\}r$	$\{13,2,1\},\{12,8,2\}^*rJ$
$\{13,3,1\},\{10,9,8\}s$	$\{13,3,1\},\{12,8,3\}^*sJ$
$\{13,4,1\},\{12,8,4\}^*J$	$\{13,4,2\},\{10,10,7\}$
$\{13,4,3\},\{12,9,1\}$	$\{13,5,1\},\{12,8,5\}^*J$
$\{13,6,1\},\{10,10,8\}t$	$\{13,6,1\},\{12,8,6\}^*tJ$
$\{13,6,2\},\{11,11,1\}$	$\{13,6,3\},\{11,9,8\}$
$\{13,6,4\},\{12,8,7\}q$	$\{13,6,4\},\{12,10,1\}q$
$\{13,6,6\},\{11,10,8\}u$	$\{13,7,1\},\{12,8,7\}^*qJ$
$\{13,7,1\},\{12,10,1\}^*qK$	$\{13,7,1\},\{13,6,4\}^*qA$
$\{13,7,2\},\{12,10,2\}^*K$	$\{13,7,3\},\{12,9,6\}n$
$\{13,7,3\},\{12,10,3\}^*nK$	$\{13,7,4\},\{12,10,4\}^*K$
$\{13,7,5\},\{12,10,5\}^*K$	$\{13,7,6\},\{10,10,10\}p$
$\{13,7,6\},\{12,10,6\}^*pK$	$\{13,7,6\},\{12,11,1\}p$
$\{13,7,7\},\{12,11,4\}$	$\{13,8,2\},\{13,6,6\}^*uB$
$\{13,8,7\},\{12,10,8\}^*vK$	$\{13,9,1\},\{12,9,8\}^*J$
$\{13,9,6\},\{13,10,3\}^*D$	$\{13,9,7\},\{12,10,9\}^*wK$
$\{13,9,7\},\{12,12,3\}w$	$\{13,10,1\},\{12,10,8\}^*vJ$
$\{13,10,1\},\{13,8,7\}^*vC$	$\{13,10,2\},\{12,11,6\}$
$\{13,10,3\},\{13,9,6\}^*D$	$\{13,10,4\},\{12,12,1\}$
$\{13,10,6\},\{12,11,8\}y$	$\{13,10,7\},\{12,12,6\}x$
$\{13,11,1\},\{12,11,8\}^*yJ$	$\{13,11,1\},\{13,10,6\}^*yE$
$\{13,11,4\},\{12,10,10\}x$	$\{13,11,4\},\{12,12,6\}x$
$\{13,11,4\},\{13,10,7\}^*xH$	$\{13,11,7\},\{12,11,10\}^*zK$
$\{13,12,2\},\{13,11,7\}^*zF$	$\{13,12,3\},\{13,10,9\}^*L$
$\{13,12,6\},\{13,10,10\}^*G$	$\{14,1,1\},\{12,10,8\}$
$\{14,2,1\},\{11,10,10\}$	$\{14,3,2\},\{11,11,9\}$
$\{14,4,1\},\{12,12,1\}^*M$	$\{14,4,2\},\{12,11,7\}aa$
$\{14,4,2\},\{12,12,2\}^*aaM$	$\{14,4,3\},\{12,10,9\}w$
$\{14,4,3\},\{12,12,3\}^*wM$	$\{14,4,3\},\{13,9,7\}w$
$\{14,5,4\},\{12,12,5\}^*M$	$\{14,6,1\},\{12,11,8\}y$
$\{14,6,1\},\{13,11,1\}^*yN$	$\{14,6,2\},\{11,11,10\}bb$
$\{14,6,2\},\{13,11,2\}^*bbN$	$\{14,6,3\},\{13,11,3\}^*N$

Table **3.7.1.** (Continued)

$m = 30$

$\{14, 6, 4\}, \{12, 12, 6\}^*ccM$ $\{14, 6, 4\}, \{13, 10, 7\}cc$
$\{14, 6, 4\}, \{13, 11, 4\}^*ccN$ $\{14, 6, 5\}, \{13, 11, 5\}^*N$
$\{14, 7, 1\}, \{12, 10, 10\}cc$ $\{14, 7, 1\}, \{12, 12, 6\}cc$
$\{14, 7, 1\}, \{14, 6, 4\}ccA$ $\{14, 7, 3\}, \{12, 11, 9\}$
$\{14, 7, 4\}, \{12, 12, 7\}^*M$ $\{14, 7, 6\}, \{12, 11, 10\}z$
$\{14, 7, 6\}, \{13, 11, 7\}^*zN$ $\{14, 7, 6\}, \{13, 12, 2\}z$
$\{14, 8, 2\}, \{13, 11, 6\}$ $\{14, 8, 2\}, \{14, 6, 6\}^*B$
$\{14, 8, 4\}, \{12, 12, 8\}^*M$ $\{14, 8, 4\}, \{13, 12, 1\}$
$\{14, 8, 6\}, \{13, 11, 8\}^*N$ $\{14, 8, 7\}, \{13, 10, 10\}dd$
$\{14, 9, 4\}, \{12, 12, 9\}^*M$ $\{14, 9, 6\}, \{13, 11, 9\}^*N$
$\{14, 9, 6\}, \{14, 10, 3\}$ $\{14, 9, 8\}, \{13, 13, 3\}$
$\{14, 10, 1\}, \{13, 12, 6\}dd$ $\{14, 10, 1\}, \{14, 8, 7\}^*ddC$
$\{14, 10, 2\}, \{12, 11, 11\}$ $\{14, 10, 3\}, \{13, 11, 9\}$
$\{14, 10, 3\}, \{14, 9, 6\}^*D$ $\{14, 10, 4\}, \{12, 12, 10\}^*iiM$
$\{14, 10, 4\}, \{13, 12, 7\}ii$ $\{14, 10, 6\}, \{13, 11, 10\}^*eeN$
$\{14, 10, 6\}, \{13, 13, 2\}ee$ $\{14, 10, 7\}, \{12, 12, 11\}ff$
$\{14, 10, 8\}, \{13, 13, 6\}$ $\{14, 10, 10\}, \{13, 12, 11\}gg$
$\{14, 11, 1\}, \{14, 10, 6\}^*eeE$ $\{14, 11, 4\}, \{12, 12, 11\}^*ffM$
$\{14, 11, 4\}, \{14, 10, 7\}^*ffH$ $\{14, 12, 1\}, \{13, 13, 7\}$
$\{14, 12, 2\}, \{14, 11, 7\}^*F$ $\{14, 12, 3\}, \{14, 10, 9\}^*L$
$\{14, 12, 6\}, \{13, 12, 11\}^*ggN$ $\{14, 12, 6\}, \{14, 10, 10\}^*ggG$
$\{14, 12, 8\}, \{13, 13, 10\}hh$ $\{14, 13, 1\}, \{14, 12, 8\}^*hhJ$
$\{14, 13, 2\}, \{14, 11, 10\}^*$ $\{14, 13, 4\}, \{13, 12, 12\}^*M$
$\{14, 13, 7\}, \{14, 12, 10\}^*K$

$m = 36$

$\{8, 4, 1\}, \{7, 5, 3\}$ $\{11, 2, 1\}, \{10, 5, 3\}$
$\{11, 7, 6\}, \{10, 10, 2\}$ $\{12, 2, 1\}, \{10, 8, 1\}^*A$
$\{12, 3, 2\}, \{10, 8, 3\}^*A$ $\{12, 4, 1\}, \{11, 7, 1\}^*B$
$\{12, 4, 2\}, \{11, 7, 2\}^*B$ $\{12, 4, 3\}, \{11, 7, 3\}^*B$
$\{12, 5, 2\}, \{10, 8, 5\}^*A$ $\{12, 5, 3\}, \{11, 8, 1\}$
$\{12, 5, 4\}, \{11, 7, 5\}^*B$ $\{12, 7, 2\}, \{10, 8, 7\}^*A$
$\{12, 8, 4\}, \{11, 8, 7\}^*B$ $\{12, 9, 2\}, \{10, 9, 8\}^*A$
$\{12, 9, 4\}, \{11, 9, 7\}^*B$ $\{12, 10, 4\}, \{11, 10, 7\}^*B$
$\{12, 11, 2\}, \{11, 10, 8\}^*A$ $\{13, 4, 1\}, \{12, 7, 3\}$
$\{13, 10, 5\}, \{12, 12, 2\}$ $\{13, 12, 2\}, \{13, 10, 8\}^*A$
$\{13, 12, 4\}, \{13, 11, 7\}^*B$ $\{14, 2, 2\}, \{13, 6, 5\}$
$\{14, 7, 3\}, \{13, 10, 1\}$ $\{14, 11, 7\}, \{12, 12, 10\}$
$\{14, 12, 4\}, \{14, 11, 7\}^*B$ $\{14, 13, 5\}, \{14, 12, 8\}^*$
$\{15, 2, 1\}, \{14, 7, 5\}$ $\{15, 3, 2\}, \{13, 10, 5\}$
$\{15, 3, 2\}, \{14, 8, 4\}$ $\{15, 3, 3\}, \{13, 11, 1\}$

Table 3.7.1. (Continued)

$m = 36$

$\{15, 6, 3\}, \{14, 10, 2\}$	$\{15, 10, 3\}, \{14, 11, 7\}$
$\{15, 12, 2\}, \{15, 10, 8\}^*A$	$\{15, 12, 4\}, \{15, 11, 7\}^*B$
$\{15, 13, 5\}, \{15, 12, 8\}^*$	$\{15, 14, 4\}, \{15, 12, 10\}^*$
$\{16, 4, 2\}, \{15, 10, 3\}$	$\{16, 5, 4\}, \{15, 11, 1\}$
$\{16, 7, 5\}, \{15, 12, 1\}$	$\{16, 8, 4\}, \{15, 12, 3\}$
$\{16, 8, 7\}, \{15, 13, 1\}$	$\{16, 10, 8\}, \{15, 14, 3\}$
$\{16, 12, 4\}, \{16, 11, 7\}^*B$	$\{16, 13, 5\}, \{16, 12, 8\}^*$
$\{16, 15, 3\}, \{16, 12, 12\}^*C$	$\{17, 2, 1\}, \{14, 12, 12\}$
$\{17, 2, 1\}, \{15, 14, 3\}$	$\{17, 3, 2\}, \{14, 13, 11\}$
$\{17, 5, 3\}, \{16, 11, 8\}$	$\{17, 5, 4\}, \{15, 12, 11\}$
$\{17, 6, 1\}, \{14, 14, 10\}$	$\{17, 7, 2\}, \{15, 13, 10\}$
$\{17, 7, 3\}, \{16, 13, 4\}$	$\{17, 7, 5\}, \{15, 15, 3\}$
$\{17, 8, 4\}, \{15, 13, 11\}$	$\{17, 8, 7\}, \{15, 13, 12\}$
$\{17, 10, 5\}, \{15, 14, 11\}$	$\{17, 12, 2\}, \{17, 10, 8\}^*A$
$\{17, 12, 3\}, \{16, 13, 11\}$	$\{17, 12, 4\}, \{17, 11, 7\}^*B$
$\{17, 13, 5\}, \{17, 12, 8\}^*$	$\{17, 14, 4\}, \{17, 12, 10\}^*$
$\{17, 15, 3\}, \{17, 12, 12\}^*C$	$\{17, 16, 2\}, \{17, 14, 12\}^*$

The problem is that in the discussions in Sections 3.5, 3.6, and (so far) in 3.7 we did not worry about possible *unintended incidences* of points and lines. However, such incidences may well happen, as is shown by the example in Figure 3.7.4. The reason is easily discerned from the symbol of the configuration, as first pointed out in [**24**]. In the language of the characteristic paths this happens when a segment of the path (or the line it determines) passes through a point that is in the same orbit as the endpoint of another segment of the path but is not itself a vertex of the path. This is illustrated in Figure 3.7.4(b), where the characteristic path starts at the red point of the middle orbit and goes towards the innermost orbit of points, but the second segment contains the blue point of the orbit of the starting point. That causes each point of this orbit to be on six lines. This description is easily translated into the language of the configuration symbols: As we step from one entry of the general symbol $m\#(s_1, t_1; s_2, t_2; \ldots ; s_k, t_k)$ to the next, a consecutive string of entries needs to be changed in only its first or else its last terms in order to obtain a valid symbol for an h-astral configuration with $h < k$ and the same m. In the 3-astral example in Figure 3.7.4, its symbol $12\#(5, 4; 1, 5; 4, 1)$ contains the string $5, 4; 1, 5$. If the last entry is changed to 4, the resulting symbol $12\#(5, 4; 1, 4)$ corresponds to the 2-astral configuration we have seen in Figure 3.6.2. Since the configuration in Figure 3.7.4 is selfdual, it is clear that there necessarily are lines that meet six of

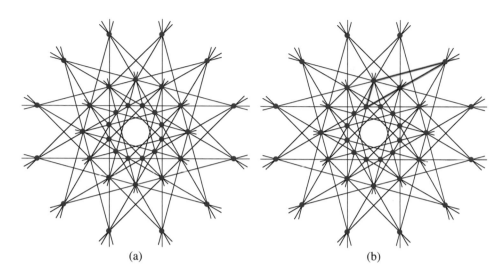

Figure 3.7.4. (a) The trivial 3-astral configuration $12\#(5,4,1,5,4,1)$ is not a configuration at all—it is a prefiguration due to the presence of an orbit of points each incident with six lines and an orbit of lines each incident with six points. (b) The explanation for this situation is detailed in the text.

its points. We formulate this in the general case of k-astral configurations by the following requirement:

(A8) The symbols of a k-astral configuration $m\#(s_1,t_1;s_2,t_2;\ldots;s_k,t_k)$ should not contain a string such that changing one of the ends of the string results in a valid symbol for an h-astral configuration with the same m and with $h < k$.

Exercises and Problems 3.7.

1. Find the symbols for the two configurations in Figure 3.7.5 and decide whether it is trivial, systematic, or sporadic. Find all the other configurations that are in the same cohorts.

2. Find other examples of unintended incidences like the ones in Figure 3.7.4.

3. Find all systematic configurations with $m = 20$ and draw three of them.

4. The configuration in Figure 3.7.5(a) has its lines parallel in sets of six, three on each side of the center. Find a 3-astral 4-configuration in which the points appear in collinear sets of six.

5. Is there a trivial 3-astral configuration in which the points are in collinear sets of six?

6. Find the symbol of the configuration in Figure 3.7.6 and explain your findings.

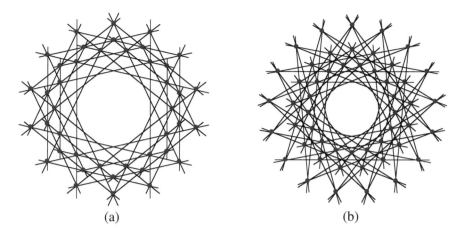

(a)   (b)

Figure 3.7.5. Identify these configurations.

Figure 3.7.6. An interesting configuration.

7. Prove that the configuration (21_3) in Figure 3.7.1 is the only k-astral configuration (21_3) for any k.

3.8. *k*-Astral configurations for $k \geq 4$

As in the preceding three sections, configurations with 4 or more orbits of points that are known in greatest detail are those k-astral configurations in which a dihedral symmetry group acts transitively on the points (and lines) of k different orbits and each orbit has the same number of points. We shall discuss these first. Although the definitions are completely analogous to the ones in previous cases, striking differences in properties led us to separate the present case from the cases of 2- and 3-astral configurations. The main change is in the possibility of various unintended incidences not encountered earlier; hence it is clear that in a number of cases we can speak only of representations and not of realizations. But before we get to that, let us review the definitions.

A **k-astral** 4-configuration (n_4), with $n = k \cdot m$, is a configuration with a dihedral symmetry group d_m that operates transitively on each of k orbits of points situated at vertices of a regular m-gon and k orbits of lines, provided all orbits have the same number of elements and each element is incident with two elements of the other kind from each of two orbits. As detailed in Section 3.5, we can attach to each k-astral configuration a well-determined set of mutually equivalent symbols, derived from the consideration of the characteristic paths possible in the configuration.

This is illustrated in Figure 3.8.1, where a characteristic path starts at P_0 and goes on to P_1 and other points that are not labeled to avoid clutter. Since P_1 has symbol $(2//1)$ and the following points of the characteristic path shown have symbols $(5//3)$, $(4//5)$, $(1//2)$, and $(3//4)$, while the orbits have size 11, the symbol of this configuration is $11\#(2, 1; 5, 3; 4, 5; 1, 2; 3, 4)$. Hence it is a 5-astral configuration.

In the case of the symbol $9\#(2, 1; 4, 2; 1, 3; 2, 3; 1, 3)$ a similar procedure leads to the diagram shown in Figure 3.8.2(a). Here we come face to face with a serious problem: Our graphics, which frequently show only (slightly elongated) segments that are necessary to connect all points that are incident according to the symbol, are misleading. Configurations consist of *lines*, not segments—and if the segments we used are extended to the rim of the diagram, additional incidences become evident; thus, we do not have a configuration at all. Instead, we have a prefiguration that can be interpreted as a *representation* of the abstract configuration $9\#(2, 1; 4, 2; 1, 3; 2, 3; 1, 3)$.

As mentioned already in Section 3.7, the explanation of the problem is that the characteristic path essentially crosses itself at a configuration point. This is detectable from the symbol of the configuration and leads to the following condition we repeat here from Section 3.7 in a slight reformulation; the condition was signaled by Boben and Pisanski in [**24**]:

(A8) (Reformulated) The symbols of a k-astral configuration $m\#(s_1, t_1; s_2, t_2; \dots ; s_k, t_k)$ should not contain a string of even length such that changing at most one of the ends of the string results in a valid symbol for an h-astral configuration with the same m and with $h < k$.

If condition (A8) is not fulfilled, the symbol $m\#(s_1, t_1; s_2, t_2; \dots ; s_k, t_k)$ encodes for a *representation* by a prefiguration and not for a *realization* by a configuration. In the example of Figure 3.8.2, the symbol can be written in the equivalent form $9\#(1, 3; 2, 1; 4, 2; 1, 3; 2, 3)$. Then the string

Figure 3.8.1. Depiction of a 5-astral configuration (55_4) with symbol $11\#(2,1;5,3;4,5;1,2;3,4)$ and symmetry group d_{11}.

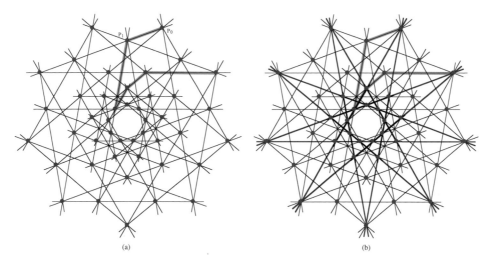

Figure 3.8.2. Problems with realization of $9\#(2,1;4,2;1,3;2,3;1,3)$.

$3,2,1,4,2,1$ can be replaced by $4,2,1,4,2,1$, which leads to the trivial 3-astral configuration $9\#(4,2;1,4;2,1)$.

The change in (A8) consists of the words "at most", which really means that the string itself should be usable in a configuration symbol. This could

Figure 3.8.3. Given the symbol $12\#(5,4;1,4;1,4;5,4)$ for a 4-configuration (48_4), the software produced this result which has three orbits of points and three orbits of lines; it would be a (36_4) configuration if it were not for the middle orbit of points: Each appears to be on *six* lines! In fact, these are two coinciding points of the configuration, and the lines with span 5 also represent two lines of the configuration.

not have happened with 3-astral configurations, but it can happen in the situation considered here.

As an example we show in Figure 3.8.3 the result of drawing the configuration that corresponds to the symbol $12\#(5,4;1,4;1,4;5,4)$. The expected (48_4) configuration did not come about. Instead, we obtained a prefiguration that looks like it has three orbits of points and lines, but points in one orbit are incident with six lines while lines of one orbit are incident with six points.

The explanation for the (mis)behavior of the symbols in these cases is the rather obvious failure of (A8): The characteristic path returns to one orbit three times; in other words, a proper string of the symbol already codes for a 4-configuration and so does the remaining part; hence there are six or more lines through each point of the appropriate orbit. In fact, the points of the middle orbit are doubled up and so are the span-5 lines. Other examples are given in Figures 3.8.4 and 3.8.5.

The example in Figure 3.8.6 shows that if the relative interior of a segment of the characteristic path contains the endpoint of another segment, unexpected incidences occur as well.

Figure 3.8.4. For the symbol $18\#(8,6;1,7;2,5;7,6)$ we obtain the pre-figurations shown. It does have four orbits of lines, each incident with two points from each of two point orbits; but there appear to be only three point orbits, one of which consists of points incident with eight lines. These actually represent pairs of coinciding points of the configuration.

There is one more set of circumstances in which unexpected incidences of a different kind occur; it was also signaled by Boben and Pisanski in [**24**]. It is illustrated by Figure 3.8.7, in which points of one orbit are on five lines while lines of one orbit contain five points. To avoid such incidences, the following condition is imposed by Boben and Pisanski beyond the ones we already require:

(**A9**) The symbol of a k-astral configuration $m\#(s_1,t_1;s_2,t_2;\ldots;s_k,t_k)$ should not contain a string of odd length, such as $s_i,t_i;s_{i+1},t_{i+2};\ldots;s_j$, such that $s_i + t_i + s_{i+1} + t_{i+2} + \cdots + s_j$ is an even integer and

$$\prod_{i\le g\le j} \cos(s_g \cdot \pi/m) = \prod_{i\le g\le j-1} \cos(t_g \cdot \pi/m).$$

Clearly, the 5-astral configuration $12\#(3,2,3,4,2,3,1,3,4,1)$ in Figure 3.8.7 violates this condition: The sum of the string of the first five entries is 14, and

$$\cos\pi/4 \cdot \cos\pi/4 \cdot \cos\pi/6 = \cos\pi/6 \cdot \cos\pi/3.$$

Hence we have the unintended incidences.

Figure 3.8.5. The symbol $18\#(8,6;7,5;2,7;1,6)$ leads to a prefiguration (polar to the one in Figure 3.8.4) in which each line of one orbit is
incident with eight points (in four different orbits).

Figure 3.8.6. The characteristic path of the prefiguration corresponding to the 5-astral symbol $14\#(5, 1; 3, 2; 4, 3; 2, 5; 1, 4)$ contains the string $3, 2, 4, 3, 2, 5$; replacing its last entry by 4, we get the symbol of the 3-astral trivial configuration $14\#(3, 2; 4, 3; 2, 4)$. The blue point is a vertex of the characteristic path, but it is also in the relative interior of a different segment of this path, hence an orbit of points that are on six lines each. The line of that segment is clearly incident with six points, as are all lines in its orbit.

Figure 3.8.7. $12\#(3,2,3,4,2,3,1,3,4,1)$ is not a 4-configuration.

Exercises and Problems 3.8.

1. The symbol of the prefiguration in Figure 3.8.7 contains many pairs of equal entries. Explain why canceling any such pair would not yield an example violating condition (A9).

2. The string s_1, t_1, s_2 with $m = 12$ is in a sense the only *known* source of examples violating condition (A9). By this is meant that one can obviously add the same numbers to the even and odd positions (2 was added in the example of Figure 3.8.7), and one can use multiples of the string with the appropriate multiples of m. Decide whether there are any essentially different strings.

3. Prove that no 4-astral configuration can violate condition (A9).

4. The symbol that yielded the example in Figure 3.8.2 belongs to the cohort $m\#\{\{4,2,2,1,1\}, \{3,3,3,2,1\}\}$ with $m = 9$, while the one in Figure 3.8.1 corresponds to $m = 11$. Why are the results different in the two cases?

5. Does the cohort $9\#\{\{4,2,2,1,1\}, \{3,3,3,2,1\}\}$ contain any geometrically realizable configurations?

6. List all the configuration symbols for 4-astral configurations (28_4).

7. Draw the (potential) configuration $7\#(3, 2; 1, 3; 1, 3; 2, 1; 3, 1)$ and describe what happens.

8. Find some systematic families for 4-astral configurations other than the ones that arise from an h-astral configuration with $h < k$ by insertion of matched pairs.

3.9. Open problems

Among the most intriguing open problems concerning 4-configurations are the following.

1. Is there *any* analog for 4-configurations of Steinitz's theorem (Theorem 2.6.1)? This theorem can be interpreted as saying that every connected combinatorial 3-configuration can be represented geometrically in the plane if one incidence is disregarded. How much of a combinatorial 4-configuration can be realized?

2. Can any of the cyclic configurations (n_4) with generating lines $\{0, 1, 6, -3\}$ or $\{0, 1, 5, -2\}$, described in Section 3.1, be geometrically realized (for any $n \geq 18$)? Can *any* cyclic 4-configurations be geometrically realized?

3. It is clear that there are some 4-configurations that can be geometrically realized in the *rational* plane; as an example we may take the configuration $LC(4)$ introduced in Section 1.1 and other similarly built configurations. Can any astral (or k-astral) configuration that can be geometrically realized (or represented) in the Euclidean plane be realized (or represented) in the rational plane? (It is well known that this cannot be done in a k-astral way.)

4. The astral configuration (24_4) in Figure 3.6.2 and the 3-astral configuration (21_4) in Figure 3.7.1 have the property that the underlying combinatorial configurations have groups of automorphisms that act transitively on the flags of the configuration. (A *flag* consists of a line and a point, incident with each other.) The configuration $LC(4)$ mentioned above has the same property. Do there exist any other 4-configurations with a single orbit of flags (under automorphisms)?

Chapter 4

Other Configurations

4.0. Overview

We devoted long chapters to each of 3-configurations and 4-configurations. In contrast, this shorter chapter covers all other configurations. The reason for this difference in extent of coverage is a direct and inevitable consequence of the paucity of knowledge about configurations that are neither 3- nor 4-configurations.

Despite the generality of the definition of configurations proposed a century and a quarter ago, strikingly little effort was devoted to the study of the k-configurations for $k \geq 5$ and the related unbalanced configurations.

In Section 4.1 we review the information that is available about 5-configurations. The first images are barely a decade old, and there is still great uncertainty concerning what is possible and what is not possible regarding 5-configurations.

Section 4.2 nominally deals with all k-configurations for $k \geq 6$. In fact, it is mostly devoted to 6-configurations. I am indebted to L. Berman for permission to include the recently found (and not previously published) images of (110_6) and (120_6). These are the first 6-configurations to appear in print anywhere.

The unbalanced configurations $(12_4, 16_3)$ and $(16_3, 12_4)$ have enjoyed a measure of popularity, but other $[4, 3]$- and $[3, 4]$-configurations have fared much less well. The material about these is presented in Section 4.3.

Unbalanced $[k_1, k_2]$-configurations with $\{k_1, k_2\} \neq \{3, 4\}$ are considered in Section 4.4.

Section 4.5 deals with a recently discovered class of configurations, the "floral" configurations. They are characterized by their hierarchical construction rather than by the particular incidence parameters.

In Section 4.6 we collect results on topological configurations. These have been investigated in some detail only very recently, and the topic abounds in open questions.

Several kinds of unconventional configurations are presented in Section 4.7. We briefly touch on configurations of circles and on two kinds of configurations involving infinite sets of points and lines.

The concluding Section 4.8 presents a few open problems that have not been mentioned in the earlier sections.

4.1. 5-configurations

The history of 5-configurations is even shorter than that of 4-configurations, and the knowledge is also much skimpier. However, there are several interesting aspects that do not appear in 3- and 4-configurations.

From the obvious necessary conditions it follows that any (n_5) configuration must satisfy $n \geq 21$. The buildup of a combinatorial configuration (21_5) using the "greedy" approach (as for (7_3) in Table 2.2.2) can probably be carried out without undue effort. However, it seems more interesting to note that (21_5) is the cyclic configuration based on $(0, 3, 4, 9, 11)$. As noted by Gropp [79], while it is obvious that this cyclic basis works for all $n \geq 2 \cdot 11 + 1 = 23$, its validity for $n = 21$ is unexpected but easily verified. The configuration is presented in Table 4.1.1. Gropp [103] also seems to be the first to discover that $(0, 1, 4, 9, 11)$ is a cyclic basis for (n_5) for all $n \geq 23$ as well; but it does not yield (21_5). Gropp establishes a connection of these bases with the "Golomb rulers"—combinatorial objects interesting in their own right; for some details see [89], [74], [75][1].

Table 4.1.1. The cyclic combinatorial configuration (21_5) generated by the basis $(0, 3, 4, 9, 11)$. This basis also works for each $n \geq 23$ to yield a configuration (n_5).

0	1	2	3	4	5	6	7	8	9	10	11	12	13	14	15	16	17	18	19	20
3	4	5	6	7	8	9	10	11	12	13	14	15	16	17	18	19	20	0	1	2
4	5	6	7	8	9	10	11	12	13	14	15	16	17	18	19	20	0	1	2	3
9	10	11	12	13	14	15	16	17	18	19	20	0	1	2	3	4	5	6	7	8
11	12	13	14	15	16	17	18	19	20	0	1	2	3	4	5	6	7	8	9	10

So far we avoided mentioning the configuration (22_5). It is a particularly interesting one because—in contrast to the situation we encountered for 3-

---

[1]Much additional information can be found on the Internet. See, for example, [44], [206], and, in particular, "Golomb ruler" in Wikipedia.

and 4-configurations—this configuration does not exist even combinatorially. The proof of this requires tools that are outside the scope of this text; see [88], [93], [102].

Except for the existence of two non-isomorphic cyclic combinatorial configurations (23_5) there seems to be no information available regarding the numbers of distinct (n_5). It is easy to construct, for all $n \geq 25$, additional cyclic bases such as $(0, 3, 4, 10, 12)$, $(0, 1, 4, 10, 12)$, or $(0, 1, 6, 10, 12)$; but neither their number nor possible isomorphisms nor the existence of non-cyclic configurations seems to have been investigated.

For 4-configurations we have seen in Section 3.4 that one needs to increase the number of points only slightly from the minimal value $n = 13$ to reach values for which topological or geometric configurations exist—$n = 17$ for the former and $n = 18$ for the latter. Moreover, all these are best possible values. In the case of 5-configurations the information available is far less satisfactory.

It is obvious that the configuration $LC(5)$ (see definition of $LC(k)$ in Section 1.1) is geometrically realizable; however, with $5^5 = 3,125$ points and lines there is no intelligible planar realization. The first description of a graphically presentable 5-configuration appeared in [122]; it is a (60_5) configuration that is 3-astral in the extended Euclidean plane and is also shown in [117] and as Figure 4.1.1. (By the convention adopted in Section 1.5, we may call such 3-astral 5-configurations **astral**.) The construction is based on the idea that many 4-configurations have quadruplets of points aligned on diameters and are such that these diameters are parallel to quadruplets of lines. Then the addition of the diameters gives $[5, 4]$-configurations, for which the addition of points-at-infinity results in 5-configurations. This construction is also illustrated in Figure 4.1.2 in the case of a (50_5) configuration, which is the smallest such configuration known. Another (50_5) configuration is shown in Figure 4.1.3.

All these configurations are symmetric only in the extended Euclidean plane since they include points at infinity. Switching to their polars is no remedy, due to the lines through the center.

The smallest 5-configuration discovered so far is the (48_5) configuration, found by L. Berman and shown in Figure 4.1.4. It has cyclic symmetry c_{12}; moreover, it is 4-astral in the Euclidean plane.

As mentioned above, the configuration (60_5) illustrated in Figure 4.1.1 has the advantage of being *astral*—but only in the *extended Euclidean* plane E^{2+}. One of the long-standing conjectures (see [117], [9]) is

Conjecture 4.1.1. *There are no 5-configurations 3-astral in the Euclidean plane E^2.*

Figure 4.1.1. Deleting the 12 lines (green) through the center yields the astral (48_4) configuration (2) $12\#(5,4;1,4)$. With these lines it is a $(48_5, 60_4)$ configuration. Adding 12 points-at-infinity, in the directions of the ten green lines, results in a (60_5) configuration that is astral in the extended Euclidean plane.

Figure 4.1.2. Deleting the ten lines (green) through the center yields the 4-astral configuration $10\#(4,3,2,3,1,2,1,2)$. With these lines it is a $(40_5, 50_4)$ configuration. Adding ten points-at-infinity, in the directions of the ten green lines, results in a (50_5) configuration that is 5-astral in the extended Euclidean plane.

The existence of certain types of astral 5-configurations in the Euclidean plane has been ruled out in the recent paper [**14**], but the more general question is still open.

Figure 4.1.3. The addition of ten diameters (green) to the 4-astral configuration $10\#(4, 3, 1, 2, 3, 4, 2, 1)$ (black lines) together with the inclusion of ten points at infinity in the direction of quintuplets of parallel lines, yields a 5-astral (50_5) configuration.

One of the basic differences in the knowledge about 5-configurations compared to 3- and 4-configurations is our lack of knowledge about whether geometric configurations (n_5) exist for **all** n that are greater than some fixed bound. On the other hand, a similarity appears to exist: Among the known 5-configurations, there are topological configurations that are smaller than the smallest known geometric configuration. One of several topological (42_5) configurations is shown in Figure 4.1.5. This is to be compared with the result mentioned in the proof of Theorem 3.2.1 to the effect that any topological (n_5) configuration must satisfy $n \geq 25$. Although the gap from 25 to 42 is still large, it is not unexpected: There has been no investigation of 5-configurations—topological or geometric—till very recently, and no systematic approaches have been developed so far.

Figure 4.1.4. The smallest 5-configuration known is this 4-astral (48_5) configuration. (L. Berman, private communication)

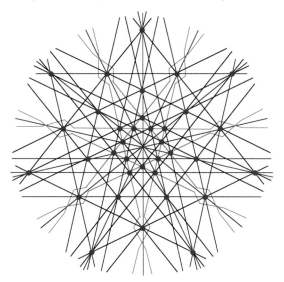

Figure 4.1.5. The geometric configuration $7\#(2, 1; 2, 1; 3, 2; 1, 2; 1, 3)$ has unintended incidences and is just a prefiguration. If these incidences are avoided by using pseudolines, we obtain a topological (35_4) configuration formed by the black lines and green pseudolines. Adding the seven blue lines yields a $(35_5, 42_4)$ configuration, and also adding the seven points-at-infinity (in the directions of the quintuplets of lines/pseudolines) results in a topological (42_5) configuration.

Exercises and Problems 4.1.

1. Determine the 4-configurations that can be turned into 5-configurations by adding lines and by adding points-at-infinity. It seems that all 4-astral 4-configurations can be used, admitting duplication if necessary. Are there any others?

2. The configuration in Figure 4.1.1 was constructed in an obvious way from two copies of the astral (24_4) configuration. Can this method be applied to all astral 4-configurations?

3. Are there any 4-astral 5-configurations in the *Euclidean* plane that have dihedral symmetry?

4. Decide whether any of the configurations in Figures 4.1.1 to 4.1.3 are selfpolar.

5. Decide whether there are geometric (n_5) configurations for any $n < 48$.

6. Decide whether there are topological (n_5) configurations for any $n < 42$.

7. Find a useful and convenient way of encoding symmetric 5-configurations.

8. Show that the 4-astral configuration $10\#(4, 3; 1, 2; 3, 4; 2, 1)$ can be used to construct a configuration (50_5). Determine all 3-astral configurations (40_4) that can be used for that purpose.

9. The constructions we have seen can be generalized. Determine criteria on 4-astral configurations $((4m)_4)$ that make it possible to obtain configurations $((5m)_5)$, and similarly for $((5m)_4)$ configurations to yield $((6m)_5)$ configurations.

4.2. k-configurations for $k \geq 6$

As justification for general existence statements for k-configurations with $k \geq 6$ we recall the configurations $LC(k)$ introduced in Section 1.1. They illustrate the possibility of geometric configurations $((k^k)_k)$ for all k. Naturally, one may be interested in smaller examples, and there are systematic ways to find them, even though they yield configurations that are neither stimulating to look at nor very small.

The first such construction, during the "prehistory period" of configurations, is due to Cayley [41] in 1846. Reflecting the spirit of the times, Cayley writes (in French, in a paper published in a German journal!):

"...sans recourrir à aucune notion métaphysique à l'égard
de la possibilité de l'éspace à quatre dimensions, ..."[2]

and proceeds to define configurations of flats of various dimensions spanned by families of points in general position; intersecting these with suitable planes, he devises (for $k \geq 2$) configurations (n_{k+1}) where

$$n = (2k+1)!/k!(k+1)!$$

Thus, what Cayley describes are geometric configurations $(35_4), (126_5),$ $(462_6), (1716_7),$ and so on. He also describes various unbalanced configurations, about which we shall report in Section 4.3.

Although Cayley's constructions yield smaller configurations than the $LC(k)$, there are better construction methods that are easy generalizations of the ones we detailed in Section 3.3, considered there for the 4-configurations.

The $(5m)$ construction which in Section 3.3 led from a configuration (m_3) to $((5m)_4)$ generalizes immediately: From any (m_k) configuration, taking $k+1$ copies that all intersect at the same points of a suitable line and then adding m appropriate lines connecting corresponding points in these copies, we obtain a $(((k+2)m)_{k+1})$ configuration. Using the smallest configurations available, this construction leads from (9_3) to (45_4), from (18_4) to (108_5), from (48_5) to (336_6), from (110_6) to (880_7), and so on. Except for the last one, these are not the known minimal configurations—but in the last case this is the best available. Carrying out only the first step, with only k copies of the starting configuration, leads to a $(k, k+1)$-configuration. Taking a stack of k configurations and adding the lines connecting the corresponding points leads to a $(k+1, k)$-configuration. Some of the other methods in Section 3.3 generalize as well.

For 6-configurations we can do better than for general k. Figure 4.2.1 shows a 10-astral (110_6) configuration, and Figure 4.2.2 shows a 4-astral configuration (120_6); both were discovered by L. Berman (private communication).

On the other hand, there is negative information available concerning astral (that is, 3-astral) 6-configurations. As proved by Berman [8], no such configurations exist, nor do any astral $[2k, 2h]$-configurations for $k \geq 3$, $h \geq 3$.

The paucity of information on the topic of this section is clearly evidenced by its brevity and the absence of references beyond [41] and [8]. Notice that these are separated by more than a century and a half!

---

[2] "...without resorting to any metaphysical idea concerning the possibility of four-dimensional space..."

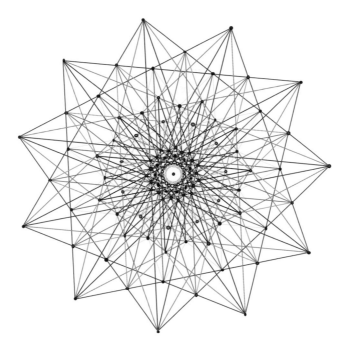

Figure 4.2.1. A 10-astral (110_6) configuration with symmetry group d_{11}, found by L. Berman.

Figure 4.2.2. A 4-astral (120_6) configuration with symmetry group d_{12}, found by L. Berman.

Exercises and Problems 4.2.

1. Decide whether there exist any other 4-astral 6-configurations.

2. Find some "small" topological 6-configurations.

3. Is there some systematic construction for 6-configurations that is analogous to the passage from 4-configurations to 5-configurations mentioned in Exercise 9 of Section 4.1.

4. Find a visually intelligible 7-configuration.

4.3. $[3, 4]$- and $[4, 3]$-configurations

In the present section we shall survey the known facts concerning combinatorial and geometric $[3, 4]$- and $[4, 3]$-configurations.

The parameters of any combinatorial (p_3, n_4) or (n_4, p_3) configuration must satisfy the conditions $3p = 4n$, $p \geq 1 + 3 \cdot 3 = 10$, and $n \geq 1 + 4 \cdot 2 = 9$. Thus p must be divisible by 4 and n must be divisible by 3, so that the only possible configurations are those of the form $((4r)_3, (3r)_4)$ or $((3r)_4, (4r)_3)$, respectively, for $r = 3, 4, 5, \ldots$. For combinatorial as well as geometric configurations, the existence of $((4r)_3, (3r)_4)$ implies by duality, resp. polarity, the existence of $((3r)_4, (4r)_3)$, and conversely. Hence it is sufficient in the following result to limit our attention to one of the two cases.

Theorem 4.3.1. *For each integer $r \geq 3$ there exists a combinatorial configuration $((4r)_3, (3r)_4)$; topological and geometric $((4r)_3, (3r)_4)$ configurations exists for each $r \geq 4$.*

Proof. We start with a combinatorial $(12_3, 9_4)$ configuration, given in Table 4.3.1.

Table 4.3.1. A configuration table for a $(12_3, 9_4)$ configuration.

1	2	3	4	5	6	7	8	9
A	*A*	*A*	*L*	*L*	*L*	*M*	*M*	*M*
B	*G*	*K*	*B*	*G*	*K*	*B*	*G*	*K*
C	*F*	*J*	*J*	*C*	*F*	*F*	*J*	*C*
D	*E*	*H*	*E*	*H*	*D*	*H*	*D*	*E*

In order to complete the proof in the case $r = 3$, we have to prove that no combinatorial configuration $(12_3, 9_4)$ can be realized by points and lines. For that we recall the result known as "Sylvester's problem", which we mentioned in Section 2.1 as Lemma 2.1.1.

To apply the Sylvester result to the question at hand, we note that in any combinatorial configuration $(12_3, 9_4)$ the 36 pairwise intersections of the 9

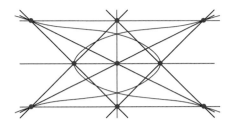

Figure 4.3.1. A realization of a $(9_4, 12_3)$ configuration, dual to the one in Table 4.3.1. Two of the "lines" are not straight. With a slight modification these two "lines" could have been chosen as circles.

lines have to occur in 12 triplets—three intersections at each of the 12 points of the configuration. However, since (by Sylvester) in every topological or geometric configuration at least one such intersection is an "ordinary" one (which is therefore not a point of the configuration), there are not enough pairwise intersections to form 12 triplets.

On the other hand, it is possible to give a geometric realization of the dual configuration but with two of the "lines" neither straight lines nor pseudolines. An example is shown in Figure 4.3.1.

For the remaining part of the proof of Theorem 4.3.1 we only have to exhibit appropriate geometric configurations of points and lines. The literature contains a number of papers devoted to the $(16_3, 12_4)$ configurations, or to the $(12_4, 16_3)$ configurations dual to them; several examples of the former kind are shown in Figure 4.3.2.

There appears to be no published mention of geometric $((4r)_3, (3r)_4)$ configurations with $r \geq 5$. However, examples of such configurations are very easy to produce. One method (see Figure 4.3.3) starts by placing $2r$ points equidistributed on a circle. Each of these points is connected to the one diametrically opposite to it, as well as to the two points separated from it by two other points. Adjoining the $2r$ triple intersections (whose existence is clear by the symmetry of the diagram) yields a $((4r)_3, (3r)_4)$ configuration, as required. □

Other $((4r)_3, (3r)_4)$ configurations may be constructed by slight variations of this method; several are shown in Figure 4.3.4. In all these cases, the geometric existence of the configurations is an obvious consequence of the high degree of symmetry involved.

Although the configurations $(16_3, 12_4)$ and/or $(12_4, 16_3)$ have been studied for at least 150 years (starting not later than Hesse [**124**] in 1848, in the "prehistoric" era of configurations), there still are many unresolved questions. It has been shown (or claimed—there seems to have been no independent verification) that there are precisely 574 combinatorial configurations

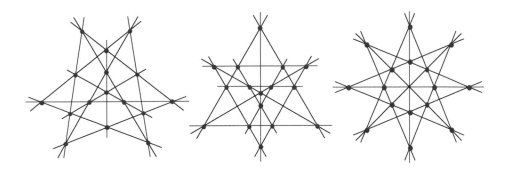

Figure 4.3.2. Three examples of configurations $(16_3, 12_4)$.

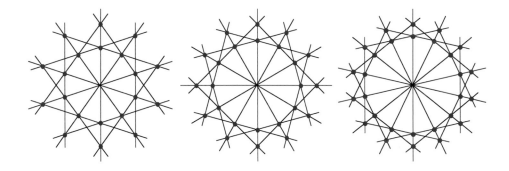

Figure 4.3.3. Examples of configurations $(20_3, 15_4)$, $(24_3, 18_4)$, and $(28_3, 21_4)$.

$(12_4, 16_3)$; see Gropp [85], [87]. The large number of such configurations helps explain why there is no clarity on the question of which (or whether all) configurations $(12_4, 16_3)$ have geometric realizations in the Euclidean plane. Two additional aspects probably contribute to the lack of clarity: On the one hand, most of the relevant papers have been published in journals that are not well known or widely available, many in Czech, which is not too widely spoken; a large number of references is listed below. On the other hand, from the very beginning, these configurations have been studied in close connection with the theory of cubic curves. This connection, in turn, is not too well known these days and also makes it hard to know which parts of the claims of possibility of realization rely on configurations in the complex plane and which claims of impossibility are due to the restriction of attention to configurations with vertices on cubic curves. See below for some relevant ideas.

From the duality in the projective plane it follows that geometric configurations $((3r)_4, (4r)_3)$ exist if and only if $r \geq 4$. One example of a $(12_4, 16_3)$

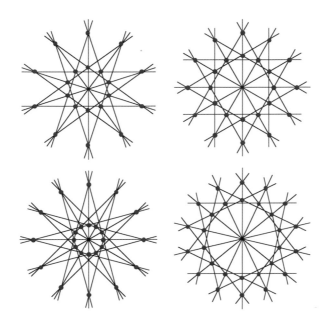

Figure 4.3.4. Depiction of additional examples of configurations $(20_3, 15_4)$, $(24_3, 18_4)$, and $(28_3, 21_4)$.

configuration is shown in Figure 4.3.5. In contrast to the very symmetric diagrams representing the $(16_3, 12_4)$ configurations, the diagrams of the $(12_4, 16_3)$ configurations shown in most publications are far from symmetric.

The reason for the difference is that projective duality does not in general preserve Euclidean symmetries—unless one considers the configurations in the extended Euclidean plane. In particular, all examples in Figures 4.3.3 and 4.3.4 have lines passing through the center of symmetry (taken at the origin) which have to be mapped to points-at-infinity in order to preserve symmetry. If this is accepted, then it is easy to produce very symmetric $(16_3, 12_4)$ configurations, such as the one in Figure 4.3.6.

Additional examples of quite symmetric $(12_4, 16_3)$ configurations are shown in Figure 4.3.7. These have vertices on cubic curves.

In order to give a feeling for the relation of cubic curves to configurations, we show another example in Figure 4.3.8. This is a geometric configuration $(12_4, 16_3)$ on a cubic curve, from the paper by V. Metelka [**168**]. The equation of this cubic curve in homogeneous coordinates (x, y, z) is

$$z(x^2 + y^2) + x(x^2 - 3y^2) = 0$$

and the points are

$$M = (1, 1, 1), \qquad N = (0, 1, 0), \qquad O = (1, -1, 1), \qquad P = (1, -t, 2),$$
$$Q = (1, t, 2), \qquad R = (t, 1, 0), \qquad S = (-t, 1, 0), \qquad T = (1, t - 2, 1 - t),$$
$$U = (1, 2 - t, 1 - t), \quad V = (1, t + 2, t + 1), \quad W = (1, -t - 2, t + 1), \quad X = (1, 0, -1),$$

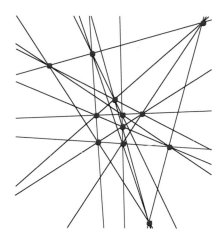

Figure 4.3.5. An example of a geometric configuration $(12_4, 16_3)$.

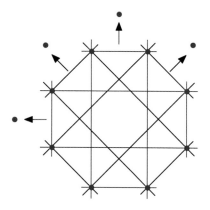

Figure 4.3.6. An example of a geometric configuration $(12_4, 16_3)$ that is astral in the extended Euclidean plane.

where $t = \sqrt{3}$.

As is well known, an easy way to see whether three points given in homogeneous coordinates are collinear is by checking whether the determinant formed by their coordinates is 0. Thus the assertions about which triplets are collinear (as indicated by Figure 4.3.8) can be algebraically verified.

As Metelka observed (this is the reason he considered the configuration "special"), there are three additional lines that pass through three of the points. These three lines are indicated by the green lines in Figure 4.3.9. It is worth noting that the maximal number of collinear triplets determined by 12 points is 19—this is one of the frequently raised "orchard problems"; see more details in [**37**].

As a clarification of what was briefly mentioned above regarding the use of cubic curves in looking for the construction of configurations and related objects, in Figure 4.3.10 we show a diagram of a cubic curve on which are marked several values of the "degree" parameter. The following explanations are taken from the old paper [37], from which data for the curve in Figure 4.3.10 was taken as well. References to texts that establish the properties in question are given in [37]; the notation is the one that seems traditional in the literature.

A suitable projective image of each real non-singular cubic curve has an equation of the form

$$(1) \qquad\qquad y^2 = 4x^3 - g_2 x - g_3$$

where g_2 and g_3 are real constants. The curve C given by equation (1) may be parametrized by

$$(2) \qquad\qquad x = \wp(u), \qquad y = d\wp(u)/du,$$

where $\wp(u)$ is the Weierstrass elliptic function defined by

$$u = \int_{\wp(u)}^{\infty} (4x^3 - g_2 x - g_3)^{-1/2} dx.$$

The Weierstrass elliptic function $\wp(u)$ is a doubly-periodic meromorphic function of the **complex** variable u, and for real g_2, g_3 it has a real period that we shall denote 2ω (as well as a purely imaginary period $2\omega'$). The parametrization (2) yields for real u the "odd circuit" (branch) of the cubic C. In the case $D = g_2^3 - 27g_3^2 < 0$ this is the only real part of the curve C ("unipartite cubic"), while in the case $D > 0$ the curve C also has an "even circuit" corresponding to the values $u = v + w'$, where v is real. (We shall be interested only in the "odd circuit".)

The importance of cubic curves for the present concerns is based on the following result of N. H. Abel:

Denoting by $P(u) = (\wp(u), d\wp(u)/du)$ the point on the cubic C given by (1), (2) and corresponding to the real parameter u, a necessary and sufficient condition for the collinearity of the points $P(u)$, $P(u')$, $P(u'')$ on the odd circuit of C is

$$u + u' + u'' \equiv 0 \ (\mathrm{mod}\, 2\omega).$$

The curve we use is given by the equation $y^2 = 4x^3 - 1$, and by consulting appropriate tables or software, we find that $\omega = 1.529954037\ldots$. As in much of the numerical work on the elliptic functions, we replace 2ω by $360°$; in Figure 4.3.10 we denote the points simply by their parameter value in "degrees".

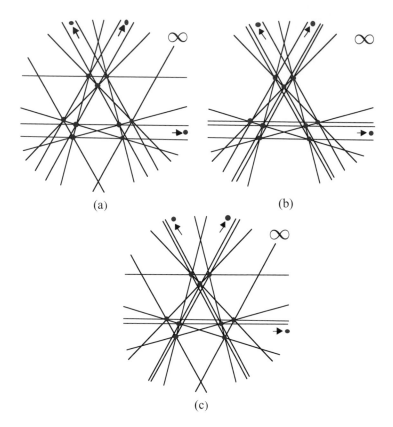

(a)   (b)

(c)

Figure 4.3.7. Depiction of three examples of quite symmetric configurations $(12_4, 16_3)$. The ∞ symbol is meant to indicate that the line-at-infinity is a line of the configuration.

A practical weakness of the method is an inconvenient bunching of the points of interest. The situation can be improved by using a suitable projective transformation of the curve C; this goes back to W. K. Clifford in 1865. The "odd circuit" of C contains the three collinear points of inflection $P(0)$, $P(2\omega/3)$, $P(4\omega/3)$. If we choose the line determined by these points as the line-at-infinity and the points themselves to be in equiinclined directions, there results a very convenient and symmetric representation of C.

We are using the curve C with equation $y^2 = 4x^3 - 1$, for which the Clifford transformation may be achieved by

$$x = (2x^* + 1)/(2x^* - 2), \qquad y = 3y^*/(x^* - 1).$$

This results (on omitting the asterisks) is the equation

$$(x - 1)(3y^2 - (x + 2)^2) = \text{const}.$$

For better visibility we choose the constant as -300, yielding the curve in Figure 4.3.10. This curve is used in some of the exercises below.

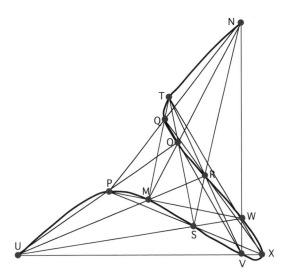

Figure 4.3.8. A configuration $(12_4, 16_3)$ with points on a cubic curve.

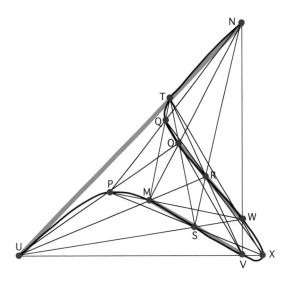

Figure 4.3.9. The points of the $(12_4, 16_3)$ configuration in Figure 4.3.8 determine three additional lines each incident with three of the points.

The following is an extensive list of papers that the author is aware of that deal with $(16_3, 12_4)$ or $(12_4, 16_3)$ configurations: [**38**], [**58**], [**85**], [**124**], [**160**], [**161**], [**162**], [**163**], [**164**], [**165**], [**166**], [**167**], [**168**], [**156**], [**157**], [**158**], [**193**], [**231**], [**232**], [**233**], [**236**]. Some of them contain additional references to earlier papers.

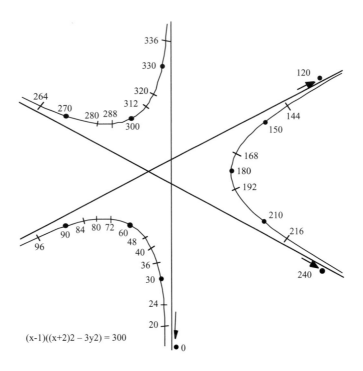

Figure 4.3.10. A cubic curve, with a parametrization derived from the Weierstrass $\wp(u)$ function, as explained in the text.

A configuration $(15_4, 20_3)$ was described by Cayley [**41**] in 1846. There seems to be no other discussion in the literature of $((4r)_3, (3r)_4)$ configurations with $r \geq 5$ (or their duals).

$$* \; * \; * \; * \; * \; *$$

The introduction of k-astral configurations helped develop the study of 3- and 4-configurations. It seems reasonable that investigations of $[3, 4]$-configurations and similar objects would be advanced by moving from the concentration on the smallest cases to more general situations. As examples capable of various generalizations, we show in Figures 4.3.11 and 4.3.12 configurations $(20_3, 15_4)$ and $(15_4, 20_3)$ with cyclic symmetry group c_5, and in Figure 4.3.13 we show a configuration $(18_4, 24_3)$ with symmetry group c_6. It is clear that such configurations fit into infinite families for which the systematic investigation and notation still need to be developed.[3]

---

[3]Added in proof (February 2009): For some steps in this direction see [**121**].

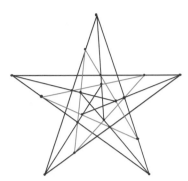

Figure 4.3.11. A $[4, 3]$-astral $(20_3, 15_4)$ configuration with symmetry group c_5.

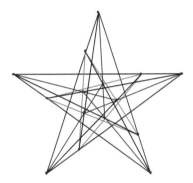

Figure 4.3.12. A $[3, 4]$-astral $(15_4, 20_3)$ configuration with symmetry group c_5.

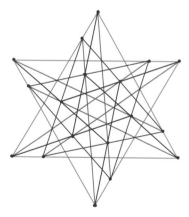

Figure 4.3.13. A $[3, 4]$-astral $(18_4, 24_3)$ configuration with symmetry group c_6.

Exercises and Problems 4.3.

1. Show that each of the permutations (described by their cycle decompositions) $(A)(L)(M)(BGK)(CFJ)(DEH)$ and $(ABCD)(LGJHMKFE)$ maps the combinatorial configuration $(12_3, 9_4)$ of Table 4.3.1 onto itself. Deduce that the automorphisms of the configuration act transitively on its points as well as on its lines. Decide whether the configuration is flag-transitive? (**Flag** = pair consisting of a "point" and a "line" incident with it.)

2. Decide whether all combinatorial $(12_3, 9_4)$ configurations are isomorphic, that is, whether the configuration $(12_3, 9_4)$ is unique. (Hint: Delete a line and all its points.)

3. Prove that any geometric realization of the $(12_3, 9_4)$ configuration must contain at least two "lines" that are not straight.

4. Set up the configuration table of the configuration $(9_4, 12_3)$ dual to the configuration in Table 4.3.1. Decide whether this configuration can be geometrically realized with straight lines or with pseudolines.

5. Describe the configuration table of the $((4r)_3, (3r)_4)$ configuration constructed in the proof of Theorem 4.3.1.

6. Show that the two $(20_3, 15_4)$ configurations shown in Figures 4.3.3 and 4.3.4 are not isomorphic.

7. Decide whether any among the three configurations $(16_3, 12_4)$ in Figure 4.3.2 are isomorphic and whether the two configurations $(24_3, 18_4)$ in Figure 4.3.4 are isomorphic.

8. Starting with 12 points equidistributed on a circle, how many $(24_3, 18_4)$ configurations can you construct that have different appearance? Are any two among them isomorphic?

9. For general r, starting with $2r$ points equidistributed on a circle, how many $((4r)_3, (3r)_4)$ configurations can you construct that have different appearance? Are any among them isomorphic?

10. Draw symmetric realizations in the extended Euclidean plane of the polars of the configurations in Figure 4.3.2.

11. Decide whether any among the three configurations $(12_4, 16_3)$ in Figure 4.3.7 are isomorphic.

12. Draw the polar configurations of the configurations in Figure 4.3.7.

13. Verify that those triplets shown as collinear in Figures 4.3.8 and 4.3.9 that contain the point U are, in fact, collinear.

14. Find in Figure 4.3.9 a configuration $(12_4, 16_3)$ that contains the dashed lines, and decide whether it is isomorphic with the configuration in Figure 4.3.8.

15. Determine the group of automorphisms of the configuration in Figure 4.3.8.

16. On the cubic curve in Figure 4.3.10, find a configuration $(9_2, 6_3)$ and a configuration (12_3). Can you find any other configurations?

17. Decide whether the configurations in Figures 4.3.11 and 4.3.12 are duals of each other? If so, find a duality map. If not, find their duals.

18. Find the dual of the configuration in Figure 4.3.13.

19. Develop a theory—similar to the ones in Chapters 2 and 3—of the [4, 3]-configurations.

4.4. Unbalanced [q, k]-configurations with [q, k] ≠ [3, 4]

Very little has been published about geometric [q, k]-configurations with $q \neq k$ and $\{q, k\} \neq \{3, 4\}$. As mentioned in Section 4.2, a few results were found by Cayley [41], using planar sections of configurations of flats of various dimensions generated by families of points in general position. Specific instances will be mentioned below. Some of these methods have been used (mostly in special cases) by later writers.

There is more information about the corresponding combinatorial configurations, much of it due to H. Gropp. Here is a survey of what is known.

For combinatorial [3, 5]-configurations (p_3, n_5) the necessary conditions for existence are $3p = 5n$, $p \geq 13$, and $n \geq 11$. Therefore we must have $p = 5r$ and $n = 3r$ for some integer r, so that we are looking at $((5r)_3, (3r)_5)$ configurations with $r \geq 4$. A combinatorial configuration $(20_3, 12_5)$ is shown in Table 4.4.1. From results on the "orchard problem" (see [37]) it is known that 12 lines determine at most 19 triple points; it follows that no geometric $(20_3, 12_5)$ configuration is possible. Unfortunately, the author does not know of any simple proof of the orchard problem result.

Table 4.4.1. A $(20_3, 12_5)$ combinatorial configuration.

1	1	1	2	2	3	3	4	4	5	5	8
2	6	10	6	9	7	13	8	9	6	7	13
3	7	11	14	10	11	17	12	11	10	12	15
4	8	12	15	16	14	18	14	18	17	15	16
5	9	13	18	19	16	19	17	20	20	19	20

There are interesting connections between combinatorial configurations $(12_5, 20_3)$ and Steiner triple systems $S(2, 3, 13)$. We recall that a Steiner triple system $S(2, 3, v)$ is a collection of triplets from a v-element set, such that each pair of elements occurs in one and only one triplet. It is well known that a Steiner triple system $S(2, 3, v)$ exists if and only if $v \equiv 1$ or

Table 4.4.2. The Steiner triple system $S(2,3,13)_1$.

1	1	1	1	1	1	2	2	2	2	2	3	3
2	4	6	8	10	12	4	5	8	9	11	4	5
3	5	7	9	11	13	6	7	10	12	13	8	12

3	3	3	4	4	4	5	5	5	6	6	7	7
6	7	9	7	10	11	6	8	9	8	9	8	10
10	11	13	9	13	12	13	11	10	12	11	13	12

Orbit: $\{1,2,3,4,5,6,7,8,9,10,1,12,13\}$.

Automorphisms group has order 39.

Generators: (1 2 13 5 3 11 6 12 7 9 8 10 4)(1 2 3)(4 10 7)(9 13 12)

Table 4.4.3. The Steiner triple system $S(2,3,13)_2$.

1	1	1	1	1	1	2	2	2	2	2	3	3
2	4	6	8	10	12	4	5	8	9	11	4	5
3	5	7	9	11	13	6	7	10	12	13	8	12

3	3	3	4	4	4	5	5	5	6	6	7	7
6	7	9	7	10	11	6	8	9	8	9	8	10
13	11	10	9	13	12	10	11	13	12	11	13	12

Orbits: $\{1,2,5,6,8,13\}\{3,9,10\}\{4,11,12\}\{7\}$.

Automorphisms group has order 6.

Generators: (1 2 8)(3 10 9)(4 11 12)(5 13 6)(1 5)(2 6)(3 10)(8 13)(11 12)

3 (mod 6). (For general information about Steiner triple systems see, for example, [**33**, Section 10.3] or [**192**, pp. 388–390].) The unique system $S(2,3,7)$ is one of the incarnations of the combinatorial configuration (7_3) (the Fano plane), which we discussed in Section 2.1. There is a unique system $S(2,3,9)$. There are two (and only two) non-isomorphic systems $S(2,3,13)$, which are of interest here. Information about them is presented in Tables 4.4.2 and 4.4.3, taken from [**155**]. For a history of the system $S(2,3,13)$ see Gropp [**83**].

One interesting property of the Steiner systems $S(2,3,13)$ is that the deletion of one point and the triplets containing it yields a combinatorial configuration $(12_5, 20_3)$. It is clear that the deletion of different points from the same orbit yields isomorphic configurations. As it happens, deleting points from different orbits of the Steiner systems $S(2,3,13)$ yields non-isomorphic configurations. Hence there are five such configurations $(12_5, 20_3)$. This result is due to Novák [**175**]; see also Gropp [**88**].

* * * * * *

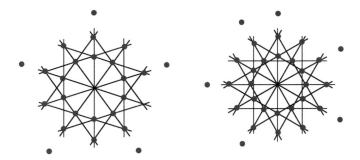

Figure 4.4.1. Typical examples of $[3, 5]$-configurations astral in the extended Euclidean plane. The two examples correspond to $r = 5$ and $r = 6$.

Concerning values of $r \geq 5$ we shall show that there exist geometric $((5r)_3, (3r)_5)$ configurations for all $r \geq 5$. By duality and polarity, the same is true for configurations $((3r)_5, (5r)_3)$.

Theorem 4.4.1. *There exist geometric $((5r)_3, (3r)_5)$ configurations for all $r \geq 5$; moreover, they can be chosen as astral in the extended Euclidean plane.*

Proof. The validity of this statement follows at once from the family of configurations illustrated in Figure 4.4.1; clearly, analogous configurations exist for all $r \geq 5$.   □

Additional examples of geometric $[3, 5]$-configurations are shown in Figures 4.4.2 and 4.4.3.

Cayley [**41**] described a $(21_5, 35_3)$ configuration.

$$* * * * * *$$

For combinatorial $[3, 6]$-configurations (p_3, n_6) the necessary conditions for existence are $p = 2n$ and $n \geq 13$. A combinatorial configuration $(26_3, 13_6)$ is shown in Table 4.4.4. It can also be shown (see [**91**]) that combinatorial configurations $((2n)_3, n_6)$ exist for all $n \geq 13$. Gropp [**88**] states that there are exactly 787 distinct $(28_3, 14_6)$ combinatorial configurations.

There seems to be no geometric $(26_3, 13_6)$ configuration, but the author is not aware of any proof. Also, there is a large difference between the case of $[3, 6]$-configurations and the $[3, 5]$-configurations considered above. In the latter case, for all values of n that satisfy the necessary conditions and are beyond a certain limit (in fact $n \geq 15$), an astral configuration is possible in the extended Euclidean plane. For $[3, 6]$-configurations this is not the case. We have the following theorem.

Figure 4.4.2. Examples of astral [3, 5]-configurations in the Euclidean plane. These are clearly representatives of an infinite family, and several variants are possible.

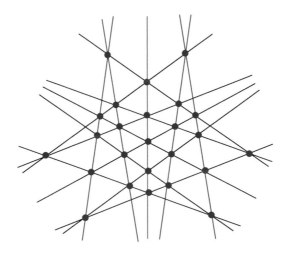

Figure 4.4.3. Another $(25_3, 15_5)$ configuration.

Table 4.4.4. A $(26_3, 13_6)$ combinatorial configurations (found in 1999 by Xin Chen, at the time a student in one of the author's classes). The number of distinct $(26_3, 13_6)$ combinatorial configurations seems not to be known.

1	1	1	2	2	3	3	4	4	5	5	6	6
2	7	12	7	8	7	8	9	10	10	11	9	11
3	8	13	12	16	15	13	14	12	16	13	15	14
4	9	14	17	20	23	17	17	22	19	18	18	21
5	10	15	18	25	24	19	20	24	21	20	19	22
6	11	16	21	26	26	22	23	25	23	24	25	26

Theorem 4.4.2. *For all $r \geq 5$ there exist astral $((6r)_3, (3r)_6)$ geometric configurations in the Euclidean plane.*

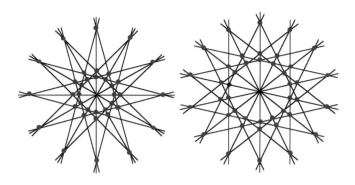

Figure 4.4.4. Configurations $(36_3, 18_6)$ and $(42_3, 21_6)$.

Proof. In Figure 4.4.4 we show the two typical configurations of this kind for $r = 6$ and 7. The only known configuration $(30_3, 15_6)$ is not typical; it is shown in Figure 4.4.5, and we have already seen it in Figure 1.6.8. $\square$

Thus, except for small values of n, there exist geometric configurations $(2n_3, n_6)$ for all n that are multiples of 3. For no other values of n are any geometric $[3, 6]$-configurations known.

It is clear that geometric $[3, k]$-configurations and $[k, 3]$-configurations can be constructed for all $k \geq 7$ in analogy to the configurations in Figures 4.4.1, 4.4.2, and 4.4.4. As these are not really interesting and no additional information seems to be available, we shall not pursue this topic any farther. Instead, we turn now to $[4, k]$-configurations and their duals.

For $[4, 5]$-configurations (p_4, n_5) the necessary conditions are $p \geq 17$, $n \geq 16$, $5n = 4p$. Therefore the configurations are necessarily of the form $((5r)_4, (4r)_5)$ for $r \geq 4$. According to Gropp [**91**], combinatorial configurations with these parameters exist for all $r \geq 4$. There seems to be no information available concerning the number of distinct configurations for each value of r.

Concerning topological or geometric $[4, 5]$-configurations, there is an elegant family of geometric configurations $((5r)_4, (4r)_5)$ for $r \geq 9$. (The author does not know whether or not there are any for $r \leq 8$.) Its two smallest members are shown in Figure 4.4.6, and their construction can be explained as follows: Starting for $r \geq 9$ from the 4-astral 4-configurations $((4r)_4)$, such as the ones denoted in Section 3.8 by $9\#(3, 1; 2, 4; 3, 2; 3, 2)$ or $r\#(3, 1, 2, 4, 1, 3, 4, 2)$, additional r points are added at infinity (in the directions of the quadruplets of parallel lines of the 4-configuration). This yields a $((5r)_4, (4r)_5)$ configuration with five orbits of points and four orbits of lines. Polars of these configurations are $[5, 4]$-configurations. Other $[5, 4]$ configurations can be obtained by adding r mirrors to the same 4-configurations

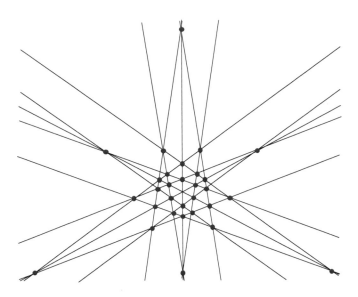

Figure 4.4.5. The only known $(30_3, 15_6)$ configuration.

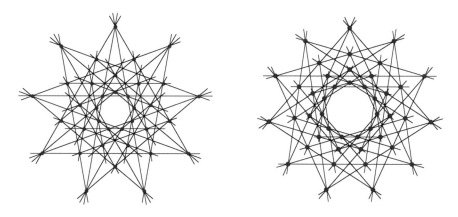

Figure 4.4.6. Depiction of typical $((4r)_4)$ configurations with symbols $9\#(3, 1; 2, 4; 3, 2; 3, 2)$ and $10\#(3, 1, 2, 4, 1, 3, 4, 2)$. Addition of r points-at-infinity to each yields configurations $((5r)_4, (4r)_5)$. Addition of r mirrors gives $((4r)_5, (5r)_4)$.

$((4r)_4)$, but only for odd $r \geq 9$. The two cases analogous to the ones in Figure 4.4.6 are illustrated in Figure 4.4.7. We used a combination of these methods in Section 4.1.

There is very little information available about small $[q, k]$-configurations with still larger values of q and k. Some examples, similar to those above and possible for some particular parameter values, are shown in Figures 4.4.8 and 4.4.9. For other examples see [**115**].

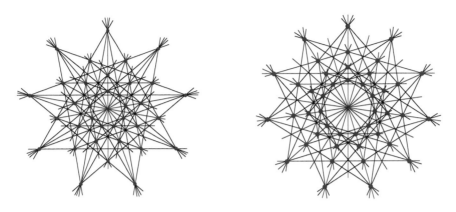

Figure 4.4.7. For odd r, adding r mirrors to 4-configurations such as $9\#(3,1;2,4;3,2;3,2)$ and $11\#(3,1,2,4,1,3,4,2)$ yields $[5,4]$-configurations $((4r)_5, (5r)_4)$.

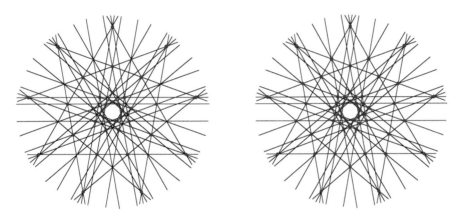

Figure 4.4.8. Depicted on the left is a 4-configuration (36_4) with symbol $9\#(3,1,4,2,1,3,2,4)$. Adding 18 points yields a $(4,6)$-configuration $(54_4, 36_6)$ shown on the right.

Many other 4-astral 4-configurations can be used in constructions similar to the ones illustrated in Figures 4.4.7 to 4.4.10.

A complete determination of astral $[6,4]$-configurations (and their polars) was carried out by L. Berman [8]. These are configurations in which each point is on six lines and each line contains four points, there being two orbits of points and three orbits of lines. As demonstrated in [8], there are precisely five connected astral $(60_6, 90_4)$ configurations and no other connected astral $[6,4]$-configurations. One of these is shown in Figure 4.4.11. This configuration can be understood as the superposition of three astral (30_4) configurations: the sporadic $30\#(12, 10; 6, 10)$ and the systematic $30\#(12, 10; 3, 9)$ and $30\#(10, 6; 3, 9)$, and similarly for the other four. Some other results on $[q,k]$-configurations can also be found in [8].

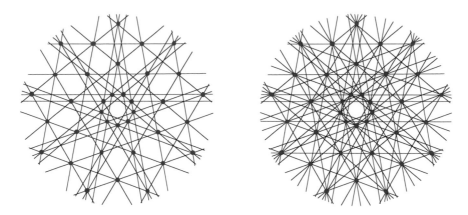

Figure 4.4.9. Depicted on the left is a 4-configuration (36_4) with symbol $9\#(4,3;1,4;2,1;3,2)$. Adding 18 lines yields a $[6,4]$-configuration $(36_6, 54_4)$ shown on the right. Adding to it the nine lines of mirror symmetry yields a $(36_7, 63_4)$ configuration. Adding instead the nine points-at-infinity leads to a configuration $(45_6, 54_5)$.

The material we have presented in this section exhausts the knowledge available to us. As in most other sections, there are lots of obvious questions and open problems for which we have no guesses as to the correct answers. The hope is that some readers will take it as a challenge to enlarge the compass of known facts.

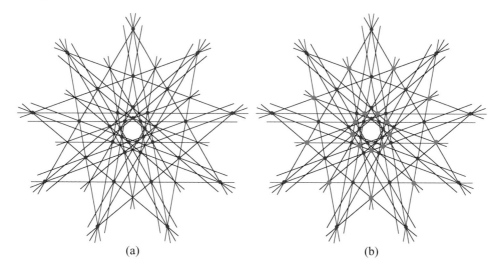

(a)   (b)

Figure 4.4.10. (a) The 4-astral configuration $9\#(3,1;4,2;1,3;2,4)$ has quadruplets of lines concurrent at points that are not configuration points. (b) Adding these 18 points (green) yields a $(54_4, 36_6)$ configuration. Adding nine points-at-infinity (in the direction of quadruplets of parallel lines) yields a $(63_4, 36_7)$ configuration. Adding instead the nine mirrors results in a $(54_5, 45_6)$ configuration.

Figure 4.4.11. An astral $(60_6, 90_4)$ configuration, taken from [8].

Exercises and Problems 4.4.

1. Decide whether any combinatorial configuration $(26_3, 13_6)$ can be realized geometrically or topologically.

2. Do there exist any geometric configurations $(2n_3, n_6)$ with n not a multiple of 3?

3. Draw a configuration $(45_4, 36_5)$.

4. Draw a configuration $(54_4, 36_6)$.

5. Draw a configuration $(49_3, 21_7)$.

6. Draw as small a configuration of type (p_8, n_3) as you can find.

7. By consulting the lists in Section 3.6, describe the other four astral $(60_6, 90_4)$ configurations.

8. Find configurations $(60_6, 90_4)$ that are not astral.

4.5. Floral configurations

Floral configurations provide a means of visualizing some rather large configurations in a pleasing and visually accessible way. The topic was initiated by J. Bokowski late in 2006 in an email message asking whether the configuration attached to the message had already been found by anyone. The configuration in question is shown in Figure 4.5.1. It was a completely new type of configuration, and the curiosity it engendered quickly led to a wealth of configurations analogous in some sense. Collectively they became known as "floral configurations". The results of the early investigations of these configurations have been presented in [15]; most will be reviewed here, together with new developments. Many of the latter arose in discussions with the coauthors of [15], and the present author owes them sincere gratitude.

Loosely speaking, a floral configuration is a connected configuration that has a number of parts, called florets, arranged within the configuration in a symmetric way. Reasonable people may (and do) differ regarding what level of generality is reasonable in the context of floral configurations. For our purposes the following approach seems most appropriate; it is more general than the approach in [15], with which we shall compare it near the end of this section.

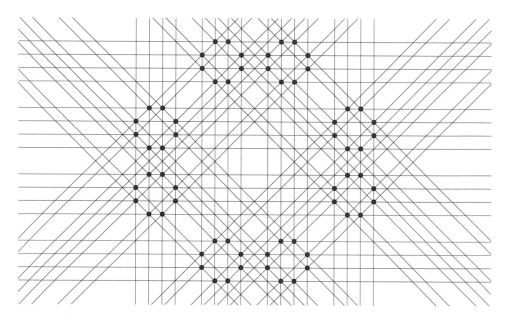

Figure 4.5.1. The first "floral" configuration, from an email by J. Bokowski on October 28, 2006, reproduced in [15].

Definition 4.5.1. A **floret** is a collection of points and lines with prescribed incidences. A **floral configuration** is a configuration that consists of a collection of florets such that the symmetry group of the configuration acts transitively on the florets.

In view of the generality of the concept, it is not surprising that one may distinguish several varieties of floral configurations. To begin with, there is the question of what symmetry group is being considered. It turns out that dihedral groups are much more productive in this context, and we shall devote most of the section to them. The cyclic groups will be considered briefly afterwards.

Even before discussing methods for the construction of floral configurations, a limitation of their appeal needs to be discussed. It is quite clear that the florets in Bokowski's configuration are easily picked out, as are the ones in parts (a) and (b) of Figure 4.5.2. However, this becomes increasingly more difficult in the other parts of that illustration, and the question arises whether it is appropriate to call all of them "floral". One may wish to restrict consideration to only those configurations in which each floret is contained in a single sector determined by the mirrors of the symmetry group or in two such sectors—but there really is no obvious and natural delimitation. Hence we shall not make any such restriction, although we shall endeavor to present examples in which the florets can be readily discerned.

Four construction methods seem to furnish all known examples of floral configurations. Since they depend on the mirrors of the dihedral groups, we denote them as constructions $(\mathcal{M}_1)$, $(\mathcal{M}_2)$, $(\mathcal{M}_3)$, and $(\mathcal{M}_4)$.

Construction $(\mathcal{M}_1)$. Let S be a family of s concurrent lines, equiinclined to each other, so that they represent the s mirrors of a dihedral group d_s. Let the **protofloret** F be a $[q, t]$-configuration such that each of the lines in F is perpendicular to one of the mirrors in S and no mirror in S is a mirror for F. Then images of the protofloret F under all reflections in the s mirrors of S create *in general* a floral $(q, 2t)$-configuration with $2s$ florets.

The "in general" part refers to two possibilities of failure of the construction:

• The resulting configuration may be disconnected, hence cannot qualify as a floral configuration.

• There may be some accidental incidences, which make this a representation of the underlying combinatorial configuration—but not a realization of it.

It should be stressed that neither the points of an individual floret nor its lines are required to have any non-trivial symmetries—although in many cases they do have them.

An illustration of the construction $(\mathcal{M}_1)$ is provided in Figure 4.5.3. There $s = 3$ (in parts (a) and (b)) or $s = 6$ (in parts (c) and (d)), with $q = 3$ and $t = 2$. The lines of S are shown in green. The protofloret F is shown with black points and red lines, while the other points are red and lines are black. Bokowski's original floral configuration, shown in Figure 4.5.1, can also be obtained by the $(\mathcal{M}_1)$ construction.

The color coding in Figure 4.5.3 will be used throughout the present section for all constructions with method $(\mathcal{M}_1)$.

It should also be stressed that all floral configurations obtained by the construction $(\mathcal{M}_1)$ have at least two degrees of freedom. This is most easily

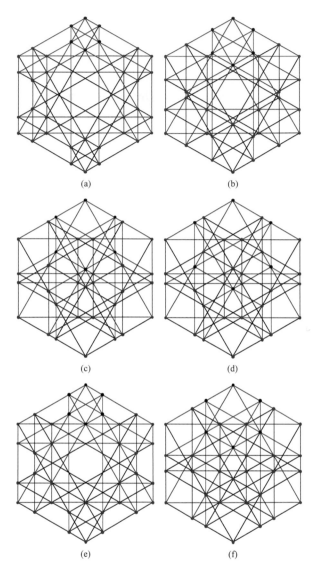

Figure 4.5.2. The first four floral (36_4) configurations are isomorphic and differ only in the size of the florets. The florets of the configuration are clearly distinguishable in (a) and (b) but less easily in (c) and especially in (d). In fact, in all four cases the situation is complicated by the fact that two different sets of florets can be picked out. When constructing the configurations, the top florets in (a), (b), and (c) had their top three points on the upper half of the top sides of the overall hexagon, while in (d) they reached beyond the top half. However, with a bit of contemplation it is easy to reverse the perception in all four of the configurations. The diagrams in (e) and (f) show two representations of the same configuration—with some unintended incidences.

seen by observing that if the distance of the center of symmetry from the centroid of the protofloret is kept fixed, the size of the protofloret can be changed continuously, as can its position with respect to the mirrors. The configuration in Figure 4.5.2 shown earlier was constructed by the $(\mathcal{M}_1)$ method as well, and the influence of the size of the protofloret can be discerned easily.

Construction $(\mathcal{M}_2)$. As before, let S be a family of s concurrent lines, equiinclined to each other, so that they represent the s mirrors of a dihedral group d_s. Let the protofloret F be a $[q, t]$-configuration such that each of the lines in F is perpendicular to one of the mirrors in S and precisely one mirror M in S is a mirror for F. Moreover, let no line of F be perpendicular to M. Then images of the protofloret F under all reflections in the s mirrors of S create in general a floral $[q, 2t]$-configuration with s florets.

An example of construction $(\mathcal{M}_2)$ is provided in Figure 4.5.4. There $s = 3$, $q = 2$, and $t = 2$. The result is a floral $[2, 4]$-configuration $(18_2, 9_4)$. Figure 4.5.2 also shows configurations obtained by method $(\mathcal{M}_2)$. As with $(\mathcal{M}_1)$, unintended incidences may occur. This happened, for example, in Figure 4.5.2, parts (e) and (f).

The floral configurations constructed using $(\mathcal{M}_2)$ have at least one degree of freedom for continuous changes—the size of the protofloret relative to its distance from the center of symmetry. This is illustrated in parts (a) and (b) of Figure 4.5.4. In some cases the floret itself may have continuous changes in shape; this is shown in part (c).

Construction $(\mathcal{M}_3)$ starts with a floral $[q, k]$-configuration C constructed by method $(\mathcal{M}_2)$. Assuming, for ease of formulation, that the mirror M of the protofloret F is vertical, we look for the uppermost points on florets F' and F'' symmetric with respect to M. If each of these two florets has a single highest point X' and X'', respectively, we focus on the corresponding points Y' and Y'' in the protofloret F and on a pair of lines, symmetric with respect to M, which go to points Z' and Z'' on F that are lower than Y' and Y''. We create a new protofloret F^* by deleting from F the two lines just mentioned and introducing a horizontal line H containing the two points Z' and Z''. The protofloret F^* is not a configuration, since it has two points (Y' and Y'') that are incident with only $q - 1$ lines. Mirroring the changes we made to get F^* from F to all the florets of C, we now utilize the presence of a degree of freedom in floral configurations constructed like C by method $(\mathcal{M}_2)$ and change the size of the protofloret F^* until the line H passes through the highest points on F' and on F''—if this is possible. Then, if there are no unintended incidences, this renders the whole collection of points and lines into a floral $[q, k]$-configuration.

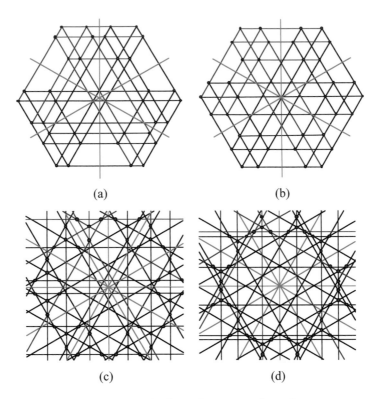

Figure 4.5.3. Four floral (3,4)-configurations formed using construction ($\mathcal{M}_1$). In (a) and (b) the protofloret F is a $(6_3, 9_2)$ configuration, with points at the vertices of an isogonal hexagon. In each case the result is a $(36_3, 27_4)$ configuration with symmetry group d_3; the two configurations are isomorphic. In (c) and (d) the configurations are $(48_3, 36_4)$ with symmetry group d_6, and the protofloret F consists of the vertices of an equilateral triangle and its center, the sides of the triangle, and its mirrors; hence it is a $(4_3, 6_2)$ configuration. In all parts the lines incident with F are shown in red, and the mirrors in S are shown green.

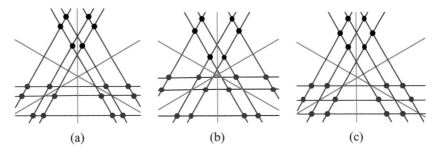

Figure 4.5.4. Three isomorphic floral configurations $(18_2, 9_4)$ obtained by method ($\mathcal{M}_2$). The florets in (a) and (b) differ only in size; those in (c) differ from them in shape.

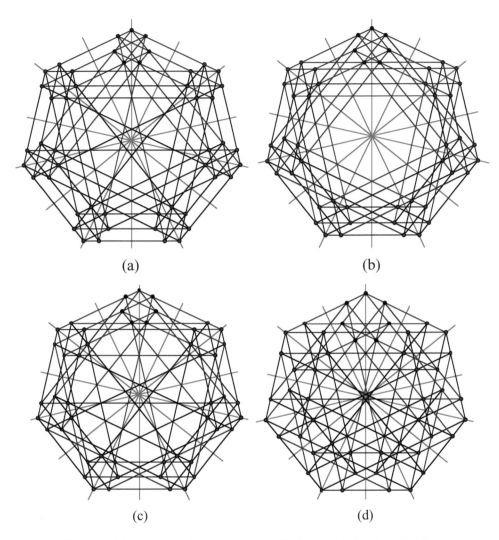

(a)   (b)

(c)   (d)

Figure 4.5.5. A floral configuration (49_4) obtained using $(\mathcal{M}_2)$ is shown in (a). From it three distinct configurations (49_4) are obtained by construction $(\mathcal{M}_3)$. The special lines resulting from the construction are shown in blue.

An illustration of the $(\mathcal{M}_3)$ construction is shown in Figure 4.5.5. In part (a) we have a (49_4) floral configuration constructed by the $(\mathcal{M}_2)$ method, and in the other parts are three floral configurations obtained from it by $(\mathcal{M}_3)$. The newly introduced lines in these configurations are shown in blue. In distinction from the other floral configurations we have seen, configurations obtained by $(\mathcal{M}_3)$ have lines that are incident with three florets each. It also should be noted that although in many cases the florets F' and F'' are the ones nearest to F, this is not always the case. In Figures 4.5.6 and 4.5.7 we show floral configurations (56_4) in which the points of the protofloret are

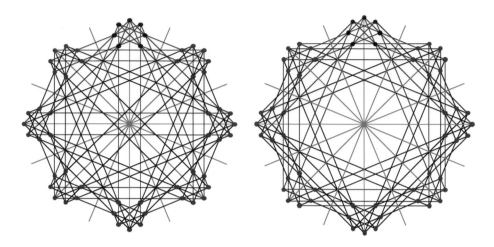

Figure 4.5.6. A floral (56_4) configuration obtained by the ($\mathcal{M}_2$) construction and another (56_4) configuration resulting from it by the ($\mathcal{M}_3$) method. The points of the protofloret are seven of the eight vertices of a regular octagon. The protofloret has d_1 symmetry, and the configurations have d_8 symmetry.

seven of the eight vertices of a regular octagon. In the second of these the special lines go to non-adjacent florets.

Construction ($\mathcal{M}_4$) is analogous to ($\mathcal{M}_3$), but it starts with a floral $[q, k]$-configuration C constructed by either method ($\mathcal{M}_1$) or method ($\mathcal{M}_2$). Assuming, for ease of formulation, that a mirror M mapping the protofloret F to its adjacent floret F° is vertical, we look for the uppermost points on florets F' and F'' symmetric with respect to M and adjacent to F and F°, respectively. If each of these two florets has a single highest point X' and X'', respectively, we focus on the corresponding points Y' and Y'' in the protofloret F and the floret F° and on a pair of lines, symmetric with respect to M, which go to points Z' and Z'' on F and F° that are lower than Y' and Y''. We create a new protofloret F^* by deleting from F the line just mentioned and omitting its companion from F° and introducing a horizontal line H containing the two points Z' and Z''. The protofloret F^* is not a configuration, since it has a point (namely Y') that is incident with only $q-1$ lines. Mirroring the changes we made to get F^* from F to all the florets of C, we now utilize the presence of a degree of freedom in floral configurations constructed like C by method ($\mathcal{M}_1$) and change the size of the protofloret F^* until the line H passes through the highest points on F' and on F''—if this is possible. Then, if there are no unintended incidences, this renders the whole collection of points and lines into a floral $[q, k]$-configuration.

Construction ($\mathcal{M}_4$) is illustrated in Figures 4.5.8 and 4.5.9 by floral configurations (36_4) and (128_4). It should be noted that in contrast to the

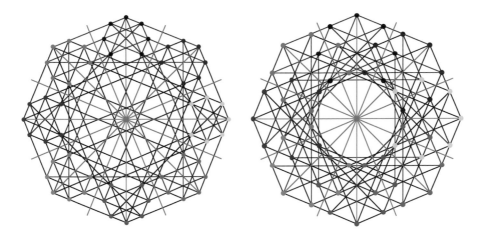

Figure 4.5.7. The same procedure as in Figure 4.5.6, except that the special line determined by two points of the protofloret was aimed not at adjacent florets but at a pair of more distant florets. To achieve the intended incidence, the size of the protofloret had to be increased, leading to a situation similar to that in Figure 4.5.2(d). As a visual aid, the points of each floret were given distinct colors, matched in the two parts.

other constructions, $(\mathcal{M}_4)$ leads to configurations in which some lines are incident with four florets. It is also worth mentioning that, as in construction $(\mathcal{M}_3)$, instead of "adjacent" florets, in some cases florets lying farther away may be used.

In order to illustrate the esthetic appeal of floral configurations and their great variety, we shall now present a number of examples. Most deal with the more versatile $(\mathcal{M}_1)$ construction.

Figure 4.5.10 shows a floral $[5, 4]$-configuration $(120_5, 150_4)$, while Figure 4.5.11 has a floral $[3, 4]$-configuration $(72_3, 54_4)$, both obtained using construction $(\mathcal{M}_1)$. A floral configuration (72_4) obtained by the same method is shown in Figure 4.5.12. Figure 4.5.13 shows a (98_4) floral configuration in which the points of the protofloret are at the vertices of a regular heptagon and the lines are diagonals of that heptagon. In contrast, the protofloret in Figure 4.5.14 is a $(6_4, 12_2)$ configuration without any symmetry, used to construct by method $(\mathcal{M}_1)$ a floral (108_4) configuration with d_9 symmetry.

Construction method $(\mathcal{M}_2)$ is illustrated by Figure 4.5.15 that shows a floral $(150_3, 75_6)$ configuration with d_5 symmetry, in which the protofloret is a (15_3) configuration with d_5 symmetry; analogs of this configuration are the only example found so far of floral configurations in which each line is incident with six points. Construction method $(\mathcal{M}_2)$ also yields the (72_4)

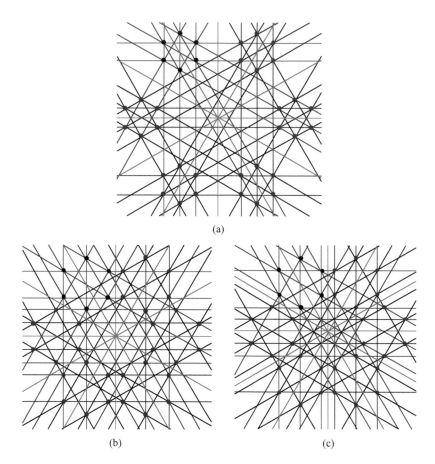

(a)

(b)   (c)

Figure 4.5.8. Floral configurations (36_4). The protofloret in (a) has
points at the vertices of a regular hexagon and symmetry group d_2; the
configuration is obtained by the $(\mathcal{M}_2)$ method and has symmetry group
d_6. The other two configurations are obtained from it by the $(\mathcal{M}_4)$
construction, which involves changing the size of the protofloret and
deleting different lines from the original.

configuration in Figure 4.5.16, in which the points of the protofloret are at
the vertices of an isogonal dodecagon; the shape of the protofloret is variable.

Another method of constructing floral configurations starts with a floral $[q, k]$-configuration F_2, with protofloret F_1. Using F_1 as protofloret, we
can construct a **3-strata** floral configuration F_3. With this terminology, F_2
would be a 2-strata configuration and F_1 a 1-stratum configuration. This
construction method is illustrated in Figure 4.5.17. It shows a 3-strata floral $(4, 8)$-configuration $(250_4, 125_8)$, in which the protofloret F_1 has points
at the vertices of a regular pentagon. The second stratum is a (25_4) floral configuration obtained by the $(\mathcal{M}_3)$ method; with it as protofloret the
complete configuration is obtained by the $(\mathcal{M}_1)$ construction.

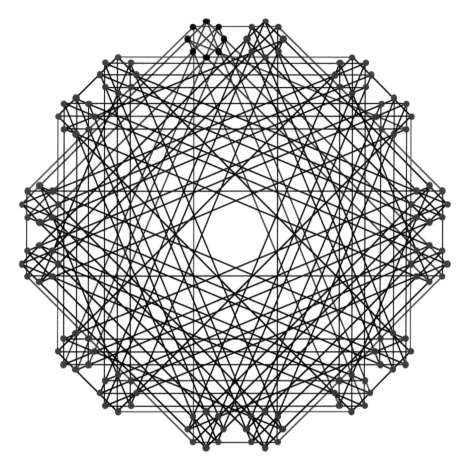

Figure 4.5.9. A floral configuration (128_4) with 16 florets and symmetry group d_8, obtained by construction $(\mathcal{M}_4)$ from a (128_4) configuration reached through the $(\mathcal{M}_1)$ method. The special lines are again shown in blue, but the mirrors are not shown.

Clearly, this procedure could be repeated to 4-strata configurations, and so on; however, the diagrams become far too crowded for visual comprehension.

Here is a brief comparison of our floral configurations with the presentation in [**15**]. The main and fundamental difference is that in [**15**] the protoflorets are considered as only sets of points that are restricted to coincide with either the vertices of a regular polygon or with the vertices of an isogonal but not regular polygon; a further restriction related the symmetries of the protofloret to those of the configuration. The main exposition in [**15**] is restricted to 4-configurations, with more general types mentioned only briefly as "generalized floral configurations". The lines of the protofloret are left to be examined in each case. With these differences in mind, the classification of floral configurations into five varieties can be explained in our terminology as follows. Varieties (A) and (C) of [15] are obtained by the $(\mathcal{M}_1)$ construction, (B) and (D) by the $(\mathcal{M}_2)$ method. The protoflorets in (A) and (B) have points coinciding with vertices of isogonal, non-regular polygons, those in (C) and (D) at vertices of regular polygons. Variety (E) consists of configurations obtained by the $(\mathcal{M}_3)$ method; they have no degrees of freedom beyond similarities.

Many of the configurations illustrated in this section do not fit into the classification of [15], since they do not satisfy the definition of floral configurations adopted there. Those that do are Figure 4.5.1 of variety (A), Figure 4.5.16 of variety (B), Figures 4.5.12 and 4.5.13 of variety (C), Figures 4.5.2, 4.5.5(a), and 4.5.8(a) of variety (D), and Figure 4.5.5(b), (c), (d) of variety (E). Figure 4.5.17 is analogous to the generalized floral configuration $(512_4, 256_8)$ in [15].

We conclude with a brief description of chiral floral configurations. We have already encountered a wide class of these, when investigating the chiral astral 3-configurations in Chapter 2. In Figure 4.5.18 we show two examples of such configurations. Many additional examples can be found in Sections 2.7 (a protofloret consists of two points and two lines) and 2.9 among the multiastral 3-configurations.

A different example of a chiral floral 4-configuration was first presented in [15]; it is reproduced in Figure 4.5.20(b). A general method of generating such floral configurations is based on the $(\mathcal{M}_1)$ construction and illustrated in a simple case in Figure 4.5.19. The crucial step is the replacing of one-half of the florets in the dihedral configuration by their mirror images.

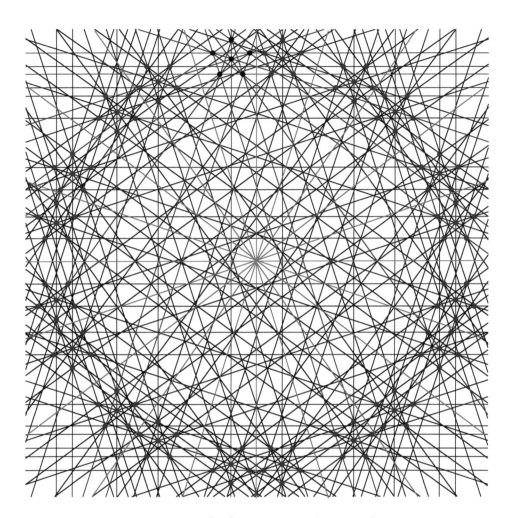

Figure 4.5.10. A floral $[5, 4]$-configuration $(120_5, 150_4)$, constructed by method $(\mathcal{M}_1)$. The protofloret is a $(6_5, 15_2)$ configuration consisting of the vertices of a regular pentagon and its center and all the lines determined by these six points. The protofloret has symmetry group d_5, and the configuration has symmetry group d_{10}.

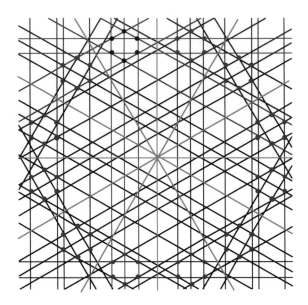

Figure 4.5.11. A floral [3, 4]-configuration $(72_3, 54_4)$ obtained using construction $(\mathcal{M}_1)$ and with symmetry group d_6. The protofloret is a $(6_3, 9_2)$ configuration, with symmetry group d_2; there are 12 florets. The color conventions are the same as in Figure 4.5.3.

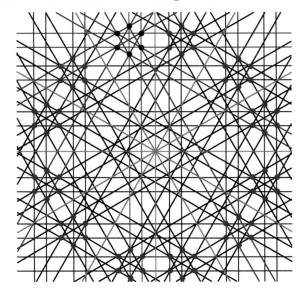

Figure 4.5.12. A floral configuration (72_4) obtained by method $(\mathcal{M}_1)$. The protofloret has points at vertices of a regular hexagon and symmetry group d_3. The configuration has twelve florets and symmetry group d_6.

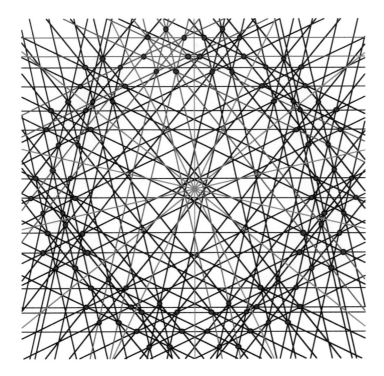

Figure 4.5.13. A floral configuration (98_4) obtained by method $(\mathcal{M}_1)$. Both the protofloret and the configuration have symmetry group d_7.

Figure 4.5.14. A floral (108_4) configuration with protofloret a $(6_4, 12_2)$ configuration devoid of any symmetry. The configuration was constructed using method $(\mathcal{M}_1)$ and has 18 florets and symmetry group d_9.

Figure 4.5.15. A $(150_3, 75_6)$ floral configuration with d_{10} symmetry, resulting from the $(\mathcal{M}_2)$ construction using as protofloret a (15_3) configuration with d_5 symmetry.

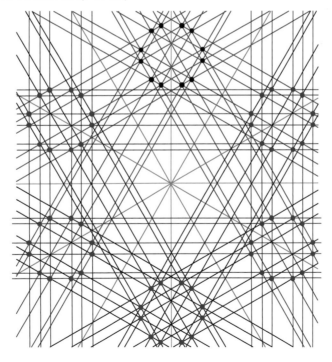

Figure 4.5.16. A floral configuration (72_4) obtained by method $(\mathcal{M}_2)$. The points of the protofloret are at the vertices of an isogonal 12-gon, and the protofloret has symmetry group d_2.

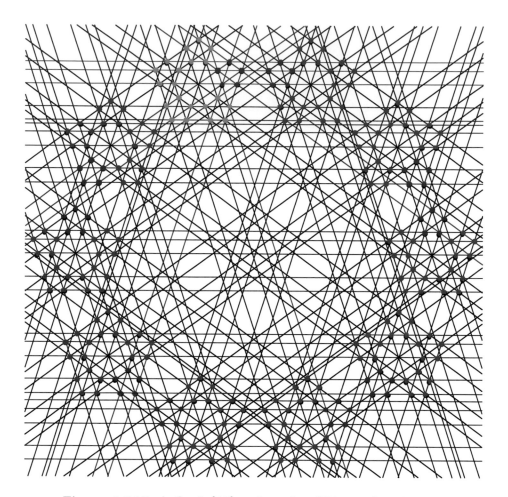

Figure 4.5.17. A floral [4,8]-configuration $(250_4, 125_8)$ with three strata. The protofloret F_0 of the first stratum has points at the vertices of a regular pentagon. Using method $(\mathcal{M}_2)$, five copies of the protofloret form a second stratum F_1, which is a floral configuration (25_4). By method $(\mathcal{M}_1)$, ten copies of F_1 form the third stratum F_2. The protoflorets of the first two strata have symmetry group d_5, as does the complete configuration. The protofloret of the first stratum in each second stratum floret is shown in blue, and one second stratum protofloret is shown in green (and blue). The lines incident with one floret of the first stratum are shown in red.

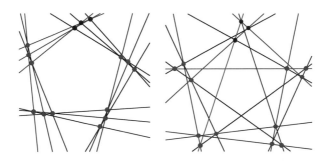

Figure 4.5.18. Two 3-astral configurations (15_3) that are also chiral floral configurations.

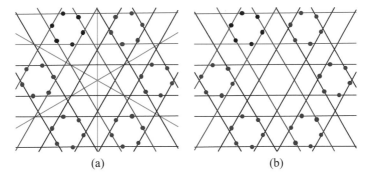

(a)   (b)

Figure 4.5.19. (a) An application of the ($\mathcal{M}_1$) construction to a (6_1) configuration (that is, a protofloret consisting of six points and six lines). (b) Replacing one half of the florets by their mirror images yields a chiral floral configuration.

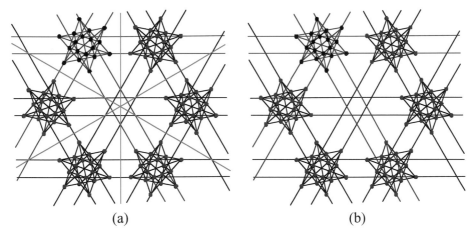

(a)   (b)

Figure 4.5.20. (a) An adaptation of the ($\mathcal{M}_1$) construction to the case in which not all lines in a protofloret are perpendicular to one of the mirrors. (b) Replacing half of the florets by their mirror images yields a chiral floral configuration.

Exercises and Problems 4.5.

1. Construct your own floral configurations using each of the four methods $(\mathcal{M}_i)$.

2. Using the Martinetti "module" shown in Figure 2.4.2, show that a chiral floral (geometric) configuration (n_3) can be constructed for every $n = 10m$, $m \geq 3$.

4.6. Topological configurations

Studies of topological configurations have begun only in the very recent past. While in many ways analogous to geometric configurations, there are significant differences that deserve to be investigated in more detail. Here we will try to present the material that is available at this time.

The distinction between geometric and topological configurations became evident long ago, through Schröter's proof [**199**], [**201**] that one of the ten combinatorial configurations (10_3) cannot be geometrically realized; see Section 2.1 for more details. The fact that it can *almost* be realized geometrically (as in Figure 1.2.2, with lines just a bit bent) means that it is topologically realizable. However, neither this nor the fact that it is not known whether there exist geometrically non-realizable 3-connected (n_3) configurations with $n > 10$ that are topologically realizable resulted in any consistent effort to find clarification. It took almost forty years after Schröter's discovery for Levi [**145**] to even define the appropriate concepts.

Another rather frustrating aspect of the situation concerning topological 3-configurations comes about through Steinitz's theorem (see Section 2.6). In the case of *topological* 3-configurations unintended incidences pose no problem, and one may formulate the resulting statement as follows:

Theorem 4.6.1. *Every connected combinatorial 3-configuration with $n \geq 7$ can be realized by pseudolines if the incidence of an arbitrary point-line pair is disregarded.*

Naturally, just as in the case of the Steinitz theorem itself, the unfulfilled incidence can always be restored by allowing a curve of degree at most 2. But there is no guarantee that this curve can be chosen in such a way that we obtain a topological configuration. As we have seen in Section 2.1, for $n = 7$ or 8, this is, in fact, impossible, and there is no topological realization of these configurations.

A separate question is whether in certain families of 3-configurations (such as astral or 3-astral or others) there exist topological configurations that cannot be realized by geometric ones of the same character. An affirmative answer to one of these questions arises from the examples in Section 2.7 (in particular, see Figure 2.7.6). However, the full extent of such situations for connected astral 3-configurations has not been determined. More precisely, in Figure 4.6.1 we show four different astral 3-configurations of pseudolines which arise from unintended incidences in geometric astral configurations—all four resulting in the same astral 4-configuration (24_4).

A different situation happens with the astral 3-configuration $12\#(5, 1; 3)$. Its drawing does not produce either the intended (24_3) nor a (24_4) configuration. Instead, the resulting family of points and lines has some points on three lines and some on four, while some line are incident with just three points and some with 4. This is illustrated in Figure 4.6.2(a). Again it is possible to avoid unintended incidences by replacing one orbit of lines by pseudolines, as indicated in Figure 4.6.2(b).

In all these cases it is not known whether actual geometric realizations of the 3-configurations can be obtained if one does not impose symmetry restrictions.

Concerning topological 4-configurations, we have already discussed in Section 3.2 the non-existence of topological (n_4) configurations for $n \leq 16$ and the fact that for every $n \geq 17$ there exist topological (n_4) configurations. Very recently, L. Berman [12] determined the conditions for the existence of astral (that is, 2-astral) configurations of pseudolines with dihedral group of symmetries. The main result of [12] is the following:

Theorem 4.6.2. *Astral topological configurations (n_4) exist if and only if n is even and $n \geq 22$.*

For the existence part of the proof it is sufficient to provide examples. An astral (22_4) configuration of pseudolines was first shown in [122] and has been reproduced in several other publications; see Figure 4.6.3. Applying the notation we used in Sections 3.5 and 3.6 to topological configurations, this is $11\#(5, 4; 1, 4)$. It can be used as a template for all even $n = 2m \geq 22$: For each $m \geq 11$, the symbol $m\#(5, 4; 1, 4)$ represents such a configuration. An example (with $m = 17$) is provided in Figure 4.6.4. To establish the inequality for n, it is necessary to first notice that due to the requirements for topological astral configurations, one can assume the configuration to be connected, to have its points coincide with the vertices of two concentric regular m-gons, and to have the concept of "span" of diagonals available— just as for geometric configurations. Then it is easy to verify that the shorter span must be at least 4, hence the larger span at least 5, and therefore m greater than twice 5. (This is an abbreviated version of the detailed arguments in [6].) □

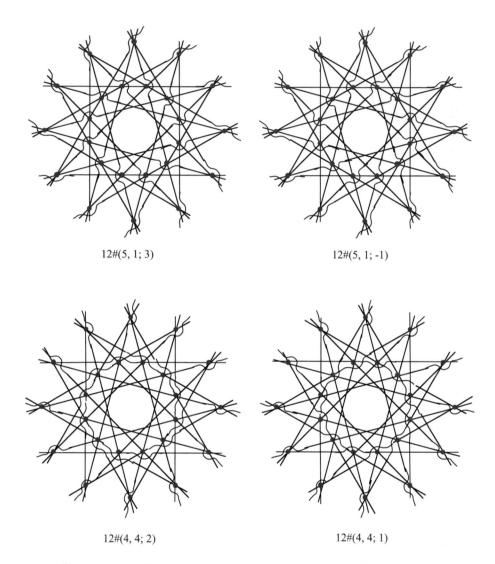

12#(5, 1; 3)   12#(5, 1; -1)

12#(4, 4; 2)   12#(4, 4; 1)

Figure 4.6.1. Four instances where an astral geometric 3-configuration (12_3) leads to the astral 4-configuration (24_4). The pseudolines can avoid the unintended incidences.

A more detailed description of astral topological 4-configurations is given in [**12**] as well. It concentrates on those with dihedral symmetry. With a slight modification of the notation in [**12**], we may summarize the results as follows. Using the symbol $m\#(b, c; d, e)$ with the same meaning as explained in Sections 3.5 and 3.6, we note the following:

- The configuration points of the inner orbit can be situated on a circle of a radius that can vary between certain limits.

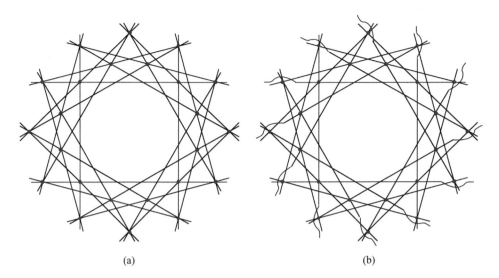

(a)   (b)

Figure 4.6.2. A drawing (a) of the astral 3-configuration $12\#(5, 1; 3)$ produces no geometric configuration but can be modified to a topological configuration (b).

Figure 4.6.3. A topological astral configuration (22_4) that can be described as $11\#(5, 4; 1, 4)$.

- The points of the inner orbit are either aligned with those of the other orbit (Type 1) or else they are situated at positions that enclose with them angles that are odd multiples of π/m (Type 2).

- $m\#(b, c; d, e)$ and $m\#(d, e; b, c)$ are equivalent. Moreover $c \neq d$ and $b \neq e$; we conventionally assume that $b < e$.

- It follows that $c < b$ and $d < e$ and that $b - c > e - d$.

- For Type 1 configurations we have $b - c \equiv e - d \equiv 0 \bmod 2$, and for Type 2 configurations we have $b - c \equiv e - d \equiv 1 \bmod 2$.

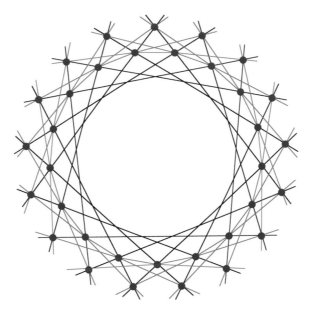

Figure 4.6.4. An astral topological configuration (34_4) of Type 2. It can be specified as $17\#(5, 4; 1, 4)$.

While these conditions impose many restraints on astral topological configurations, it is also clear that most of them cannot be "straightened" or "stretched" into geometric astral configurations. The reason is that the geometric $m\#(b, c; d, e)$ configurations exist only if m is a multiple of 6, while no such restriction holds in the topological case.

The smallest topological astral configuration is $11\#(4, 1; 4, 5)$ shown in Figure 4.6.3. It is the only astral configuration (22_4) and is of Type 2. The smallest astral topological configuration of Type 1 is $13\#(5, 1; 4, 6)$, shown together with $17\#(5, 1; 4, 6)$ in Figure 4.6.5.

Even when m is divisible by 6, there are topological astral configurations $m\#(b, c; d, e)$ that are not stretchable. The smallest such configuration is $18\#(6, 1; 5, 8)$, shown in Figure 4.6.6.

Berman's paper [**12**] contains a number of other results that we cannot get into here. It should only be mentioned that there are examples of *essentially* chiral configurations, that is, configurations that are astral under a cyclic symmetry group but are not even isomorphic to an astral configuration with mirror symmetries. Such configurations do not exist for geometric

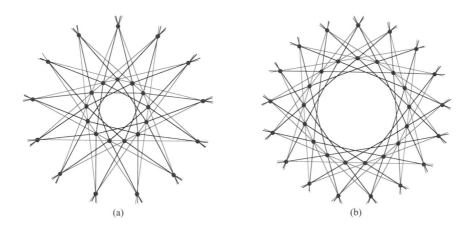

(a)   (b)

Figure 4.6.5. Two astral topological configuration of Type 1. (a) The configuration $13\#(5, 1; 4, 6)$, the smallest such configuration. (b) Another (34_4) topological astral configuration that can be specified as $17\#(5, 1; 4, 6)$.

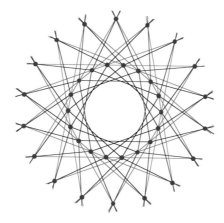

Figure 4.6.6. The astral topological configuration $18\#(6, 1; 5, 8)$, the smallest configuration with m divisible by 6 that is not a geometric astral configuration.

astral configurations. An example of an essentially chiral astral configuration is shown in Figure 4.6.7. A complete description of such configurations is still lacking, as is also any treatment of k-astral topological configurations for $k \geq 3$.

An interesting conjecture in [**12**] can be formulated as follows:

Conjecture 4.6.1. *If the outer orbit of points in an astral topological configuration $m\#(b, c; d, e)$ with dihedral symmetry is on a circle of radius 1, then the inner orbit is on a circle of radius r, where*

$$0 < r < \cos((b - c - 1)\pi/m)/\cos(\pi/m).$$

For a study of simplicial arrangements of pseudolines see [**11**].

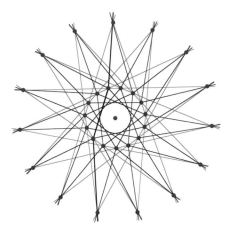

Figure 4.6.7. A chiral astral configuration (30_4) that is not isomorphic to any astral configuration that admits mirror symmetries. One of the pseudolines is drawn by heavy segments. It can be labeled $15\#(6, 1; 5, 7)$, and the fact that $6 - 1 \equiv 1 \bmod 2$ while $7 - 5 \equiv 0 \bmod 2$ shows that it cannot be dihedral of either Type 1 or Type 2.

Exercises and Problems 4.6.

1. Justify the claims that the configurations in Figures 4.6.5(a) and 4.6.6 are the smallest of their kind.

2. What is the smallest topological 5-configuration you can find?

3. How many distinct astral topological configurations (26_4) and (30_4) can you find?

4. What are the smallest topological 3-astral 4-configurations you can find?

5. Generalize the statement (in the proof of Theorem 4.6.2) that the symbol $m\#(5, 4; 1, 4)$ describes a valid topological astral 4-configuration for each $m \geq 11$. What about analogous statements for 3-astral configurations?

4.7. Unconventional configurations

In this section we shall consider several families of objects that we shall call "configurations" even though they do not fit the definition of that word accepted in all the other sections of this book.

The first of these families are "configurations of points and circles". Some examples are shown in Figures 4.7.1 and 4.7.2. In analogy to configurations of points and lines we may denote them by a symbol such as (p_q, n_k), where p, n are the numbers of points and of circles and q, k are the number of circles incident with each point and the number of points incident with each circle; in case the numbers are equal, we use the notation (n_k). Hence the three configurations shown are (4_3), $(8_3, 6_4)$, and (10_4).

Several aspects of configurations of points and circles deserve notice.

First, such configurations are generalizations of configurations of points and lines in a very direct way: Every configuration of points and lines in the projective plane can be shown as a configuration of antipodal pairs of points and great circles in the model of the projective plane on the sphere; a stereographic projection then maps this into a configuration of points and circles in the plane. However, these are only very special cases of such configurations—none of those in Figures 4.7.1 and 4.7.2 are of this kind.

Second, in all but name, configurations of points and circles made their appearance before configurations of points and lines. For example, the configuration in Figure 4.7.1(b) is an illustration of a theorem of A. Miquel [**171**] dating to 1844, asserting that if four pairwise intersections of four circles are concyclic, the other four intersections of the same pairs are concyclic as well. This is one of several results of Miquel, some of which have been greatly generalized by many writers, starting with Clifford [**42**] in 1871 and de Longchamp [**147**] in 1877. One of the achievements is the so-called "chains of theorems" bearing the names of Clifford and de Longchamps. The former establishes the existence of configurations of points and circles $((2^{n-1})_n)$ for all $n \geq 1$. The cases $n = 1$ or 2 are trivial, and $n = 3$ is shown in Figure 4.7.1(a). For more recent works on this and related topics see, for example, Ziegenbein [**239**], Rigby [**190**], Longuet-Higgins [**148**], Longuet-Higgins and Parry [**149**], and references given therein to other works.

Third and last—why is there no greater activity regarding these configurations? We venture to guess that the preoccupation with just a few specific results (such as the "chains of theorems") tended to discourage more general inquiries. There are various subclasses of circle configurations that may well be worth investigating: Are pairs of circles required to intersect twice, are touching circles allowed, can disjoint circles appear, are straight lines admitted, does one wish to consider symmetries in the inversive plane—the choices and possibilities are very wide and almost entirely unexplored. (The inversive plane seems to be an appropriate setting for many of the considerations of symmetries of configurations of points and circles; see, for example, Coxeter [**47**], Eves [**67**], Yaglom [**230**].)

The configuration (10_4) in Figure 4.7.2 is an example of configurations $((2n)_{n-1})$ that exist for all $n \geq 5$ and exhibit remarkable symmetry in the inversive plane. The (10_4) configuration has a single orbit of points and a single orbit of circles under inversive transformations. The author does not know what other configurations are as symmetric, but probably there are many additional ones.

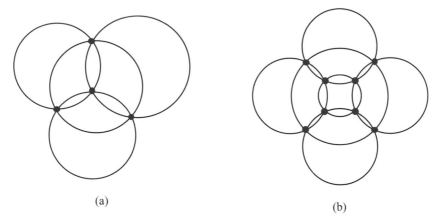

(a)

(b)

Figure 4.7.1. Configurations of points and circles. (a) An (4_3) configuration. (b) An $(8_3, 6_4)$ configuration.

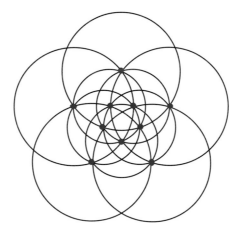

Figure 4.7.2. A (10_4) configuration of points and circles.

The second family of "unconventional configurations" is illustrated by the examples in Figures 4.7.3 and 4.7.4. The objects in this family are the traditional points and lines of the Euclidean plane, and the configurations satisfy all the conditions assumed throughout the book—except the requirement that there are only finite numbers of points and of lines. More precisely, we are now looking at infinite families of points and lines such that, for some finite k, each point is incident with k lines, each line is incident with k points, and the family is discrete in the sense that every point [line] has a neighborhood that contains no other point [line] of the family. We shall call a family of this kind an *infinite k-configuration*.

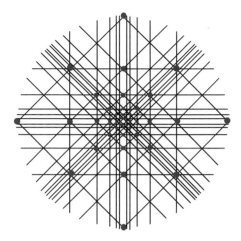

Figure 4.7.3. An infinite 3-configuration with 4-fold dihedral symmetry and single transitivity classes of points and of lines under the group of similarity transformations.

While many different kinds of infinite k-configurations (or of analogously defined infinite $[q, k]$-configurations) can be contemplated, the two examples we show have few orbits of points and of lines under similarity transformations.

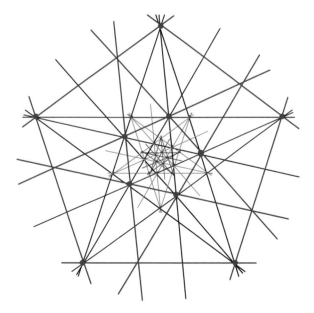

Figure 4.7.4. An infinite 5-configuration obtained by repeated inscribing/circumscribing of copies of the astral configuration $5\#(2, 2; 1)$. The copies are distinguished by colors.

These configurations can be interpreted as an iterative analogue of the $(4m)$ construction we considered in Section 3.3. The infinite 3-configuration in Figure 4.7.3 arises by repeatedly inscribing (4_2) configurations in each other. A construction of this type can be performed starting with any regular m-lateral, leading to an infinite 3-configuration with m-fold dihedral (or cyclic—with a suitable placement of the m-laterals) symmetry. Such configurations can therefore be considered as infinite analogues of the families of inscribed/circumscribed multilaterals we shall consider in Section 5.3.

The infinite 5-configuration in Figure 4.7.4 arises in the same way from repeated inscription/circumscription of copies of the astral configuration (10_3) shown in Figures 1.3.3 and 1.5.4. It is the only example of this kind that the author found in the literature; it is explicitly mentioned in van de Craats's paper [**219**]. It is clear that this type of construction can be carried out with other astral 3-configurations.

The third (and last) family of unconventional configurations is illustrated by the remaining figures of this section. In these configurations the roles of points and lines are different, although there are infinitely many of both: Each point is on precisely k lines for some finite k, but each line contains infinitely many points. We call such configurations *infinite $[k]$-configurations*. To avoid complications, we also require that there be no accumulation points or lines. It is again convenient to consider configurations with a high degree of symmetry under the group of isometric maps of the plane. It is easy to verify that infinite $[k]$-configurations exist for all $k \geq 1$.

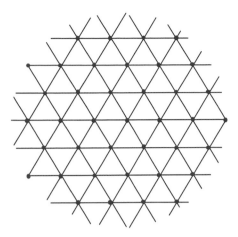

Figure 4.7.5. An infinite [3]-configuration with a single orbit of points and of lines under isometric symmetries of the plane.

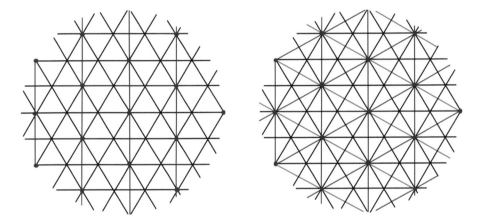

Figure 4.7.6. Examples of infinite [4]- and [6]-configurations.

Figure 4.7.7. Examples of infinite [5]- and [4]-configurations.

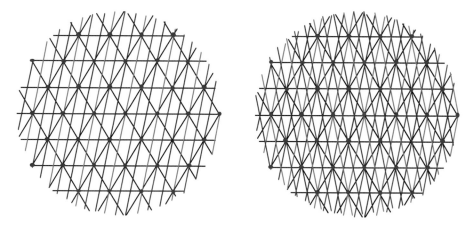

Figure 4.7.8. Additional examples of infinite [4]- and [5]-configurations.

Exercises and Problems 4.7.

1. Construct the (12_4) and (14_4) analogues of the configuration of points and circles in Figure 4.7.2.

2. Decide whether there are configurations (n_k) of points and circles for arbitrarily large k.

3. Modify the construction in Figure 4.7.3 to obtain a *chiral* infinite 3-configuration (that is, with cyclic symmetry group) and with a single orbit of points and one of lines.

4. Justify the claim that the van de Craats construction can be carried out for other astral 3-configurations.

5. Is there any infinite k-configuration such that its points have no accumulation point?

6. Find infinite [k]-configurations that differ in some essential aspect from the ones shown in Figures 4.7.5 to 4.7.8.

7. Construct infinite configurations of points and circles that share some features with the infinite configurations of points and lines described in this section.

4.8. Open problems

There is so little known about the various kinds of configurations described in this chapter that it seems presumptuous to propose specific problems about any of them. But let us try to present a few that would seem capable of being solved within our lifetime.

1. Are any cyclic 5-configurations geometrically realizable? Any cyclic k-configurations for $k \geq 6$?

2. Develop a theory of k-astral 5-configurations for some $k \geq 4$.

3. Determine whether there exist k-configurations (n_k) for all sufficiently large n, that is, for $n \geq N(k)$, where $N(k)$ depends on k only, and similarly for unbalanced configurations, taking into account the divisibility properties resulting from the symmetry of the incidence relation.

4. Clarify the relation between the configurations $((4r)_3, (3r)_4)$ for $r \geq 5$ and cubic curves in the real plane. Can such curves contain all vertices of configurations of this kind for all r? Are all such configurations realizable with all vertices on suitable cubic curves? If not, what are the smallest ones that are not realizable in that manner?

5. Consider geometric configurations of points and lines realized in 3-dimensional Euclidean or extended Euclidean space and spanning it. Find some that are astral in that setting but have no astral realization in the plane.

6. There seems to be no information whatsoever available concerning k-astral 5-configurations for $k \geq 3$.

7. Develop some concept and some results on *configurations of curves—* that is, objects that can be described as "topological configurations of points and circles" in the same sense that configurations of pseudolines are "topological configurations of points and lines".

8. Is it possible to use astral 4-configurations to construct infinite k-configurations with an accumulation point, for some k?

Chapter 5

Properties of Configurations

5.0. Overview

This chapter differs in character from the preceding ones. While each of them deals with particular kinds of configurations—most concerning their existence and constructions—here we consider certain properties of configurations that have attracted the interest of researchers at various times.

We begin by presenting in Section 5.1 the information available concerning the connectivity properties of configurations. While some of the first papers ignored the issue, it quickly became apparent that restricting the discussion to connected configurations leads to considerable simplification in formulation of results. However, the importance of the degree of connectedness has been slow to emerge. Even today, it is not clear to what extent various properties depend on whether the configuration is 2-connected or has higher degree of connectivity.

Section 5.2 deals with Hamiltonian multilaterals—the analogs of Hamiltonian circuits in graphs. The recent result that there exist 3-connected 3-configurations with Hamiltonian multilaterals is presented; this solves negatively a long-standing conjecture that 3-connectedness is sufficient for Hamiltonicity.

The next section deals with the related concept of multilateral decomposition of a configuration. By this is meant a family of multilaterals that include all lines and all points of the configuration, each just once. The topic goes back to the "prehistory" of configurations and leads to many still open questions.

Section 5.4 is devoted to the presentation of the known facts about configurations that have no trilaterals, or, more generally, no k-laterals for $k = 3, 4, \ldots, h$. This is an old topic that has recently been revived and has turned out to be related to interesting questions about graphs.

In Section 5.5 we consider the configurations for which every points is incident with the same number of trilaterals—another old topic in configurations theory.

Section 5.6 is concerned with the recently proposed question of what is the largest dimension that a configuration isomorphic to a given configuration can span. The few results are supplemented by many open questions.

Section 5.7 deals with a topic that has attracted attention only recently—configurations that can be continuously modified while keeping their combinatorial structure and keeping a sizable part of the configuration unchanged. The latest available results are presented, but there is a wide array of open questions to occupy geometers in the future.

Duality and selfduality and the special cases of polarity and selfpolarity are considered in Section 5.8. The material available shows only too clearly how much is still to be discovered.

The final section, Section 5.9, is devoted to a list of open problems. These are only a few specific questions, meant to illustrate the topics considered. By no means is the list supposed to be exhaustive.

5.1. Connectivity of configurations

We start by recalling from Section 1.4 the concept of the Levi graph $L(C)$ of a configuration C. This is a bipartite graph (that is, there are two sets of nodes—black and white) with no edge connecting vertices of the same color. Usually the black nodes of $L(C)$ correspond to the vertices of C, while the white ones correspond to the lines of C. A black node is connected to a white node by an edge of $L(C)$ if and only if the corresponding vertex of C is incident with the corresponding line. The advantage of $L(C)$ is that the graph $L(C)$ represents the configuration *faithfully*—that is, knowing the Levi graph of a configuration enables one to determine the (combinatorial) configuration uniquely. An example of the Levi graph of a (12_3) configuration is presented in Figure 5.1.1.

For a given configuration C, the Levi graph $L(C)$ is uniquely determined. However, $L(C)$ may admit various presentations, with different properties. For example, another rendition of the Levi graph of the configuration (12_3) in Figure 5.1.1 is shown in Figure 5.1.2; it can be understood as an imbedding of the Levi graph in the torus. This presentation shows that all vertices of this configuration form one orbit under automorphisms of the configuration

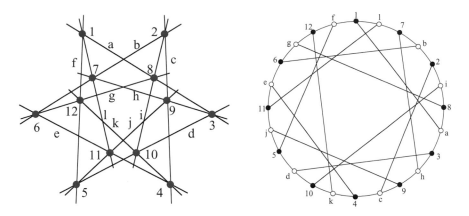

Figure 5.1.1. A configuration (12_3) and its Levi graph. Since the graph admits an incidence-preserving color reversal, which yields a graph isomorphic to the original (by reflection in the line bisecting the segments $7h$ and $12k$, for example), the configuration is selfdual.

and that all lines form one orbit as well. This is not easily visible from either the drawing of the configuration or its Levi graph in Figure 5.1.1.

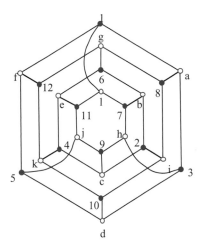

Figure 5.1.2. Another rendition of the Levi graph of the configuration (12_3) in Figure 5.1.1. It is easy to visualize this graph embedded in a torus in such a way that the combinatorial equivalence of all vertices of the configuration is obvious.

The main utility of Levi graphs comes from the fact that the graph-theoretic properties of $L(C)$ may be used to define or determine properties of the configurations involved. We shall return to this topic in the next section, in the context of multilaterals.

Levi graphs are particularly useful in connection with questions about the connectivity of configurations. Concepts such as "connected", "k-connected", etc., for a configuration are defined by asking whether its Levi graph has the property in question. It is clear that these concepts can be defined directly in the configurations, but the formulations, distinctions, relevance, and familiarity are in many cases more easily perceived on the Levi graphs.

Theorem 5.1.1. *Every connected combinatorial or geometric k-configuration C is 2-connected.*

Proof. Note that a configuration is connected but not 2-connected (that is, the deletion of a single point from the Levi graph of C disconnects the graph) if and only if the dual configuration has the same property. Hence for a proof of the theorem by contradiction we may assume that there is a line L whose removal disconnects the configuration. At least one of the connected components resulting from the removal of L has at most h points incident with L, where $1 \leq h \leq k/2$. Then the number of incidences of the m lines of this component with the p points of the component is, on the one hand, equal to km, but on the other hand, it is equal to $k(p-1) + h$, since L was incident with h points. These numbers should be equal, but as one of them is divisible by k and the other is not, a contradiction was reached. $\square$

Theorem 5.1.1 is due to Steinitz [**210**], as is the idea of its proof. We have seen earlier (Corollary 2.5.3) a different proof of this result. But that proof relied on the construction of an orderly configuration table, which was a rather deep result. The approach here provides a good example of the utility of introducing graph-theoretic concepts (in particular, the Levi graph), in considerations of configurations. Steinitz did not have such tools, and as a consequence he needed more than a page of densely printed (and clumsily formulated) arguments to state and prove Theorem 5.1.1.

As a strengthening of Theorem 5.1.1 one might conjecture that each connected k-configuration, $k \geq 3$, is 3-connected. However, this is not the case. A counterexample is shown in Figure 5.1.3. It is known that all combinatorial configurations (n_3) with $n \leq 13$ are connected and, moreover, are 3-connected. This is the best possible since for $n = 14$ there are counterexamples to both parts. The combinatorial configuration consisting of two disjoint copies of the Fano configuration is disconnected, while the configuration in Figure 5.1.3 is connected but not 3-connected. Any disconnected geometric (or topological) configuration (n_3) must have $n \geq 18$.

For 4-configurations the corresponding numbers follow.

■ There are disconnected combinatorial configurations (n_4) if and only if $n \geq 26$.

■ There are disconnected topological configurations (n_4) if and only if $n \geq 34$.

■ There are disconnected geometric configurations (n_4) if and only if $n \geq 36$.

The example in Figure 5.1.3 shows that there are 2-connected 3-configurations that are not 3-connected. This leads to the following problem: If $k \geq 4$ and $2 \leq j < k$, do there exist j-connected k-configurations that are not $(j + 1)$-connected?

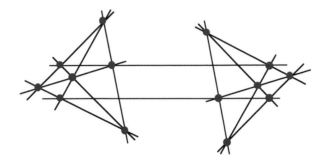

Figure 5.1.3. A connected configuration (14_3) which is not 3-connected.

An affirmative answer is given by the construction described in the proof of the following theorem:

Theorem 5.1.2. *For each $k \geq 4$ and each j with $2 \leq j < k$ there exist geometric k-configurations that are j-connected but not $(j + 1)$-connected.*

Proof. We consider first the case $j = 2$. The text deals with arbitrary k, and the illustration in Figure 5.1.4 presents in parallel the case $k = 4$. We start with copies of configurations $LC(k)$ described in Section 1.1 (see also [**186**]), with a slight modification. As described in Section 1.1, $LC(k)$ consists of an array of k^k points of the integer lattice in the Euclidean k-space E^k, with all coordinates in the range $[0, k - 1]$, together with the lines parallel to the coordinate axes through these points. The modification we need here is that in one of the directions the coordinate $k-1$ is replaced by another convenient integer. Our attention focuses on one of the lines in that direction and the k points on it. (If desired, we way think of these configurations as projected into the plane.) In Figure 5.1.4 this situation is schematically indicated; one of the gray rectangles represents the modified configuration $LC(k)$, the dashed line within the rectangle represents the chosen line, and the four dots represent the four points of $LC(k)$ on that line. We need $2k$ configurations of this type, indicated by the gray rectangles and positioned in such a way that two of the solid lines connect the k configurations on the left with the k configurations on the right, while each of the other $2k - 2$ solid lines connects the k copies in each half among themselves. Note that in each half, a single configuration is placed differently than the other $k - 1$. Finally, we delete the dotted lines, thus creating a k-configuration. It is obvious that the resulting configuration is 2-connected, but the deletion of the two lines running between the two halves disconnects it; hence the configuration is not 3-connected.

For $j > 2$ we proceed analogously, but with an additional step. The construction is illustrated in Figure 5.1.5.

The first modification of the above construction is that the left half now has $j - 1$ dots "above" the rest, and the right half has $k - j + 1$ such points. Naturally, the corresponding numbers on the "bottom" are $k - j + 1$ and $j - 1$. Also, we do not insert the bottom connecting line between the two halves. Instead, we take a stack of k copies of what we constructed so far (it is best to imagine these copies to be in parallel planes, stacked above each other) and connect the corresponding points that were originally connected by the omitted bottom line. This is the desired configuration. It is clearly j-connected, but the right part at each level can be disconnected from the rest by the omission of the $j - 1$ lines connecting it to the other levels and the line connecting it to the other half at its own level. $\qquad\square$

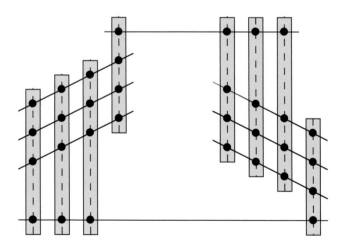

Figure 5.1.4. The construction of a geometric 4-configuration that is 2-connected but not 3-connected.

It is clear that these constructions lead to very large configurations even in the smallest cases: (2048_4) in Figure 5.1.4 and (8192_4) in Figure 5.1.5. There probably exist much smaller configurations with the same properties—but justifying the existence of the appropriate projective images to yield the alignments necessary may be more involved. Rather obviously, much smaller combinatorial configurations with the properties discussed in

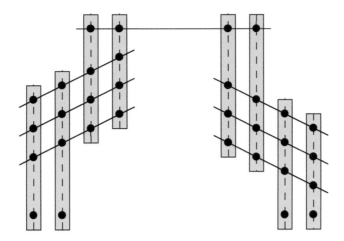

Figure 5.1.5. The construction of a geometric 4-configuration that is 3-connected but not 4-connected.

Theorem 5.1.2 may exist, but this seems of rather marginal interest. On the other hand, even though one may expect to find combinatorial configurations of this type that are smaller than the corresponding geometric configurations, no actual examples seem to be available.

$$* * * * *$$

In a course on configurations given in the 1990s the author made several conjectures that aimed at extending the result of Theorem 5.1.1 above to unbalanced $[q, k]$-configurations with $3 \leq q \neq k \geq 3$. Xin Chen, a student in that course, produced various counterexamples, among them one proving that there exist $[4, 3]$-configurations which are 1-connected but not 2-connected.

A small modification of Chen's procedure leads to

Theorem 5.1.3. *Combinatorial $[q, k]$-configurations that are 1-connected but not 2-connected exist if and only if $q \neq k$.*

Proof. As we have seen in Theorem 5.1.1, every connected k-configuration is 2-connected. For the other direction, due to duality, if $q \neq k$, it is enough to consider the case $q > k$. We start by forming a $[q, k]$-configuration, **cyclic** as far as possible. By this is meant that one uses as many cyclic sequences as necessary—in the illustration below we need two such cycles. These configurations are generalizations of the cyclic configurations $\mathscr{C}_3(n)$ we introduced in Section 2.1. (The use of "cyclic" configurations comes in only to simplify checking that the tables which will be constructed are actually configuration tables.) Then we use a Martinetti-type construction (see Section 2.4): We select some $k-1$ lines with a property specified below, add one more point, and form a fragment in which all points except the new one are incident with q lines and the new point is incident with k lines. The selected lines (which are then omitted) should be such that their points can be grouped in a way that no pair occurs in any other line; the new point is "cross-connected" to the points in these lines. As an illustration, the case $q = 4$, $k = 3$ is explicitly presented in detail. Here we take $n = 12$, the chosen lines are 1 2 4 and 7 8 10, and the additional point is 0. This way the configuration

1	2	3	4	5	6	**7**	8	9	10	11	12	1	2	3	4
2	3	4	5	6	7	**8**	9	10	11	12	1	5	6	7	8
4	5	6	7	8	9	**10**	11	12	1	2	3	9	10	11	12

yields the subfiguration

0	**0**	**0**	2	3	4	5	6	8	9	10	11	12	1	2	3	4
1	**2**	**4**	3	4	5	6	7	9	10	11	12	1	5	6	7	8
7	**8**	**10**	5	6	7	8	9	11	12	1	2	3	9	10	11	12

If three copies of this fragment are taken, distinguished by the number of dashes, and a line 0 0′ 0″ is added, we get a combinatorial $(39_4, 52_3)$ configuration which is connected but not 2-connected.

In the general case, the additional point 0 of the fragment will be on k lines only, thus having a deficit of $q - k$ lines. To supply these, we find a connected $[q - k, k]$-configuration C. Then we take as many copies of the fragment as there are points in C and identify the point 0 of one copy of the fragment with each point of the configuration C. (In the above example, C is the $[1, 3]$-configuration which consists of just three points and one line.) This clearly yields a $[q, k]$-configuration which is connected but not 2-connected since each copy of the point 0 disconnects the configuration. □

It is not known whether the $(39_4, 52_3)$ configuration is the smallest of this type.

Theorem 5.1.3 deals with combinatorial configurations, and the question arises whether there exist connected geometric $[q, k]$-configurations, with $q \neq k$, that are not 2-connected.

A partial affirmative answer is given by the following result.

Theorem 5.1.4. *For every q and k, with $\min\{q, k\} \geq 3$ and $q \neq k$, there exist geometric $[q, k]$-configurations that are connected but not 2-connected.*

Proof. We first consider the case of $[4, 3]$-configurations. We start with a tricyclic 3-configuration C_1 shown in Figure 5.1.6. (This particular configuration is (54_3), but it is likely that smaller configurations of the same general type could be used in the construction.) The significance of the two heavily drawn lines will be explained soon. As in some of the other constructions, the next step is best explained by thinking of the configuration C_1 as contained in a plane of the 3-dimensional space. By adding two congruent copies, situated perpendicularly above and below C_1, and adding vertical lines through all points of C_1, we obtain a configuration $(162_4, 216_3)$, which we designate C_2. Now we delete the two heavily drawn lines from the configuration C_1—but not from the two copies of it, which we used in C_2. Instead of these two lines we introduce three new lines, as shown in Figure 5.1.7, and a new point incident with all three of these. (The existence of such a triplet of lines depends on the variability afforded to tricyclic configurations by the presence of an arbitrary parameter.) This step leads from C_2 to a prefiguration C_3. All lines in C_3 are incident with three points, and all points of C_3 except the newly introduced point are incident with four lines.

In the final step we take two additional copies of C_3 and connect the three exceptional points by a line. This results in a $[4,3]$-configuration which is connected but obviously is not 2-connected.

An easy modification of this construction works for $q > 4$. All that is needed is to use copies of C_2 to construct (in 4-space, for greater comfort) a $[5,3]$- or $[6,3]$-, etc., configuration, before proceeding to C_3 and to the final configuration.

Clearly, the polars of these configurations (or, more precisely, of their projections into the plane) yield the appropriate connected but not 2-connected $[3,k]$-configurations, with $k \geq 4$. □

Now, if $\min\{q,k\} \geq 4$, we need an additional step in the construction. We shall assume that $q \geq k$, since otherwise we could construct the dual configurations. We start again with the (54_3) configuration C_1 shown in Figure 5.1.6. By repeatedly using the procedure we designated by $(5m)$ in Section 3.3, we generate from C_1 a $[q,k]$-configuration C^*. In Section 3.3 we went only one step, from 3- to 4-configurations. But an easy modification allows the construction of a k-configuration C^{**} that contains the original C_1 as a subconfiguration. By stacking k copies of C^{**} and connecting them with lines through corresponding points, a $[k+1,k]$-configuration is obtained; repetition of this step leads to the required C^*. Now we replace—as before—the two heavily drawn lines in Figure 5.1.6 by the three lines and a point as shown in Figure 5.1.7. The newly introduced point O is the only point that is on just $q-1$ lines. Now taking $k-1$ additional copies of C^* and connecting the k points (O and its images in the other copies of C^*) by a line gives the desired $[q,k]$-configuration which is connected but not 2-connected.

* * * * * *

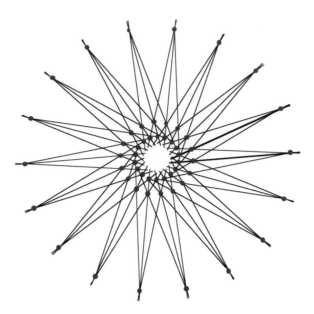

Figure 5.1.6. The tricyclic configuration C_1 used in the proof of Theorem 5.1.4.

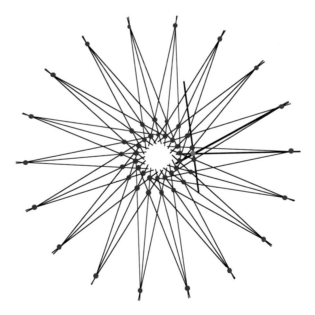

Figure 5.1.7. The prefiguration C_3 used in the proof of Theorem 5.1.4. The new point is indicated by the green dot.

Two elements (points or lines) of a configuration are said to be **independent** if they are

▷ two points that are on no line of the configuration,

▷ two lines not incident with a point of the configuration,

▷ a point and a line not incident.

A family of elements in a configuration is called **independent** if every two of its elements are independent.

A configuration C is said to be **unsplittable**[1] if the deletion of any independent family of its elements leaves a connected configuration. In other words, if for every independent family F, any two elements that do not belong to the family F are in a multilateral that does not use any element in F. Equivalently, C is **splittable** if it can be disconnected by an independent family of elements.

For example, the (12_3) configuration in Figure 5.1.8 has several 5-element independent families but no 6-element independent families. It is easy (even if tedious) to check that the configuration is unsplittable. Similarly, for the (15_3) configuration in Figure 5.1.9, the maximal number of elements in an independent family is 6, and the configuration is unsplittable. These and other examples lead to the following conjecture.

Conjecture 5.1.1. *The maximal number of elements in an independent family in a connected configuration (n_k) is $[n/k] + 1$.*

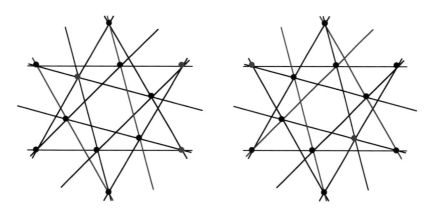

Figure 5.1.8. Two independent families of five elements each (shown in red) that do not disconnect the (12_3) configuration.

We conclude with the following theorem.

Theorem 5.1.5. *Every unsplittable 3-configuration is 3-connected.*

Proof. Assume that a connected configuration C is not 3-connected; we shall show it is splittable. As a consequence of Steinitz's Theorem 2.5.1, it

---

[1]The material on independent families and unsplittable configurations is part of an ongoing collaboration with Tomaz Pisanski; it was presented in part in [**181**].

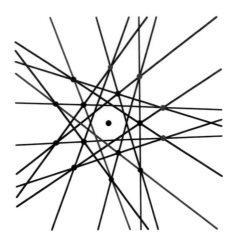

Figure 5.1.9. An independent family of six elements (red) in a (15_3) configuration.

is possible to present C in an orderly configuration table; hence it must be at least 2-connected.

If C is not 3-connected, it can be disconnected by two elements, and we have the following two possibilities:

(i) Both elements are of the same kind; without loss of generality we can assume that the disconnecting set consists of two points.

(ii) The disconnecting set consists of one point and one line.

If either of these disconnecting sets were not a splitting set, the two points must be incident with a line of C or the points and line must be incident, respectively.

In case (i), let the two points be A and B, and let L be the line incident with both. Let D be the third point of L. It is impossible that of the six lines incident with L, only two come from the same connected component. (If this were the case, either the two lines would be incident with the one of the points A, B, D—which would mean that C is not 2-connected—or else the two lines would be incident one each with A and B. Then taking two copies of this component, together with $A, B,$ and L, and attaching three such systems at a point corresponding to D in all three would again yield a configuration that is not 2-connected.) So each component has three lines incident with L. Since L is incident with only six lines and since A, B disconnect C, the arrangement must be like the one in Figure 5.1.11. But then the point B and line M are a splitting set.

In case (ii), the situation must again be as shown in Figure 5.1.10, with A and L the disconnecting elements, and again M and B are a splitting set.   □

We may note that the converse of Theorem 5.1.5 does not hold. There are 3-connected configurations that are splittable. The smallest one that the author knows of is the (15_3) configuration shown in Figure 5.1.11.

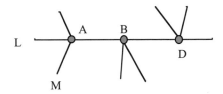

Figure 5.1.10. Schematic arrangement used in the proof of Theorem 5.1.5.

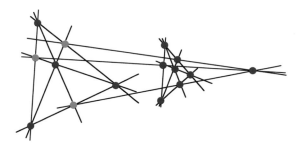

Figure 5.1.11. This 3-connected configuration (15_3) is splittable. A splitting set consists of the three green dots.

Exercises and Problems 5.1.

1. The geometric $[q, k]$-configurations constructed in the proof of Theorem 5.1.4 are quite large, even in the case of $[4, 3]$-configurations. Find smaller examples of connected but not 2-connected geometric $[4, 3]$-configurations.

2. Do there exist reasonably sized 2-connected but not 3-connected geometric $[4, 3]$-configurations? It is clear that this question can be generalized.

3. Without relying on polarity, give a detailed proof of Theorem 5.1.4 for the case of geometric $[3,4]$-configurations.

4. Prove that Conjecture 5.1.1 is true for $k = 2$, even without the connectedness assumption. Show that it is invalid for $k = 3$ if connectedness is not assumed.

5. Show that Conjecture 5.1.1 holds for all (10_3) configurations.

6. Show that the cyclic configuration $\mathscr{C}_3(n)$ (see Section 2.1 for the definition) is unsplittable.

7. Investigate which of the cyclic configurations $\mathscr{C}_3(n, a, b)$ (see definition in Exercise 2 in Section 2.1) are unsplittable.

8. The independent families of elements of a configuration C can be characterized as corresponding to independent (that is, unconnected) sets of

vertices in the **independence graph** $I(C)$ of C. (The independence graph of C can be defined as the (graph-theoretic) **square** of the Levi graph $L(C)$; this is the graph in which two vertices are connected by an edge if they are connected in the original graph or if they share a common adjacent vertex. Equivalently, $I(C)$ is the union of $L(C)$ with the edges of the Menger graph $M(C)$ described in Section 1.4 and the edges of $M(C^*)$, where C^* is the configuration dual to C.)

5.2. Hamiltonian multilaterals

As another example of the utility of Levi graphs, we consider in this section and the next some of the known results on multilaterals in configurations. We start by expanding the appropriate definitions from Section 1.3.

We call **multilateral** (or, if appropriate, r-**lateral**) any sequence of points P_i and lines L_i of a configuration that can be written as $P_0, L_0, P_1, L_1, \ldots, P_{r-1}, L_{r-1}, P_r = P_0$, with each L_i incident with P_i and P_{i+1} (all subscripts understood $\bmod r$). Thus, an r-lateral in a configuration C corresponds to a $(2r)$-circuit in the Levi graph $L(C)$. A **multilateral path** satisfies the same conditions except the coincidence of the first and last elements. Instead of 3-lateral we shall say **trilateral**, and analogously **pentalateral**, etc.[2] A **circuit decomposition** of a graph is any family of disjoint simple circuits that together include all vertices of the graph. Clearly, not every graph has a circuit decomposition, but as we have seen in Section 2.5, the Levi graph of every connected k-configuration, $k \geq 2$, has such a decomposition. The corresponding multilaterals of the configuration are said to be a **multilateral decomposition** of the configuration. A multilateral decomposition consisting of a single multilateral is a **Hamiltonian multilateral** of the configuration. In other words, a Hamiltonian multilateral of a configuration passes through all its points and uses all its lines, each precisely once.

The Hamiltonian circuit of the Levi graph in Figure 5.1.1 corresponds to a Hamiltonian multilateral of the configuration. On the other hand, from the Levi graph of the same configuration shown in Figure 5.1.2, we easily see the possibility of a circuit decomposition of the Levi graph into four 6-circuits, hence yielding a decomposition of the configuration (12_3) in Figure 5.1.1 into four trilaterals (that form a cycle of mutually inscribed/circumscribed trilaterals). It is also easy to observe a decomposition of this configuration into three quadrilaterals, such as $la8g6b71, 3d10i2c9h, 5f12k4e11j$.

---

[2]The terms "trilateral", "pentalateral", and others were used by Martinetti [**151**] in 1886, but with a slightly different meaning.

As another example, in Figure 5.2.1 we show a Hamiltonian multilateral of a configuration (10_3), as well as a decomposition of the same configuration into two pentalaterals.

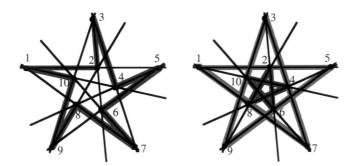

Figure 5.2.1. A configuration (10_3) with one Hamiltonian multilateral and one decomposition into two pentalaterals that are mutually inscribed/circumscribed.

The remaining part of this section is devoted to a survey of results known about Hamiltonian multilaterals. The great majority of these results deal with 3-configurations.

To begin with, here are some historical notes; in all of these papers the *circuit* terminology has been used, but we present them in our terms. Kantor in 1881 [**132**] states *as a theorem* that every (n_3) configuration has a Hamiltonian multilateral. Martinetti [**152**] in 1887, Schönflies [**195**] in 1888, and Brunel [**34**] in 1895 consider Hamiltonian multilaterals as self-inscribed/circumscribed polygons. Schröter in 1889 [**201**] states that he confirmed the existence of Hamiltonian multilaterals in all (10_3) configurations, and Steinitz in 1897 [**211**] does the same for all (11_3). Steinitz also observed that connectedness is a necessary condition, a fact not mentioned by earlier writers. He also provided a first example of a connected configuration that does not admit a Hamiltonian multilateral. The smallest configuration that would fit his description is (28_3). In 1990, Gropp [**80**] announced that all connected configurations (n_3) with $n \leq 14$ have Hamiltonian multilaterals. The statement (a1) in the paper [**137**, p. 128] by Kelmans, to the effect that every 3-valent, 3-connected bipartite graph with at most 30 vertices has a Hamiltonian circuit, implies that all connected (n_3) configurations with $n \leq 15$ have Hamiltonian multilaterals.

The example of Steinitz mentioned above was improved in [**63**] by the construction of the configuration (22_3) shown in Figure 5.2.2. It is not known whether every connected configuration (n_3) with $16 \leq n \leq 21$ admits a Hamiltonian multilateral.

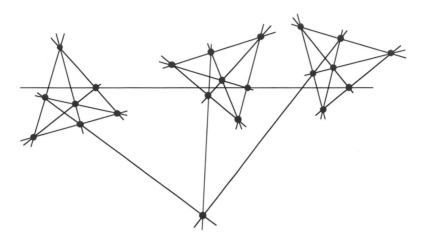

Figure 5.2.2. A (22_3) configuration that does not admit any Hamiltonian multilateral. It is obvious that such configurations (n_3) can be constructed for every $n \geq 22$.

The fact that all connected configurations (n_3) without a Hamiltonian multilateral known at the time were only 2-connected led the author to conjecture in 2002:

Conjecture 5.2.1. *All* 3-*connected* (n_3) *configurations admit Hamiltonian multilaterals.*

However, this conjecture was disproved in [**116**]:

Theorem 5.2.1. *There exists a* 3-*connected geometric configuration* (25_3) *that does not admit a Hamiltonian multilateral.*

We shall prove this by a construction, fashioned after the arguments presented in [**116**].

Our construction starts with the small bipartite graph shown in Figure 5.2.3, devised by M. N. Ellingham and J. D. Horton in [**65**]. This graph has no Hamiltonian circuit that uses both heavily drawn edges. The *proof* of this assertion is simply a follow-up of a few alternatives—the non-trivial, clever part is the *discovery* of the graph. For the next step we insert two additional vertices in each of the heavily drawn edges of Figure 5.2.3, resulting in the graph shown in Figure 5.2.4. From this the graph of Figure 5.2.5 was constructed by Georges [**71**]. It is a non-Hamiltonian, 3-connected, bipartite graph. Again, the proof of non-Hamiltonicity consists of the examination of several possibilities and showing that neither leads to a Hamiltonian circuit. After the graph is constructed, this is a matter of routine checking, as are the other relevant properties. At this step, as in the earlier, it is the construction of the graph that is ingenious. The main result of the construction that is

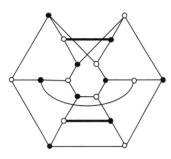

Figure 5.2.3. This bipartite graph found by Ellingham and Horton [**65**] does not admit a Hamiltonian circuit that uses both heavily drawn edges.

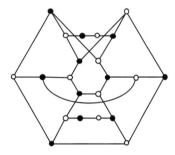

Figure 5.2.4. A modification of the graph in Figure 5.2.3.

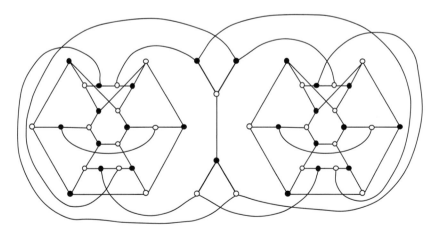

Figure 5.2.5. The Georges graph, resulting from a combination of two copies of the graph in Figure 5.2.4. The graph is bipartite, 3-connected, non-Hamiltonian, and has girth 6.

relevant to the present aim is the fact that the Georges graph has girth 6. Since it is bipartite, it follows that it is the *Levi graph* of a combinatorial configuration (25_3). This is its relevance for the present goal.

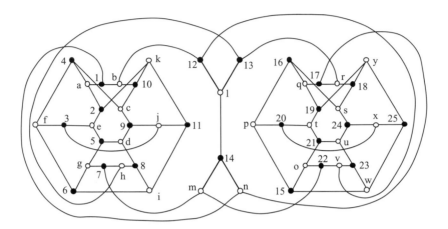

Figure 5.2.6. A labeling of the Georges graph that allows us to interpret it as the Levi graph of a combinatorial configuration. The task is simplified by the fact that there are sufficiently many characters to label all lines.

We shall now apply the method of Steinitz described in Section 2.6 to obtain first a realization of this combinatorial configuration (25_3) as a prefiguration and then apply a continuity argument to establish the possibility of its realization by a geometric configuration. To begin with, we label the Georges graph in a more or less random way, as in Figure 5.2.6. This labeling leads to the configuration Table 5.2.1. This is turned into an orderly configuration table, shown in Table 5.2.2, as required for the application of Steinitz's construction. (This table was not constructed by following the rather cumbersome Steinitz algorithm, but by a straightforward "greedy" algorithm: taking the first available choice at each step. It worked very well in the present case.)

Table 5.2.3 shows the Georges configuration with columns (that is, lines) permuted so that a decomposition of the configuration into "multilaterals" becomes obvious. This is accomplished by making the second entry in a column equal to the third entry in the preceding column, the first column being chosen arbitrarily. The exception is the last column of a multilateral, in which the last entry is the same as the second entry of the first column. In cases like the present one, where the first multilateral does not exhaust the columns, the first remaining column is used to start a new polygon. In the case of the George configuration, the first is a 22-gon, the second a triangle. Figure 5.2.7 provides an illustration of the two circuits in the Levi graph that correspond to two multilaterals.

As a last step in the Steinitz algorithm before the geometric construction, we permute the columns (lines) once more. Some vertex of a column

Table 5.2.1. A configuration table of the Georges configuration, as read off Figure 5.2.6.

a	b	c	d	e	f	g	h	i	j	k	l	m
1	1	4	5	2	3	5	7	6	3	2	12	7
2	10	9	8	3	4	6	8	8	9	10	13	14
4	12	10	9	5	6	7	13	11	11	11	14	22

n	o	p	q	r	s	t	u	v	w	x	y
1	15	15	16	13	16	19	21	12	15	20	18
14	21	16	17	17	18	20	23	22	23	24	19
17	22	20	19	18	24	21	24	23	25	25	25

Table 5.2.2. An orderly configuration table for the Georges configuration.

a	b	c	d	e	f	g	h	i	j	k	l	m
1	10	4	5	2	3	6	7	8	9	11	12	14
2	12	9	8	3	4	5	13	6	11	0	14	7
4	1	10	9	5	6	7	8	11	3	2	13	22

n	o	p	q	r	s	t	u	v	w	x	y
17	15	16	19	13	18	20	21	22	23	24	25
1	22	15	17	18	16	21	24	23	25	20	19
14	21	20	16	17	24	19	23	12	15	25	18

Table 5.2.3. A rearrangement of the columns of the Georges configuration used to show a decomposition into multilaterals. The boxed labels are explained in the text.

a	f	i	j	e	g	m	o	t	y	r	q	s
1	3	8	9	2	6	14	$\boxed{15}$	20	25	13	19	18
2	4	6	11	3	5	7	22	21	19	18	17	16
4	6	11	3	5	7	22	21	19	18	17	16	24

u	v	b	n	l	h	d	c	k	p	x	w
21	22	10	17	12	7	5	4	11	16	24	23
24	23	12	1	14	13	8	9	10	$\boxed{15}$	20	25
23	12	1	14	13	8	9	10	2	20	25	15

of the last multilateral (the second in this case) must be a vertex that appeared in a previous multilateral, since otherwise the configuration would not be connected. (In the present case, we chose the vertex labeled 15.) We place that column as the first of the last multilateral and place as the last column of the previous multilateral its column that contains the same

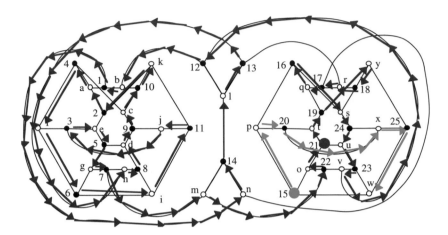

Figure 5.2.7. The two multilaterals from Table 5.2.3 that lead to a multilateral decomposition of the Georges graph.

vertex. The other columns in both multilaterals are permuted accordingly, so as to preserve the multilaterals present. The configuration table obtained in this step (resulting from the choice of 15 as the special vertex) is shown in Table 5.2.4.

The geometric realization now proceeds very simply. It can be followed in Figure 5.2.8 which was obtained using Geometer's Sketchpad® and modified by ClarisDraw®. The idea follows the explanations given in Section 2.6: Choose the vertices as arbitrary points, except when constrained to lie on one or two previously constructed lines. In this example we start with arbitrary points 21 and 19. Points 18 and 17 are also chosen freely, but the choice of 16 has to be on the line q through the previously determined points 19 and 17, and then 24 must be on the line s through 16 and 18; similarly, the point 23 must be on the line $u = [21, 24]$. The points 12 and 1 can be

Table 5.2.4. A rearrangement of the columns of Table 5.2.3, needed for the application of Steinitz's geometric construction.

t	y	r	q	s	u	v	b	n	l	h	d	c
20	25	13	19	18	21	22	10	17	12	7	5	4
21	19	18	17	16	24	23	12	1	14	13	8	9
19	18	17	16	24	23	12	1	14	13	8	9	10

k	a	f	i	j	e	g	m	o	p	x	w
11	1	3	8	9	2	6	14	$\boxed{15}$	16	24	23
10	2	4	6	11	3	5	7	22	$\boxed{15}$	20	25
2	4	6	11	3	5	7	22	21	20	25	15

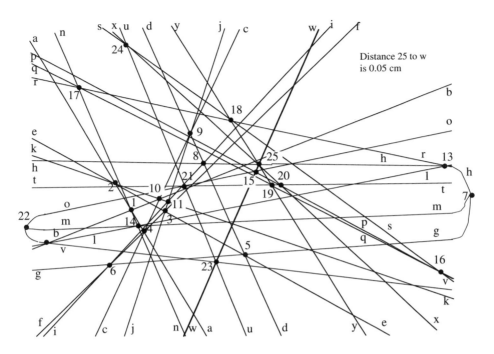

Figure 5.2.8. The construction of a 3-connected geometric configuration that does not admit a Hamiltonian multilateral. In the actual construction shown (using Geometer's Sketchpad®) the final incidence was missed by about 0.5 mm due to the discreteness of the underlying software. The curves on the left and right are meant to indicate that the triplets of lines do meet at the points they are supposed to—but too far away for inclusion in an intelligible version of the diagram.

chosen freely, but 14 must be on the line $n = [1, 17]$, and 13 must be on the intersection point of the lines $r = [17, 18]$ and $l = [12, 14]$. Next, points $8, 9, 10, 2$ are free, but 4 must be on the line $a = [1, 2]$. The point 6 is free, while 11 is the intersection point of lines $i = [6, 8]$ and $k = [10, 2]$, and point 3 is the intersection point of the lines $f = [4, 6]$ and $j = [9, 11]$. Similarly, 5 is the intersection point of $d = [8, 9]$ and $e = [2, 3]$, and 7 is the intersection point of $h = [8, 13]$ and $g = [5, 6]$, while 22 is the intersection point of lines $v = [23, 12]$ and $m = [7, 14]$. This completes the construction of the first multilateral. To start with the next (the trilateral), we select 15 on the line $o = [22, 21]$, then 20 as the intersection point of $t = [21, 19]$ and $p = [16, 15]$. The only remaining problem is the selection of point 25, which should be at the intersection of **three lines**, namely $y = [18, 19]$, $x = [24, 20]$, and $w = [23, 15]$. It is to be expected that three lines do not have a common point. This is quite general, and it is the final solution given by Steinitz, with the selections made: The last line may need to be taken as a circle (or a parabola).

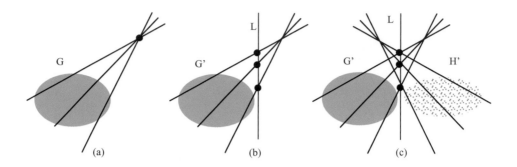

Figure 5.2.9. The construction establishing Theorem 5.2.2.

In fact in this case—just as for many other configurations—by judicious choices of the free parameters one may find selections in which the point 25 is on a certain side of the line w, as well as selections where it is on the other side. By continuity, this implies that there is a position of incidence. The final conclusion, therefore, is that the Georges configuration can be realized geometrically, by points and straight lines. Hence it is a 3-connected non-Hamiltonian geometric configuration (25_3). $\square$

The result of Theorem 5.2.1 can be extended to geometric configurations of almost all sizes:

Theorem 5.2.2. *For each $n \geq 33$ there exist 3-connected geometric configurations (n_3) that do not admit Hamiltonian multilaterals.*

Proof. Select any of the vertices of the (25_3) configuration of Theorem 5.2.1 (see Figure 5.2.9(a), where the gray oval stands for the rest of the Georges configuration (25_3), denoted here by G). We delete that vertex and replace it by the three intersection points of the lines incident with the deleted vertex and a new line L (see Figure 5.2.9(b)). We denote this truncated version of G by G'. Taking now the similarly truncated version H' of any configuration (p_3), we make its three lines incident with the three points on L (Figure 5.2.9(c)). Since both (7_3) and (8_3) have truncated versions realizable by straight lines, the above results in a configuration with $n \geq 24 + 3 + 6 = 25 + 1 + 7$ points and lines. This configuration is clearly 3-connected, but it cannot be Hamiltonian. Indeed, any Hamiltonian multilateral would have to use L and two of its points, one of which must be on a line toward G', the other towards H'. The third point on L must be on one line from G' and another from H'. But then there would be a multilateral in G' using L, and therefore—by identifying the three points of G' that are on L—we would get a Hamiltonian multilateral of G. $\square$

We make a few remarks on the background of Theorem 5.2.1. The author learned of Georges's paper [**71**] and his non-Hamiltonian graph from

Gropp [**80**]. But Gropp makes no connection between the Georges graph and configurations and, in particular, does not observe the fact that the Georges graph has girth 6 and is therefore the Levi graph of a configuration (combinatorial at least). It should also be noted that in [**71**] the rendition of the Ellingham-Horton graph, shown here in Figure 5.2.3, is missing one of the special edges.

Gropp [**80**], [**90**] also mentions that a result similar to Georges's has been found earlier by Kelmans [**136**]. This may well be the case; however, we find the presentation in [**136**] (both in the Russian original and in the translation) too confusing to be able to decide whether the graph he constructs has girth 6. Like Georges, Kelmans does not mention girth or configurations. The claim in [**80**] that Kelmans's 50-vertex graph is the same as Georges's seems quite unjustified. The expanded version of Kelmans's paper (see [**137**]) remains inscrutable to the present author. Moreover, there is no mention in [**137**] of girth 6, of Levi graphs, or of any type of configurations.

$$* \; * \; * \; * \; *$$

A natural question that can be asked in view of Theorems 5.2.1 and 5.2.2 is, *Does every geometric 4-configuration admit a Hamiltonian multilateral?*

As we shall now show, there is a negative partial answer:

Theorem 5.2.3. *There exist 2-connected geometric 4-configurations that do not admit any Hamiltonian multilaterals.*

Proof. We provide a conceptually simple construction that, unfortunately, leads to such configurations but of relatively large sizes. We use configurations (or, more precisely, fragments of configurations) such as the halves of the configurations in Figure 5.1.5. With the same conventions, we can assume that we start with an arbitrary (n_4), for example, the $C(4)$ used in Section 5.1. (We note that the configuration (18_4) in Figure 3.3.4 could be used, but at the cost of some detailed arguments about cross-ratios.) As indicated in Figure 5.2.10, we start with eight copies, delete from each a line, and take suitable projective transforms so that the points that are on three lines each are aligned as shown in Figure 5.2.10, using an additional line L and an additional point P. Since each of L and P can be used only once in any multilateral, there is no possibility for involvement of more than two of the groups of four starting configurations. Hence this non-Hamiltonian configuration is a (4087_4) configuration; if we start with an (18_4) configuration, the result is a still formidable (289_4) configuration. A somewhat smaller example can be found by replacing, in each set of four, one configuration by a single point; this would lead to a non-Hamiltonian (3077_4) configuration if using $C(4)$ or to a (221_4) configuration if using an (18_4) configuration. $\square$

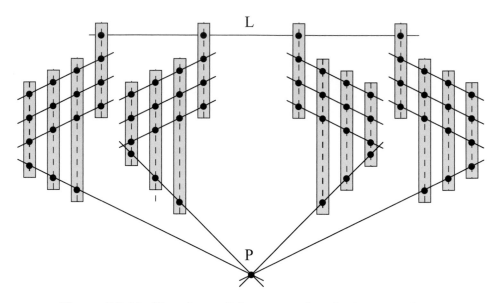

Figure 5.2.10. The scheme of the construction of a 2-connected geometric configuration (289_4) that does not admit any Hamiltonian multilaterals. Each gray rectangle represents an (18_4) configuration from which one line has been omitted.

It is easy to see that a similar construction can be applied to k-configurations for all $k \geq 5$, starting with any single geometric configuration of that kind; for example, $C(k)$ can be used. Obviously, the configurations obtained will be monstrously large.

An unsolved problem is the question of whether there are non-Hamiltonian geometric 4-configurations that are 3- or 4-connected.

A different direction in the study of Hamiltonian multilaterals in configurations is opened by the following generalization:

Definition. A $[q, k]$-configuration has a Hamiltonian multilateral if and only if it has a multilateral M such that

(i) every element of one of the two kinds (points or lines) is contained in the multilateral M;

(ii) every element (of both kinds) is incident at most once with the multilateral M.

Obviously, this definition reduces to the standard one in case of balanced configurations $(q = k)$.

Except for a few examples of type $[3, 4]$ with Hamiltonian multilaterals, there seems to be no information available on this topic.

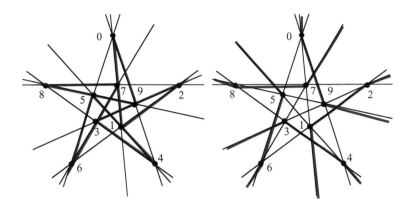

Figure 5.2.11. A symmetric Hamiltonian multilateral in the astral configurations (10_3). Both parts represent the same multilateral.

It also seems that the concept has not been studied in the context of bipartite graphs—to which it obviously applies.

$$* \; * \; * \; * \; * \; *$$

A different type of questions and results arises if we inquire about Hamiltonicity of some restricted families of configurations. For example, if we consider astral 3-configurations, one may inquire about Hamiltonian multilaterals that have the same symmetries as the configuration; we shall call them **symmetric Hamiltonian multilaterals**. In Figure 5.2.11 we show an example of a symmetric Hamiltonian multilateral in the astral (10_3) configuration. Note that both parts show the same symmetric multilateral—multilaterals are concerned with lines, not segments.

In Figure 5.2.12 we show four different symmetric Hamiltonian multilaterals on the astral (10_3) configuration. Using the notation for astral 3-configurations we introduced in Section 2.7 and which is illustrated in Figure 5.2.13, we can assert the following theorem.

Theorem 5.2.4. *The cyclic astral configuration $m\#(b, c; d)$ may have four types of symmetric Hamiltonian circuits:*

$$B_0 \to C_0 \to B_d \to \cdots$$
$$B_0 \to C_0 \to B_{d-c} \to \cdots$$
$$B_0 \to C_{-b} \to B_{d-b} \to \cdots$$
$$B_0 \to C_{-b} \to B_{d-b-c} \to \cdots .$$

A symmetric Hamiltonian circuit of one of these types exists if and only if m is relatively prime to d, $d - c$, $d - b$, or $d - b - c$, respectively.

A related result of Hladnik et al. [**127**], based in part on work of Alspach and Zhang [**1**], should be mentioned here: Every connected cyclic

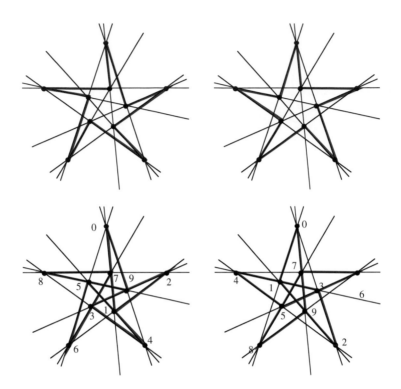

Figure 5.2.12. Four different symmetric Hamiltonian multilaterals in the astral configuration (10_3).

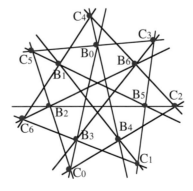

Figure 5.2.13. A reminder of the notation for astral 3-configurations, illustrated for $m\#(b, c; d) = 7\#(3, 2; 4)$.

3-configuration (combinatorial or geometric) is Hamiltonian. Unfortunately, it is not clear to the author exactly what is here meant by "cyclic 3-configuration".

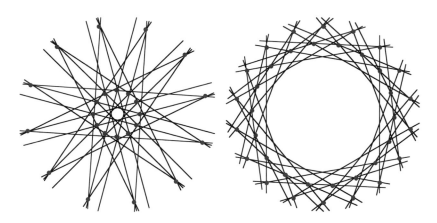

Figure 5.2.14. Depiction of selfpolar astral configurations $12\#(5,5;2)$ and $18\#(5,1;3)$ showing they have no symmetric Hamiltonians. Do they have any Hamiltonian multilaterals at all?

Exercises and Problems 5.2.

1. Decide about the validity of the following open conjecture: Every astral 3-configuration admits a Hamiltonian multilateral.

2. Find four symmetric Hamiltonian multilaterals in the (14_3) astral configuration $7\#(3,2;1)$.

3. Prove Theorem 5.2.4. Apply it to the configuration $8\#(3,2;1)$.

4. The two configurations in Figure 5.2.14 do not have any symmetric Hamiltonian multilateral. If the result of [**127**] mentioned above relates to them, they have Hamiltonian multilaterals. In any case, either find a Hamiltonian multilateral or else show that there is none.

5. Determine whether the three 3-astral configurations (9_3) in Figure 1.1.6 admit symmetric (or any) Hamiltonian multilaterals.

6. In configurations with dihedral symmetry group one cannot expect any Hamiltonian multilateral to have the same symmetry group. At most, one may look for cyclically symmetric Hamiltonian multilaterals. In Figure 5.2.15 we show three examples of this situation. Determine whether the six astral 4-configurations (36_4) shown in Figure 3.6.3 admit cyclically symmetric Hamiltonian multilaterals.

7. Following [**60**], we say that a set of points of a configuration C is a **blocking set** if it contains a point of every line of C but not all points of any of the lines. Show that the (22_3) configuration shown in Figure 5.2.2 contains no blocking set. (This example disproves two conjectures in [**60**].) For more information about blocking sets and blocking set-free configurations, see [**82**] and [**95**].

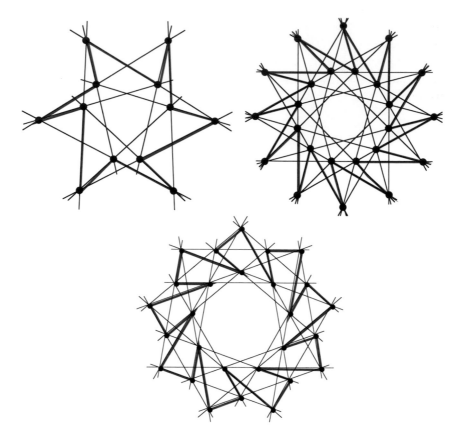

Figure 5.2.15. Examples of cyclically symmetric Hamiltonian multilaterals in three configurations with dihedral symmetry.

5.3. Multilateral decompositions

In 1828 Möbius [**172**] pointed out that it is obvious that two trilaterals cannot be mutually inscribed/circumscribed and proved the impossibility of two
quadrilaterals being in such mutual relationship; he dealt here with the real
Euclidean plane. (This seems to be the first paper dealing with this topic.)
Möbius concludes his paper by saying: "I have not extended this investigation to multilaterals with more sides" [the author's translation]. Therefore
the statement in Wikipedia [**226**] that "Möbius (1828) asked whether there
exists a pair of polygons with p sides each, having the property that the vertices of one polygon lie on the lines through the edges of the other polygon,
and vice versa" has to be taken with a grain of salt. In fact, the answer
to the question is affirmative, even if its author is lost in history. The first
mention of three mutually inscribed/circumscribed trilaterals seems to be in
Graves [**77**], where the Pappus configuration (9_3) is shown to have that property. Moreover, Graves shows that the Desargues configuration (10_3) can

be presented as a pair of mutually inscribed/circumscribed pentalaterals. He illustrated this in [**77**] by a diagram (reproduced here as Figure 5.3.1) in which one of the pentalaterals is rendered in color (this was in 1839!). Without diagrams, Cayley [**41**] describes this and several other examples of mutually inscribed/circumscribed families of multilaterals.

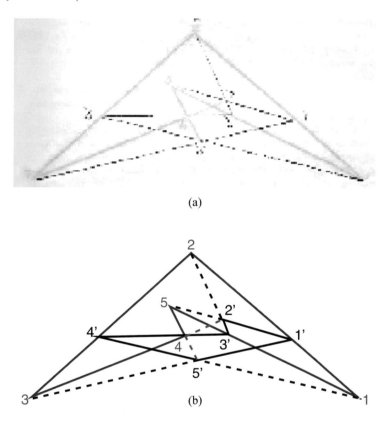

(a)

(b)

Figure 5.3.1. (a) The Desargues configuration (10_3) presented as a pair (black and color) of mutually inscribed/circumscribed pentalaterals. Scanned from a somewhat deteriorated copy of the paper [**77**] by Graves, published in 1839. (b) The same configuration, redrawn to show Graves's intention more clearly.

For any multilateral decomposition of a 3-configuration by a family F of p-laterals P_i, $1 \leq i \leq r$, we shall say that it is an **inscribed/circumscribed family** (or decomposition) provided every line of P_i contain a vertex of P_{i+1}, for all i (subscripts taken mod r). This is illustrated (for $r = 2$) by the two examples in Figure 5.3.2. (The first is clearly the same as the Graves example in Figure 5.3.1.) As we discussed previously, Hamiltonian multilaterals of 3-configurations fit this description for $r = 1$. The astral 3-configurations considered in Section 2.7 show that for $r = 2$ there exist inscribed/circumscribed families of n-laterals for all $n \geq 5$. The case $r = 2$, $n = 5$ is illustrated in

Figure 5.3.2(b). The k-astral 3-configurations discussed in Section 2.10 show that for every r, there exist inscribed/circumscribed families of n-laterals for all $n \geq 3$. However, in different forms, these results go much further back.

The Desargues configuration (which we denoted $(10_3)_1$ in Section 2.2) can also illustrate a refinement of the definition. We say that a family F is an **orderly** inscribed/circumscribed family if consecutive vertices of P_{i+1} belong to consecutive lines of P_i. Although the example in Figure 5.3.2(a) shows a pair of inscribed/circumscribed pentalaterals, this family is not orderly—in contrast to the example in Figure 5.3.2(b). In fact, it can be proved that the Desargues configuration does not admit any orderly pair of pentalaterals. It is worth stressing (as Graves [77] did long ago) that the two pentalaterals in the Desargues configuration are combinatorially in a very symmetric reciprocal relationship although this finds no reflection in the geometric rendition.

Among other early publications are the papers [194] and [195] by Schönflies. He was led to this topic by his investigation of the density of trilaterals which we shall discuss in Section 5.5. In these papers Schönflies formally introduced the families of mutually inscribed/circumscribed multilaterals and provided examples. (Minor errors in [195] are corrected in [196]. More serious shortcomings are pointed out by Steinitz in [212, p. 488], [213, p. 307].) Additional related investigations by Schönflies are reported in [197] and [198]. Brunel [34] considers the topic as well.

Other examples of orderly inscribed/circumscribed families are provided in Figure 5.3.3. A variety of examples is shown in [63]. These can be generalized to all n-laterals with $n \geq 3$ and all $r \geq 3$.

Like many other aspect of the theory of configurations, families of inscribed/circumscribed multilaterals have found a home of sorts in the theory of set configurations and more general combinatorial incidence systems. Recent publications one may wish to consult for these developments include [179], [186], [150], and especially [220]; additional references may be obtained through these, as well as from other publications of the given authors. However, despite the language used in the publications of this trend (including words such as points, lines, polygons, collinearity, Pappus and Desargues configurations, and others), in most cases the meaning is completely divorced from the accepted geometric interpretation of the terms used. In some of the publications there are diagrams, but they are only schematic tools, not configurations in the geometric (or even topological) sense.

As far as the author is aware, there has been no consideration given to any concepts analogous to inscribed/circumscribed families of multilaterals or multilateral decompositions in k-configurations, for $k \geq 4$. In fact, it is easy to come up with several meaningful interpretations. In Figure 5.3.4

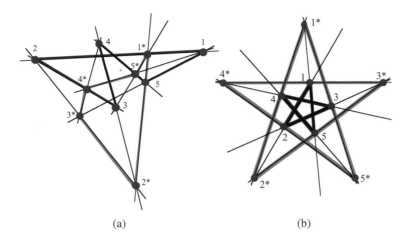

(a)   (b)

Figure 5.3.2. The configurations $(10_3)_1$ (in the labeling of Section 2.2, shown in (a)) and $(10_3)_{10}$ can both be interpreted as pairs of inscribed/circumscribed pentalaterals. However, the one in (b) has an orderly pair of inscribed/circumscribed pentalaterals: The line determined by the points i and $i + 1$ of the blue pentalateral contains the point $i + 1$ of the green one. The Desargues configuration $(10_3)_1$ does not have such an orderly pair.

we show what is possibly the simplest of these. A family of multilaterals in a 4-configuration is an inscribed/circumscribed family if each line of a multilateral is incident with two vertices of another multilateral in a cyclical arrangement of the multilaterals. This is illustrated in Figure 5.3.4 in the case of a 3-astral configuration (21_4); in both versions the red heptalateral is inscribed in the blue one, which is inscribed in the green one, which is inscribed in the original red heptalateral.

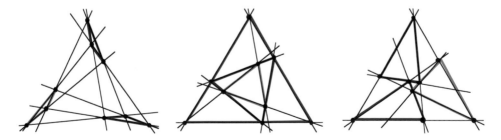

Figure 5.3.3. Examples of orderly inscribed/circumscribed families with $r = 3$ in rotationally symmetric realizations of the three configurations (9_3).

A different interpretation is illustrated in Figure 5.3.5. The multilateral decomposition of the astral configuration (24_4) consists of eight mutually inscribed/circumscribed trilaterals in a more complicated way. Each line contains points of three distinct trilaterals, and each point is incident with lines of three distinct trilaterals.

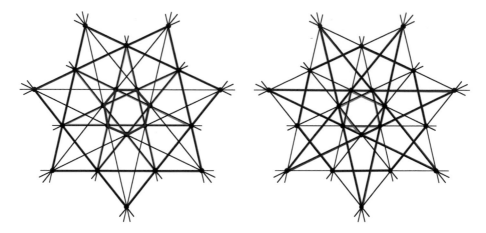

Figure 5.3.4. Two families of three mutually cyclically inscribed heptalaterals that form a multilateral decomposition of the 3-astral configuration (21_4).

Figure 5.3.5. An illustration of the alternative interpretation of inscribed/circumscribed trilaterals in the astral configuration (21_4).

Exercises and Problems 5.3.

1. Prove that there is no *orderly* pair of pentalaterals in the Desargues configuration $(10_3)_1$.

2. Show that the cyclic configuration $\mathscr{C}_3(n)$ (see Section 2.1) contains an orderly family of three inscribed/circumscribed m-laterals whenever $n = 3m$. Does it contain a family of m trilaterals?

3. Investigate the general cyclic configurations $\mathscr{C}_3(n, 1, k)$ (see Exercise 2 of Section 2.1) for the presence of orderly families of inscribed/circumscribed multilaterals.

4. Prove the result of Schönflies [**198**]: None of the three (9_3) configurations can be obtained by starting with nine points in general position in Euclidean 3-space, generating some of the lines and planes they determine, and intersecting these by a plane.

5. Can each of the (10_3) configurations be presented as an inscribed/circumscribed family of two pentalaterals?

6. Consider the astral configuration (60_4) shown in Figure 5.3.6. Find inscribed/circumscribed multilateral decompositions of this configuration into (i) twelve pentalaterals and (ii) four 15-laterals.

Figure 5.3.6. The astral configuration (60_4) used in Exercise 6.

5.4. Multilateral-free configurations

We turn now to some questions concerning trilaterals (and multilaterals) in configurations that go back to the classical period of configurations in the last quarter of the nineteenth century. It has seen new life in recent decades, mostly without any acknowledged relation to the earlier results.

The first question that will occupy us asks for configurations that contain no trilaterals. Here is what is known.

Theorem 5.4.1. *For every $k \geq 2$ there exist geometric k-configurations that are trilateral-free.*

The proof is immediate on recalling the configurations $LC(k)$ described in Section 1.1, as well as in [**182**], and utilized in Section 5.1. The only drawback of this answer is the rather large size of these configurations. The resulting trilateral-free geometric configurations are (n_k) with $n = k^k$.   □

We shall see below how smaller trilateral-free geometric configurations can be found in some cases. For some general estimates see Lazebnik et al. [**144**].

Another general result gives a lower bound on the size of trilateral-free configurations.

Theorem 5.4.2. *If an (n_k) configuration with $k \geq 2$ is trilateral-free, then $n \geq k(k-1)^2 + k$.*

The proof is straightforward on considering the situation schematically presented in Figure 5.4.1, assuming $k = 4$. Any one line (represented by the horizontal line) carries k points; through each of the points go $k - 1$ other lines of the configuration, each carrying $k - 1$ additional points. The only remark that needs to be made is that these points must all be distinct, since otherwise there would be a trilateral present in the configuration. This argumentation (or something similar) was shown to the author by J. Bokowski. Notice that the argument does not use any geometry, and hence the result holds for combinatorial configurations as well.   □

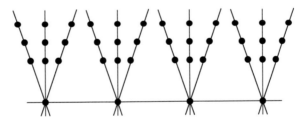

Figure 5.4.1. Schematic representation of the proof of Theorem 5.4.2.

For $k = 3$ the result was known to Martinetti [**151**] in 1886.

The cubic bound in Theorem 5.4.2 is in contrast to the exponentially large examples in Theorem 5.4.1. We shall next show that we can do much better than the exponential example for $k = 3$ and slightly better for $k = 4$.

As a consequence of Theorem 5.4.2 we see that for $k = 3$ any trilateral-free k-configuration must have at least $n \geq 15$ points. Martinetti [**151**] in 1886 seems to be the first to have raised the question of trilateral-free 3-configurations. He proved that such configurations have at least 15 points

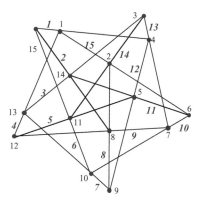

Figure 5.4.2. The Cremona-Richmond trilateral-free configuration (15₃). Point labels are in plain font; line labels are in red italics. Same digits establish a duality correspondence between points and lines.

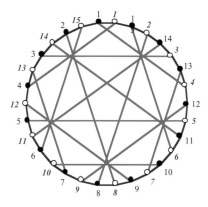

Figure 5.4.3. A Levi graph of the Cremona-Richmond configuration. It is the "Tutte 8-cage", the smallest 3-valent graph with girth 8. Similar presentations appear in [46] and many other places.

and provided a combinatorial description of the unique (15₃) configuration that is trilateral-free. It needs to be stressed that, from all that we can read in his publications, Martinetti thought at that time (as well as later) that combinatorial 3-configurations are all geometrically realizable. In fact, Martinetti's trilateral-free (15₃) configuration is, indeed, geometrically realizable; see Figure 5.4.2. Moreover, in the prehistory of configurations, traces of this (15₃) geometric configuration can be found in considerations of families of straight lines on cubic surfaces, by Schläfli in 1858 and Cremona in 1868. The configuration itself is frequently called the Cremona-Richmond configuration; see [46], [224]. More detailed historical explanations and references can be found in [22].

The Cremona-Richmond configuration is shown in Figure 5.4.2, and a Levi graph (see [**229**]) based on a Hamiltonian multilateral is shown in Figure 5.4.3. Since the configuration is trilateral-free, its Levi graph has no circuits of size smaller than 8; in other words, its *girth* is 8. In fact, it is the smallest 3-valent graph of girth 8, and it is famous as *Tutte's* $(3,8)$*-cage*, or, more simply, *Tutte's 8-cage*. Coxeter [**46**] provides an ingenious labeling of its vertices and calls it "the most regular of all graphs". Figure 5.4.3 also shows that this graph has a color-reversing symmetry, hence the Cremona-Richmond configuration is selfdual. This result can also be deduced from the fact that there is only one type of trilateral-free (15_3) configuration. On the other hand, it should be noted that there are infinitely many *projectively* inequivalent geometric realizations of this configuration. This is most easily seen by manipulations in some software such as Geometer's Sketchpad®.

Before continuing our description of the other results about trilateral-free configurations, we need to present some of the more recent definitions and results that deal with the same topic in a different language.

A $(\boldsymbol{k},\boldsymbol{g})$**-cage** is a graph with all vertices of **valence** k and of **girth** g, having the smallest possible number of vertices. For this definition and most of the known results concerning cages, see [**90**], [**229**], and the references given therein. For attractive illustrations of some of the cages see [**184**].

The Levi graph of a (combinatorial or geometric) (n_k) configuration has, as we mentioned in Section 1.5, girth $g \geq 6$; since, as we have seen in Section 2.1, the Fano configuration (7_3) has the smallest number of vertices, its Levi graph with 14 vertices is a $(3,6)$-cage—in fact, the only $(3,6)$-cage.

Trilateral-free 3-configurations have girth at least 8; hence the Levi graph of the smallest such configuration—the Cremona-Richmond (15_3) configuration—is the $(3,8)$-cage, with 30 vertices. Since the Cremona-Richmond (15_3) configuration is the unique trilateral-free (15_3) configuration, this cage is also unique: It is the Tutte $(3,8)$-cage, mentioned earlier.

Another related concept is that of "generalized quadrangles". A **generalized quadrangle** is an incidence structure in which each pair of distinct points determines at most one line and for each non-incident pair consisting of a point P and a line L, there is precisely one line L^* that is incident with P and with a point P^* of L. A (finite) generalized quadrangle is of **order** (s,t) if every line contains precisely $s+1$ points and every point is incident with precisely $t+1$ lines. The terminology is often justified by the fact that an ordinary quadrangle can be interpreted as a generalized quadrangle of order $(1,1)$. Obviously, each generalized quadrangle of order $(2,2)$ is a combinatorial 3-configuration. The smallest generalized quadrangle of order $(2,2)$ has 15 points; hence it is a (15_3) configuration, for which the definition of generalized quadrangles implies that it is trilateral-free. It follows that it

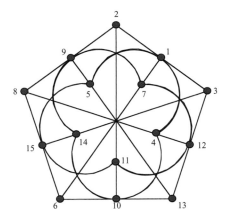

Figure 5.4.4. The "doily" of S. Payne: a geometric model of the 15-point generalized quadrangle of order $(2, 2)$—also known as the Cremona-Richmond configuration.

is isomorphic to the Cremona-Richmond configuration. Polster [**184**] shows several diagrams of this generalized quadrangle; two are particularly interesting. The first, which he attributes to Stanley Paine, is shown in Figure 5.4.4; the labels establish its isomorphism with the configuration in Figure 5.4.2.

The second interesting model shown by Polster [**184**] is by lines (actually, line segments) in 3-dimensional space; see Figure 5.4.5. The model is best understood as being spanned by a regular tetrahedron; the tetrahedron's edges are indicated by the dashed lines and are not part of the configuration. Naturally, any appropriate projection of this model into the plane provides a planar realization of the Cremona-Richmond configuration. A figure resembling such a projection illustrates the Cremona-Richmond configuration in Wells's "Dictionary" [**224**, p. 40].

Returning now to the configuration language, here are some of the additional results on trilateral-free 3-configurations, all established by Martinetti [**151**]:

• There are no trilateral-free configurations (16_3). This is not hard to show, starting with the arrangement in Figure 5.4.1 and noting that for a sixteenth point there are only relatively few possibilities of collinearities with the other points—none leading to a configuration, even in the combinatorial sense.

• There is a single trilateral-free configuration (17_3); again, the uniqueness implies that it is selfdual. It is interesting because of its very low symmetry. Under its group of automorphisms it has four point orbits: two of size 6 each, one of size 3, and one of size 2. It is geometrically realizable

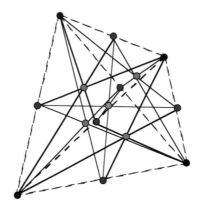

Figure 5.4.5. A realization of the 15-point generalized quadrangle of order $(2, 2)$, alias Cremona-Richmond configuration, supported in 3-space by a regular tetrahedron. (Adapted from [**184**].)

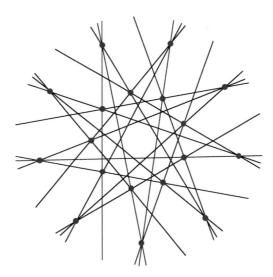

Figure 5.4.6. A realization of the trilateral-free selfpolar configuration (18_3) denoted 18-D in [**22**]; it is astral with symbol $9\#(4, 2; 3)$ and is the first of an infinite series of trilateral-free selfpolar configurations.

but with no symmetry. Details (such as a configuration table, geometric realization, Levi graph, automorphism group, orbits) can be found in [**22**].

• There are precisely four trilateral-free configurations (18_3). Two are dual to each other, and each of the other two is selfdual. Data on all four, with geometric realizations, are given in [**22**]. One of the selfdual configurations (denoted 18-D in [**151**] and [**22**]) is interesting because of its symmetry; it admits a selfpolar realization as an astral configuration $9\#(4, 2; 3)$ in the notation of Section 2.7 and is shown in Figure 5.4.6.

Figure 5.4.7. A trilateral-free configuration (24_3); it is astral with symbol $12\#(5,3;4)$.

In considering these results of Martinetti [**151**], one should bear in mind that although he uses geometrical language, there is no diagram presenting these configurations, nor is there any hint about how the corresponding geometric configurations should be constructed. The first geometric realizations seem to be the ones in [**22**].

According to the data in [**17**] (reproduced in [**22**]) there are 19 combinatorial trilateral-free configurations (19_3), 162 such configurations (20_3), and 4,713 configurations (21_3). It is not known how many are geometrically realizable.

On the other hand, we have the following:

Theorem 5.4.3. *For every $n \geq 15$ except $n = 16$ and possibly $n = 23$ and 27, there are trilateral-free geometric configurations (n_3).*

Proof. For $n = 15, 17, 18, 19, 20, 21$, trilateral-free geometric configurations are shown in [**22**]. It is easy to verify that all astral configurations $m\#(4,2;3)$ for $m \geq 9$ and $m \neq 12$ are trilateral-free; this shows that for all even $n \geq 18$, $n \neq 24$, there are trilateral-free configurations (n_3). The (18_3) and (20_3) configurations mentioned above are of this type. For the exceptional value $n = 24$ a trilateral-free geometric configuration is shown in Figure 5.4.7.

The construction of the appropriate configurations for odd n is slightly more complicated. In almost all cases, the following construction works. Starting with trilateral-free geometric configurations (p_3) and (q_3), we delete

Table 5.4.1. The configuration table of one of the trilateral-free configurations (25_3) found by Visconti [222].

1	2	3	4	5	6	7	8	9	10	11	12	13	14	15	16	17	18	19	20	21	22	23	24	25
2	3	4	5	1	8	9	10	6	7	13	14	15	11	12	18	19	20	16	17	23	24	25	21	22
6	7	8	9	10	11	14	12	15	13	16	19	17	20	18	21	24	22	25	23	1	2	3	4	5

Table 5.4.2. The configuration table of the other trilateral-free configuration (25_3) found by Visconti [222]. A geometric realization of this configuration is given in Figure 5.4.8.

1	2	3	4	5	6	7	8	9	10	11	12	13	14	15	16	17	18	19	20	21	22	23	24	25
2	3	4	5	1	8	9	10	6	7	13	14	15	11	12	18	19	20	16	17	23	24	25	21	22
6	7	8	9	10	11	14	12	15	13	16	19	17	20	18	21	24	22	25	23	1	5	4	3	2

one line in each and connect the three pairs of orphan points with an additional, new point. (The required alignment can always be obtained through suitable projective transformations.) This yields a trilateral-free geometric configuration (n_3) with $n = p + q + 1$. Starting from the trilateral-free configurations we already constructed, this yields the required geometric configurations for all odd $n \geq 31 = 15 + 15 + 1$. An alternative construction works for all $n \geq 29$: In analogy to the "deleted union" construction (DU-1) described in Section 3.3, we delete a line from a trilateral-free geometric configuration (p_3) and delete a point from a trilateral-free geometric configuration (q_3); by placing appropriate copies of the two configurations so that the lines of the latter (which are missing a point) pass through the points of the former (which are missing a line), we obtain a trilateral-free geometric configuration of $n = p + q - 1$ points. For $p = q = 15$ this yields $n = 29$.

The case of the (25_3) configuration is particularly interesting. Visconti [222] gives configuration tables for two distinct trilateral-free combinatorial configurations (25_3), each consisting of a family of five mutually inscribed/circumscribed pentalaterals. These are reproduced, in Visconti's notation, in Tables 5.4.1 an 5.4.2. In the somewhat analogous case of trilateral-free configuration (20_3) consisting of four mutually inscribed/circumscribed pentalaterals, Visconti provides a graphical representation that seems to be the first *symmetric* rendition of any multiastral configuration. However, contrasting with this is the fact that there is no indication in [222] whether the (25_3) configurations described are geometrically realizable. We have verified that at least one of these can be drawn, but not in a polycyclic manner; see Figure 5.4.8. Just as in the case of the (17_3) and (19_3) configurations investigated in [22], the configuration is asymmetric and was constructed by successive approximations. It is very likely that the same situation exists for Visconti's other (25_3) configurations.   □

Visconti [**222**] and Martinetti [**153**] provide additional examples of tri-lateral-free combinatorial 3-configurations consisting of mutually inscribed/circumscribed pentalaterals and some other multilaterals as well. It may be conjectured that these are also geometrically realizable.

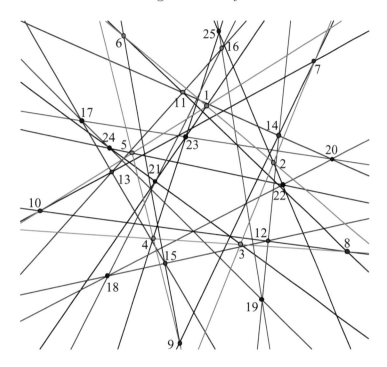

Figure 5.4.8. A geometric realization of one of the trilateral-free configurations (25_3) given by configuration tables in [**222**] and Table 5.4.2.

It is worth noting that most astral configurations $m\#(b, c; d)$ are tri-lateral-free for sufficiently large m. Exceptions (such as the $n = 24$ case mentioned above) are usually easy to spot, but there seem to be some subtler issues that have not been tackled so far. An example of such a situation is given in Figure 5.4.9.

We are turning now to quadrilateral-free configurations; this term is somewhat of a misnomer—at least in the sense we shall use it. By **quadri-lateral-free** we shall mean configurations that have neither a trilateral nor a quadrilateral. We have no example of a configuration that has no quadri-lateral but does have trilaterals; it appears to be an open question whether such configurations exist. This leads to our terminology that simplifies the locutions.

In discussing quadrilateral-free configurations, we consider only 3-config-urations, since nothing on the topic of quadrilateral-free k-configurations seems to be known for $k \geq 4$.

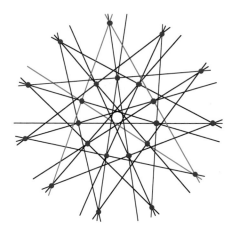

Figure 5.4.9. The astral configuration $11\#(5,4;2)$ is not trilateral-free; one trilateral is shown by green lines. For m such that $13 \leq m \neq 15$, the configuration $m\#(5,4;2)$ is trilateral-free.

The only published work that the author is aware of that deals with quadrilateral-free geometric 3-configurations is [**183**]; the configurations are studied using their Levi graphs.

Through the Levi graphs, the question of quadrilateral-free configurations is related to 3-valent bipartite graphs of girth at least 10. Such graphs have been extensively investigated; a large quantity of relevant literature can be found in [**228**] and [**229**].

The result of these graph-theoretic studies that is most relevant to our topic is that there exist exactly three 10-cages (also called $(3,10)$-cages), that is, 3-valent graphs of girth 10 with the smallest number of vertices, namely 70; all three are bipartite. The first one was found by Balaban [**4**], the other two by O'Keefe and Wong [**177**]; Wong [**228**] proved that these three are the only ones. Balaban's 10-cage has a color-interchange automorphism; the other two do not have any such automorphism.

In [**183**], Pisanski et al. describe in detail these three 10-cages and the resulting five quadrilateral-free configurations (35_3). Since Balaban's 10-cage has a color-reversing symmetry, the corresponding configuration is selfdual. The other two 10-cages yield a pair of dual configurations each. It is clear that the three 10-cages can be interpreted as Levi graphs of quadrilateral-free *combinatorial* 3-configuration (35_3); however, Pisanski et al. prove that they admit *geometric* realizations and provide in [**183**] diagrams for three of the five. These three admit polycyclic representations which the last two do not have; for them there is in [**183**] a description of the method of proof (following [**29**]) and a reference to the full set of coordinates listed at a website. In [**183**] there is also described a construction of quadrilateral-free geometric configurations (n_3) for an infinite sequence of values of n.

An improvement of this last result is the following:

Theorem 5.4.4. *For every $n = 4m \geq 40$, there exists a quadrilateral-free geometric configuration (n_3). There exists an n_0 such that for every $n \geq n_0$ there is a quadrilateral-free geometric configuration (n_3). The available estimate is $n_0 \leq 320$.*

Proof. It is easily verified that for $m \geq 10$ the astral configuration $(2m)\#(m-1, 1; 4)$ is quadrilateral-free. The author is indebted to T. Pisanski for showing him one of the two smallest of these configurations, $20\#(9, 1; 4)$; see Figure 5.4.10(a). The next members of the sequence, the pair of dual $22\#(10, 1; 4)$ configurations, are shown in Figure 5.4.11. The proof of the fact that the $(2m)\#(m-1, 1; 4)$ configurations are quadrilateral-free is easy by generalizing the argument indicated by the coloring of the points in Figures 5.4.10 and 5.4.11. Since the configuration is astral and any quadrilateral would have to contain a point of the outer ring, it is enough to show that the point marked by the large black dot is not part of any quadrilateral. The only six points at (graph-)distance 1 are the six red points, and those at distance 2 are the 24 green ones. The presence of any quadrilateral would imply that two of the green points coincide—which does not happen since this would imply that there are at most 23 green points.

In order to prove the existence of n_0, we may use the same construction as in the proof of Theorem 5.4.3 for odd n. We take two quadrilateral-free configurations (p_3) and (q_3) and, using convenient representatives, delete one line from each; an additional point and three lines through it and the points on the two deleted lines form a quadrilateral-free configuration (n_3) with $n = p+q+1$. Repeating the construction r times leads to configurations with r points more than the sum of the numbers of points of the configurations used. This yields the bound $n_0 \leq 320$. $\qquad\square$

In analogy to our convention concerning quadrilaterals, we say that a configuration is **pentalateral-free** if it contains no t-laterals for $t = 3, 4, 5$. The information available is exceedingly meager. A $(3, 12)$-cage has 126 points and happens to be bipartite; hence it can be interpreted as the Levi graph of a pair of dual (63_3) *combinatorial* configurations. Schroth [**202**] found graphic representations of these two configurations. (The title of Schroth's paper refers to the "generalized hexagons" of order $(2, 2)$; the uniqueness of the dual pair was established in [**43**].)

Schroth's diagrams naturally lead to the question of whether the two pentalateral-free configurations are geometrically realizable. The affirmative answer was provided by M. Boben and T. Pisanski, soon after the

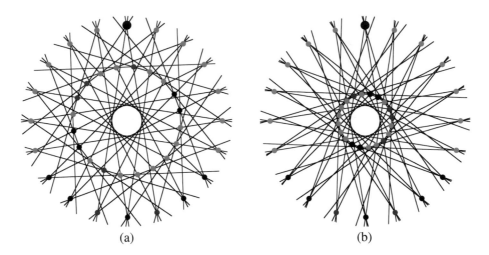

(a)   (b)

Figure 5.4.10. The two dual astral configurations $20\#(9,1;4)$. Both (40_3) configurations are quadrilateral-free.

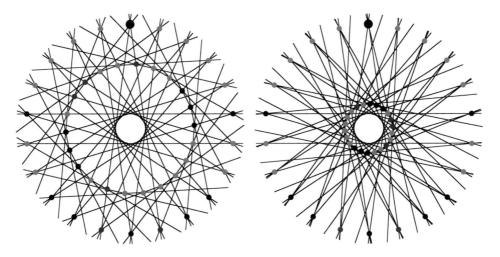

Figure 5.4.11. A dual pair of astral, quadrilateral-free (44_3) configurations $22\#(10,1;4)$.

publication of [**202**], but was never published. With their permission, one of the two geometric pentalateral-free configurations (63_4) is reproduced in Figure 5.4.12, by a diagram they supplied. In Figure 5.4.13 we show the reduced Levi diagram of this configuration, as kindly provided by Boben and Pisanski.

The Boben and Pisanski construction of the pentalateral-free (63_3) configuration is a piece of supporting evidence for Conjecture 2.6.1, according to which all 3-connected combinatorial 3-configurations can be realized by

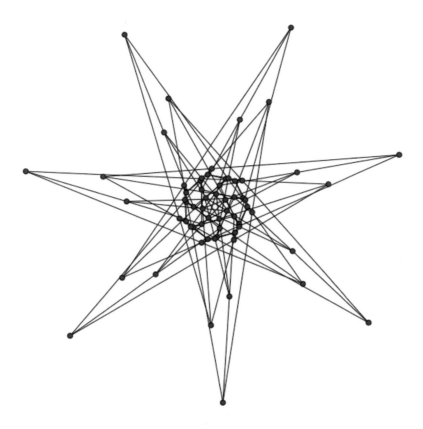

Figure 5.4.12. A geometric realization of a pentalateral-free configuration (63₃). (Courtesy of M. Boben and T. Pisanski, from unpublished work.)

points and (straight) lines. (The 3-connectedness of the 12-cage can be directly established, but it also follows from the more general result of Fu et al. [**70**] that all $(3, g)$-cages are 3-connected, or the more general result of Daven and Rodger [**57**] that for $k \geq 3$ all (k, g)-cages are 3-connected; there is a conjecture in [**70**] that all (k, g)-cages are k-connected.)

* * * * * *

Turning next to the case of trilateral-free 4-configurations, there is much less information available. A trilateral-free *combinatorial* configuration (40_4) was found recently by Hendrik van Maldeghem. Van Maldeghem's example attains the bound of Theorem 5.4.2 for $k = 4$; hence it corresponds to a $(4, 6)$-cage. The construction of this example is described by Bokowski in [**25**, pp. 263–265], where an incidence matrix is also shown (in two forms).

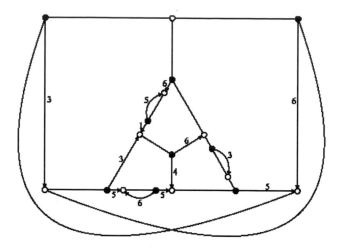

Figure 5.4.13. The reduced Levi graph (in their notation) of the pentalateral-free configuration in Figure 5.4.12. (Courtesy of M. Boben and T. Pisanski, from unpublished work.)

However, no information seems available concerning the possibility of realizing this configuration geometrically or even just topologically.

As already mentioned, the configuration $LC(4)$ provides an example of a trilateral-free geometric configuration (n_4) with $n = 4^4 = 256$. Smaller trilateral-free configurations (120_4) are the astral configurations $60\#(22, 21, 2, 9)$ and $60\#(27, 26, 3, 14)$ shown in Figures 5.4.14 and 5.4.15 and their duals. By using the $(3m+)$ and (DU-1) constructions described in Section 3.3, from the (120_4) configurations we can construct infinite families of geometric trilateral-free configurations.

However, a much better example of a trilateral-free 4-configuration is a (60_4) found very recently by M. Boben. It and its polar are shown in Figure 5.4.16.

The procedures analogous to the one describe earlier, applied to the configurations $LC(k)$ for $k \geq 5$, show that in all these cases there are infinite families of trilateral-free geometric configurations. Unfortunately, they are all far too large for intelligible graphics.

It is not known whether for $k \geq 5$ there exist trilateral-free *combinatorial* configurations (n_k) with $n = k(k-1)^2 + k$.

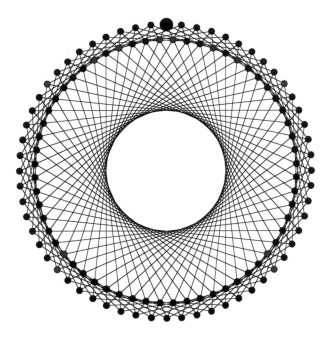

Figure 5.4.14. A trilateral-free geometric configuration (120_4). It is a sporadic astral configuration, with symbol $60\#(22, 21, 2, 9)$.

Figure 5.4.15. Configuration $60\#(27, 26, 3, 14)$

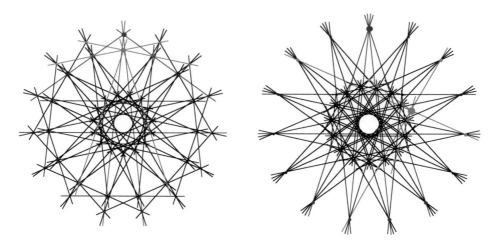

Figure 5.4.16. The trilateral-free (60_4) 4-astral configuration $15\#(1,3;7,6;4,3;2,6)$ found by M. Boben and its polar. (Courtesy of M. Boben.)

Exercises and Problems 5.4.

1. Find other astral families of quadrilateral-free 3-configurations.

2. Is there a quadrilateral-free configuration (n_3) for every $n \geq 35$? Or for all but a very small number of values of n?

3. Decide whether it is possible for a 3-configuration to have no quadrilaterals but contain some trilaterals.

Added in proof (February 2009).

Recent developments have brought about many new insights regarding multilateral-free configurations. In order to present the results obtained in [**23**] we need to change the terminology used in Section 5.4. Specifically, saying that a configuration is t-multilateral-free, for some integer t, means here that the configuration has no multilateral of size precisely t, without this implying anything about the presence or absence of multilaterals of other sizes. Let $x = (x_3, x_4, x_5)$ be an ordered triplet of zeros and ones; a configuration C is said to be of type x provided C is t-multilateral-free if and only if $x_t = 0$. The main result in [**23**] can be formulated as follows:

For each of the eight possible choices of x there is a geometric 3-configuration of type x.

5.5. "Density" of trilaterals in configurations

Another classical question is the following: Consider (combinatorial or geometric) 3-configurations for which the group of automorphisms acts transitively on the points and on the lines; for the purposes of this section (and only here) we shall follow the early writers and call such configurations **regular**. For a regular configuration we shall denote by $t = t(C)$ the number of trilaterals that meet one (hence every) point of a regular configuration C. The question of possible values of the "density" t is a topic raised by Schönflies in [**194**] and [**195**], where he provided a partial answer. We shall present his results soon, but first a word of warning. These papers, like many in that period, are far from clear regarding exactly what the configurations under discussion are; quite a few of the assertions show a degree of naïveté that one does not expect from serious mathematicians. For example, Schönflies states in [**194**] that every regular configuration is selfdual; he repeats the assertion in [**195**]. This error was noticed by Steinitz [**213**]; he mentions that the smallest counterexample is a geometric (18_3) configuration; its underlying combinatorial version is denoted 18.7 in [**17**, p. 337]. Also, although Schönflies *seems to think* that he deals with geometric configurations (in the complex plane!) and therefore does not notice the Fano plane (7_3), Schönflies does not address the question of whether his configurations can actually be geometrically realized—even in the complex plane.

Acknowledging the work of Martinetti [**151**] on configurations with no trilaterals (though giving an incorrect reference for it), Schönflies asserts in [**194**] that t must have one of the values 9, 6, 4, 3, or 2. (Note that the Fano plane has $t = 12$.) However, as noticed by Steinitz [**213**], $t = 1$ is possible as well. The first result in [**194**] is that $t(C) = 9$ happens if and only if $C = (8_3)$, the Möbius-Kantor configuration. (Unless one considers this as a combinatorial configuration, it can be "realized" only in the complex plane.) But Schönflies here clearly missed the restriction to connected configurations—if C consists of several disconnected copies of (8_3), then $t(C) = 9$ as well; he corrected this error in [**195**], as well as providing there a correct reference for [**151**]. Similar shortcomings afflict some of his other assertions. For example, another statement is that there are three possibilities for $t = 6$: (i) the Pappus configuration (9_3); (ii) the Desargues configuration (10_3); and (iii) the cyclic configurations $\mathscr{C}_3(n)$ for $n \geq 9$, with lines $(j, j + 1, j + 3)$ (which we consider in Sections 2.1 and 5.6). But as noted by Steinitz [**212**, p. 488], $t = 8$ for the cyclic (9_3). The possibilities for $t = 2, 3$, or 4 are also discussed in detail in [**194**] and in particular in [**195**]; they consist of various families of mutually inscribed/circumscribed multilaterals. Steinitz [**212**] also corrected Schönflies's erroneous assertion that $t(C) = 1$ is not possible.

Steinitz [**213**] and [**212**, p. 489] also notes that all points of a 3-configuration may be in one orbit under automorphisms of the configuration, without the lines necessarily being in one orbit. Also, even if both points and lines are in single orbits, the configuration need not be selfdual. On the other hand, all examples of this type are not relevant to cyclic configurations and their properties.

There is little motivation to discuss in more detail the results of Schönflies and others and their proofs. However, several open questions deserve mention.

- Are there any 3-configurations that are not regular but for which there exists a "density" of trilaterals—in other words, every point is incident with as many trilaterals as every other point?

Possibly some of the astral 3-configurations may provide examples, but the question seems not to have been investigated.

- Is there a reasonable classification of "regular" 4-configurations that have a "density" of trilaterals?

For example, the astral (24_4) configuration shown in Figure 3.6.2 has 16 trilaterals incident with every point.

- What about "densities" of quadrilaterals in "regular" 3- or 4-configurations? The same question may be asked for other multilaterals, including Hamiltonian multilaterals.

Questions in the same spirit have often been asked and investigated for graphs and in some cases for combinatorial configurations or more general incidence structures. As this material has little connection to geometric configurations, we do not consider it here.

Exercises and Problems 5.5.

1. Verify the claim that the configuration in Figure 3.6.2 has 16 trilaterals incident with each point.

2. Find the number of trilaterals incident with each point of the 3-astral configuration (21_4) in Figure 3.7.1. How many quadrilaterals are incident with every point?

3. Are all points of the 2-astral configuration (48_4) shown in Figure 3.6.1 incident with the same number of trilaterals?

4. Investigate the astral 3-configurations (n_3) with $10 \leq n \leq 14$ concerning the numbers of trilaterals incident with each point. Quadrilateral? Any general hypotheses?

5.6. The dimension of a configuration

In Section 1.3 we introduced the concept of "dimension of a configuration". For convenience, we repeat it here. If C is a configuration, we say that C has dimension d if this is the largest integer for which G admits a geometric representation (by points and straight lines) in some Euclidean space, such that the affine hull of the imbedding has dimension d.

Among meaningful questions that one can ask is the determination of the dimension of a given configuration, the possible dimensions of k-configurations for a given k, what criteria can be found to determine whether a given configuration (or class of configurations) has this or that dimension, and so on. The material presented in this section has not been published before; it was developed in an ongoing collaboration with Tomaz Pisanski.

Here are some examples to help develop an understanding of the issues.

First, we consider the three smallest 3-configurations, the (9_3) configurations shown in Figure 5.6.1 (which we have seen earlier, as Figure 2.2.1). Each is (obviously) drawn in the plane—but could we somehow imbed it in 3-space so that it could not be contained in a plane? The negative answer is easily established: Regardless of the dimension of the space, the plane determined by points $1, 5, 7$ of $(9_3)_1$ necessarily also contains the points $3, 4, 8$ and hence also $2, 6, 9$, and thus the whole configuration. Similarly, the plane containing the points $1, 5, 7$ of $(9_3)_2$ contains $2, 3, 8$, hence $4, 6, 9$; the plane containing $1, 5, 8$ of $(9_3)_3$ also contains $2, 6, 9$ and then $3, 4, 7$—in both cases the whole configuration.

A different situation prevails with respect to the Desargues configuration which we have denoted $(10_3)_1$; see Figure 5.6.2. Consider the four points $0, 1, 2, 3$ imbedded in any Euclidean space of dimension at least 3, in such a way that no plane contains all four; this we occasionally call "general position". Then the points $4, 5, 6$ determine another plane; we choose them so that the plane is not parallel to the plane of $1, 2, 3$ and does not contain any of the four earlier points. These two planes determine (as their intersection) a line, which contains the points $7, 8, 9$ determined, respectively, on that line by the planes of $1, 3, 6, 4$, of $1, 2, 4, 5$, and of $2, 3, 6, 5$. Hence all points (and all lines) of the configuration are in the 3-dimensional space affinely spanned by $0, 1, 2, 3$. It follows that the dimension of the configuration $(10_3)_1$ is 3.

However, it would be wrong to conclude that an increase in the number of points implies an increase in the dimension. For example, the configuration $(10_3)_2$ shown in Figure 5.6.3 is 2-dimensional. Indeed, the plane containing points $1, 3, 5$ also contains $2, 4, 9$, hence $6, 7, 8$, and 0, and thus the whole configuration.

Theorem 5.6.1. *There exist 3-configurations with arbitrarily large dimensions.*

Proof. Start with any 3-configuration in the plane. Take three copies vertically above each other in 3-space, delete copies of the same line from each, and insert three vertical lines through the points on these lines. This raised

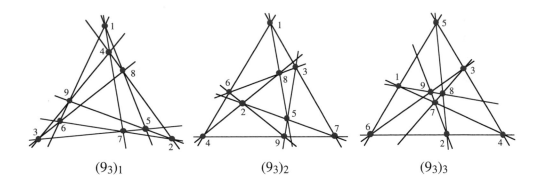

$(9_3)_1$ $(9_3)_2$ $(9_3)_3$

Figure 5.6.1. The three configurations (9_3).

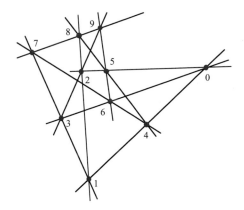

Figure 5.6.2. The configuration $(10_3)_1$—the Desargues configuration—has dimension $d = 3$.

the dimension by 1 (at least). Repeating the same procedure with three copies of this configuration, placed in suitable positions in parallel 3-spaces within a 4-dimensional space, deleting copies on one line from each and adding three transversals, raises the dimension of the resulting configuration to 4 (at least). Obviously we can continue indefinitely by the same method. $\square$

The configurations constructed in the proof of Theorem 5.6.1 are quite large. It may be of interest to find smaller examples, at least for small dimensions d. We have already seen such configurations for $d = 2$ and 3. For $d = 4$ we can use the Cremona-Richmond configuration shown in Figure 5.6.4; we encountered this configuration earlier, in Sections 1.1 and 5.4. Indeed, consider the 15 points in the 4-dimensional Euclidean space E^4, listed in Table 5.6.1 by their labels in Figure 5.6.4. Then it is easily checked that all 15 triplets that are supposed to be collinear indeed are, while it is obvious that the affine hull of the set is 4-dimensional.

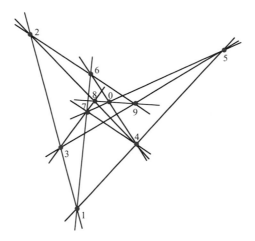

Figure 5.6.3. The configuration $(10_3)_2$ has dimension $d = 2$.

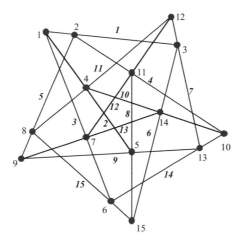

Figure 5.6.4. The Cremona-Richmond configuration (15_3) is 4-dimensional.

Table 5.6.1. Coordinates for the points of a realization of the Cremona-Richmond configuration (15_3) in the 4-dimensional Euclidean space. The names of the points refer to the labels in Figure 5.6.4.

$$1 = (0,0,0,0), \quad 2 = (1,0,0,0), \quad 3 = (-1,0,0,0), \quad 4 = (0,1,0,0),$$
$$5 = (0,-1,0,0), \quad 6 = (0,0,1,0), \quad 7 = (0,0,-1,0), \quad 8 = (0,0,0,1),$$
$$9 = (2,0,0,2), \quad 10 = (1,1,1,1), \quad 11 = (0,2,2,2), \quad 12 = (0,2,0,2),$$
$$13 = (2,2,0,2), \quad 14 = (2,0,2,2), \quad 15 = (0,0,2,2)$$

As reported in Section 2.3, there is no information available on the family of geometric configurations (n_3) for $n = 13$ or 14 (beyond some examples).

Hence it is not possible to definitely assert that the Cremona-Richmond configuration is the smallest 3-configuration of dimension $d = 4$. We venture:

Conjecture 5.6.1. *All configurations* (n_3) *with* $n \leq 14$ *have dimensions 2 or 3.*

A question that arises quite naturally is whether dual configurations have the same dimension. A negative answer is obvious from the example in Figure 5.6.5: The configuration $(6_2, 4_3)$ is clearly contained in the plane determined by any two of its lines, while the dual configuration $(4_3, 6_2)$ spans the 3-dimensional space if the four points are chosen in affinely independent positions. While it is possible to generalize this example, the situation concerning connected but not 2-connected configurations discussed in Section 5.1 makes it plausible that balanced configurations may behave differently from unbalanced configurations.

Conjecture 5.6.2. *If C is a balanced configuration, then the dimensions of C and its dual C^* are the same.*

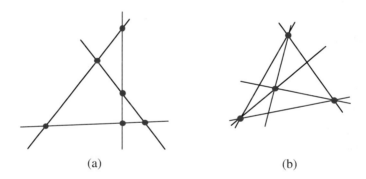

(a)   (b)

Figure 5.6.5. (a) The configuration $(6_2, 4_3)$ known as the *complete quadrilateral* is 2-dimensional. (b) Its dual $(4_3, 6_2)$, the *complete quadrangle*, is 3-dimensional.

It is easy to show that the cyclic configuration $\mathscr{C}_3(n)$ is 2-dimensional in all cases in which it is realizable by a geometric configuration, namely $n \geq 9$. Indeed, consider the typical Levi diagram of $\mathscr{C}_3(n)$, shown in Figure 5.6.6. The plane that contains the points P_0, P_1, P_2 contains the lines L_0, L_1 and hence the points P_3, P_4; then the lines L_2 and L_3 are in this plane

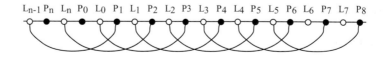

Lₙ₋₁ Pₙ Lₙ P₀ L₀ P₁ L₁ P₂ L₂ P₃ L₃ P₄ L₄ P₅ L₅ P₆ L₆ P₇ L₇ P₈

Figure 5.6.6. A stretch of the Levi graph of the cyclic configuration $\mathscr{C}_3(n)$ used to show that the configuration is 2-dimensional for all $n \geq 9$.

and therefore the points P_5 and P_6 as well. Since this pattern continues indefinitely, the whole configuration is in one plane.

So far we have dealt mainly with the dimension of 3-configurations. What is known about 4-configurations? Very little seems to be known at present. It is easy to verify that the astral configuration (24_4) shown in Figure 3.6.2 is 2-dimensional, as is the 3-astral configuration (21_4) shown in Figure 3.7.1. In the case of the six astral configurations (36_4) shown in Figure 3.6.3 the proof that all are 2-dimensional is only slightly more involved. Experimental evidence on k-astral configurations has not turned up any that are demonstrably d-dimensional with $d \geq 3$. However, it is well possible that for reasonably large n some k-astral (n_4) configurations are not 2-dimensional; it would be interesting to decide this question at least for astral 4-configurations or for 3-astral configurations.

On the other hand, the (41_4) configuration in Figure 3.3.16 is easily seen to be 3-dimensional. The two parts that are joined at the four collinear points by the four concurrent lines show how to "bend" the configuration into 3-dimensional space.

A challenging task—that may be impossible to fulfill—is finding combinatorial criteria for the dimension of a configuration.

Exercises and Problems 5.6.

1. Determine the dimensions of the remaining seven configurations (10_3), shown in Figures 2.2.3 and 2.2.5.

2. Does the analogy with the results of Section 5.1 for unbalanced configurations extend to the dimensions? Specifically, do there exist $[q, k]$-configurations (with $3 \leq q \neq k \geq 3$) such that the dimensions of C and its dual C^* are different?

3. How large can the difference between the dimensions of a dual pair of configurations in Exercise 2 be?

4. Recall from Section 2.1 that a *general cyclic configuration* $\mathscr{C}_3(n, a, b)$ consists of triples $\{j, a + j, b + j\}$, for given a, b with $0 < a < b < n$ and for $1 \leq j \leq n$, all entries taken $\mod n$. Determine the dimension of the various configurations $\mathscr{C}_3(n, 1, b)$ and possibly of the general configuration $\mathscr{C}_3(n, a, b)$.

5.7. Movable configurations

In this section we shall investigate the possibilities of changes of shape among configurations of a fixed isomorphism type. Applying any affine or projective transformation is most likely to produce a different configuration—but we shall consider such differences trivial and endeavor to find and describe more substantial modifications. In other words, we are considering equivalence classes of configurations, where members of each class are projectively equivalent to each other.

For example, all geometric realizations of (3_2) configurations are in one equivalence class, as are those of (4_2) configurations. On the other hand, configurations (5_2) have infinitely many projectively distinct forms; in fact, since any four points in general position can be projectively mapped onto any other such quadruple of points, the projective equivalence class is determined by any fifth point. It follows that the (5_2) configurations form a variety of dimension 2.

We shall say that a configuration is **rigid** if its geometric realizations form a single class under projective transformations. Both the theorem of Steinitz (as presented in Section 2.6) and practical experience suggest the following.

Conjecture 5.7.1. *There are no rigid 3-configurations.*

In view of the more stringent constraints that k-configurations with $k \geq 4$ have to satisfy, it may be tempting to believe that at least some of them are rigid. This may well be the case—however, none has been found that is demonstrably rigid. Hence we venture:

Conjecture 5.7.2. *There are no rigid k-configurations for any $k \geq 3$.*

A configuration that is not rigid shall be called **movable**. The motion itself can happen in a variety of ways. For example, with 3-configurations, in all cases that have been investigated, after having fixed a sufficient number of points and lines to eliminate projective maps, it is at least possible to either move a point arbitrarily on a line or to pivot a line about a point (or both). This is illustrated by four of the (10_3) configurations illustrated in Figure 2.2.5; after arbitrarily choosing four points, the only remaining choice is that of a point on one of the already-determined lines. In many 3-configurations (such as the ones illustrated in Figure 5.7.1) it is easy to see that they are movable even if keeping considerable parts of the configuration unchanged. However, there are analogous 3-configurations, such as the one in Figure 5.7.2, in which some of the parts have to be modified; similar examples can easily be multiplied. There are other examples in which the connecting lines between parts are neither parallel nor concurrent, and there are others in which the connections between parts are through points rather than lines.

In all the movable situations described so far there is essentially no symmetry except possibly by reflection in a mirror or by a half-turn and in some special positions. Much more interesting are movable configurations in

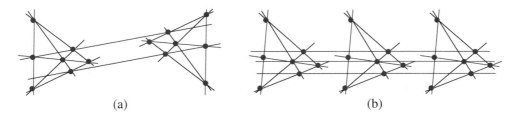

Figure 5.7.1. Two 3-configurations in which solid parts may be simply pulled apart: (a) a 2-connected (14_3) configuration and (b) a 3-connected (21_3) configuration.

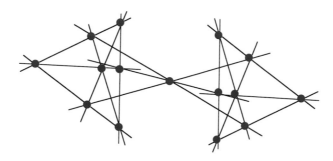

Figure 5.7.2. A 3-connected configuration (14_3) with a half-turn symmetry, in which solid parts may be separated by a greater or smaller distance.

which the configuration retains some non-trivial cyclic or dihedral symmetry throughout the motion. We have encountered such configurations in Section 2.9, when discussing dihedral astral 3-configurations.

However, it seemed rather unlikely that analogous movable k-configurations, with considerable symmetry, can exist for $k \geq 4$. But exactly this type of configuration was discovered by L. Berman in the summer of 2006 and was first published in [**10**]. We mention in passing that another kind of movable 4-configurations are the "floral configurations", the first of which was found by J. Bokowski somewhat later in 2006. We presented the relevant results in Section 4.7 and shall not dwell upon them here.

The simplest of Berman's methods of generating such configurations can be described as follows.

Starting with two 4-configurations, in one of them we omit one half of the lines of one orbit, and in the other we omit one half of the points in one orbit. If the configurations and the orbits have been chosen appropriately, it is possible to locate the deficient configurations in such a way that the points that were incident with the deleted lines slide on the lines from which a point was deleted, thus supplying the correct numbers of incidences. The new configuration has four fewer points and lines than the original ones had jointly. Naturally, the choices of the points and lines to be omitted have to be made carefully, subject to some very stringent conditions. These restrictions are made explicit in [10], and the complete characterization and proofs are given there. They are far too detailed and delicate to be included here, and the interested reader is advised to consult the original paper (which is easily accessible). A glimpse of the result of Berman's construction can be seen in Figure 5.7.3, which shows the smallest movable configuration obtainable by this method. Berman's paper [10] contains some additional constructions and developments as well.

A new paper by Berman [13] (private communication) presents additional constructions of movable configurations that retain cyclic symmetry during motion. It is more parsimonious, but the construction steps depend on the parity of the starting regular polygon. All configurations in this class have five point orbits of equal size. The smallest movable configuration that can be obtained is a (30_4) configuration. Several positions of this configuration are shown in Figure 5.7.4, adapted from [13].

Berman's construction in [13] is considerably simpler than the ones in [10], since it does not require deletion and pasting. However, it seems that the construction we shall consider next has certain advantages; it is presented here for the first time. It was discovered a very short time before this book went to print, and the author has not had the time to figure out all conditions for the applicability of the method.

Consider the following example, which is the smallest to which the construction is applicable but is typical in all other respects. We start with the tricyclic configuration $10\#(2,1;4,2;1,4)$; it has symmetry group d_{10}. However, we wish to consider it with only one of its d_5 subgroups. The situation is illustrated in Figure 5.7.5. Without the constraints imposed by the deleted mirrors and the accompanying rotations, the configuration is movable! Several stages of the motion are shown in Figure 5.7.6. The images were created with the Geometer's Sketchpad® software, the top green point being freely movable on the blue line with positive slope incident with it. An interesting and useful observation is that the points on the disregarded mirrors remain collinear throughout the motion.

This circumstance can be used to construct another family of movable configurations, using the $(5/6m)$ construction from Section 3.3. Applied to the (30_4) configuration in Figure 5.7.5(b), it yields the smallest known movable 4-configuration, a (25_4) configuration, illustrated with several snapshots of its motion in Figure 5.7.7.

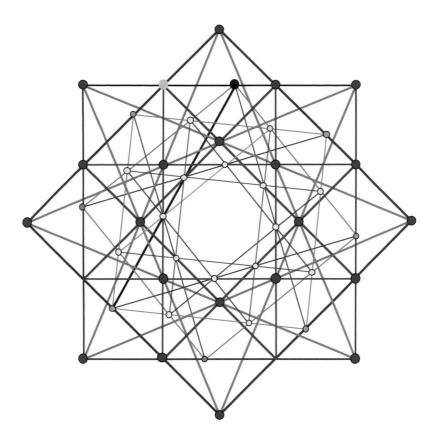

Figure 5.7.3. A movable (44_4) configuration, adapted from [**10**], Figure 7. It is constructed from two copies of the 3-cyclic configuration (24_4) with symbol $8\#(2,1;3,2;1,3]$; one is shown in heavy lines and large red dots, the other with thin lines and smaller yellow dots. From the former the gray dot and the three analogous points (not shown) have been omitted, while from the latter the dotted line and its three analogs are deleted. The missing incidences are replaced by placing the black dot on the black line and the corresponding points of its orbit on the corresponding lines; because of the choice of the parameters and orbits, the black point is freely movable on the black line, provided the sizes of the configurations are adjusted appropriately.

Figure 5.7.4. The smallest movable configuration from Berman's [**13**].
It is a 5-orbits configuration (30_4). The black point can move freely on
the black line as illustrated in the three parts.

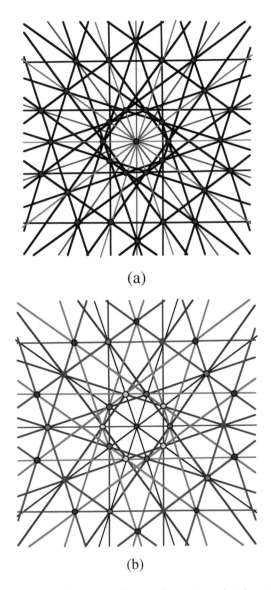

(a)

(b)

Figure 5.7.5. (a) The tricyclic configuration (30_4) with symbol $10\#(2, 1; 4, 2; 1, 4)$, shown with its ten mirrors; its symmetry group is d_{10}. (b) The same configuration but equipped with only five of the mirrors. The three orbits of lines and the five orbits of points under the symmetry group d_5 are distinguished by their colors.

(a)   (b)

(c)   (d)

Figure 5.7.6. Four snapshots of different stages in the motion of the (30_4) configuration from Figure 5.7.5(b). The purple segment indicates one of the mirrors.

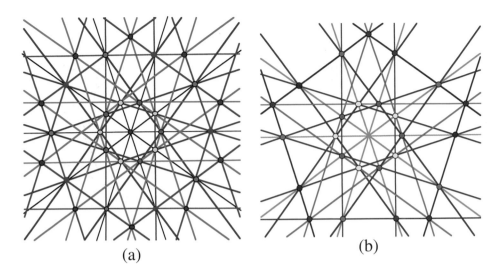

(a)   (b)

Figure 5.7.7. Part 1. (a) The (30_4) configuration with symmetry group d_5, from Figure 5.7.5(b). (b) A (25_4) configuration obtained from the configuration in (a) by omitting the points in the orbit of the lowest red point and the lines incident with these points and by adding the orange diametral lines (that go along the mirrors of the starting (30_4) configuration). This is the construction we introduced in Section 3.3 under the designation $(5/6m)$; this particular instance is the same as the first part of Figure 3.3.13.

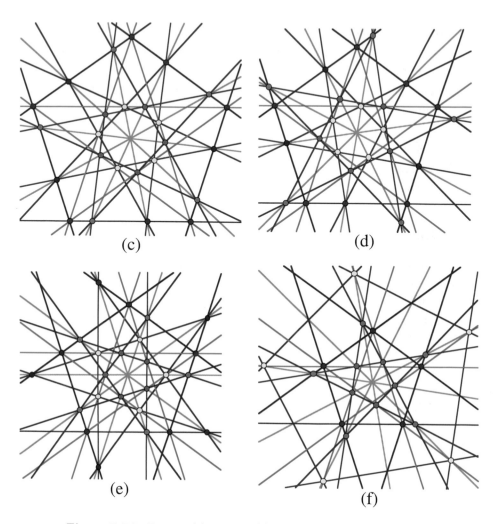

Figure 5.7.7. Part 2. (c) through (f) Four snapshots of different stages
in the motion of the (25_4) configuration from part (b).

Exercises and Problems 5.7.

1. Investigate the possible motions of the astral configurations such as (10_3), allowing for departure from astrality.

2. Describe movable examples of 4-configurations that can be obtained by using copies of $[3, 4]$-configurations. What are the smallest configurations of this kind? Apply the same construction to the superfiguration (9_3) shown in Figure 1.3.4.

3. Investigate to which configurations the methods we used in Figures 4.7.6 and 5.7.7 are applicable.

5.8. Automorphisms and duality

In Section 1.5 we introduced the combinatorial concepts of automorphism and duality in a superficial way. In Section 2.10 we presented some results on the duality of 3-configurations, in particular, the astral configurations. Here we will discuss such topics in more detail, with particular attention to the various ways in which such combinatorial properties and relations interact with geometric symmetries. The main reason for doing so is the absence of a coherent account of such interdependence of combinatorics and geometry; the examples we shall present are only brief glimpses of these relations. They lead to a large number of essentially unexplored questions.

We recall that an incidence-preserving map between the elements (points and lines) of two configurations is an **isomorphism** if points are mapped to points and lines are mapped to lines, and it is a **duality** if points are mapped to lines and vice versa. An isomorphism of a configuration with itself is an **automorphism**, while a duality of a configuration with itself is a **selfduality**. An automorphism of a configuration that is induced by an isometry of the plane is a **symmetry** of the configurations, while a selfduality that is induced by reciprocation in a suitable circle is a **selfpolarity**. Similarly, a pair of dual configurations may be **polar** to each other; this means that when situated appropriately, the reciprocation in a circle maps one onto the other. On the other hand, as we shall discuss later, the situation may be much more complicated.

We shall first illustrate the possibilities of the duality relations on a variety of examples, introducing additional classes and concepts as we are led by these experiences. In Figure 5.8.1 we show a pair of dual configurations (18_3). It is an easy exercise to find a duality mapping between the two configurations. There is no polarity between the two realizations of the dual pair. In contrast, in Figures 5.8.2, 5.8.3, and 5.8.4 we show the (24_4) configuration $12\#(5, 4; 1, 4)$ with three distinct labelings of its points and edges. The first shows that this configuration is selfdual, under the mapping

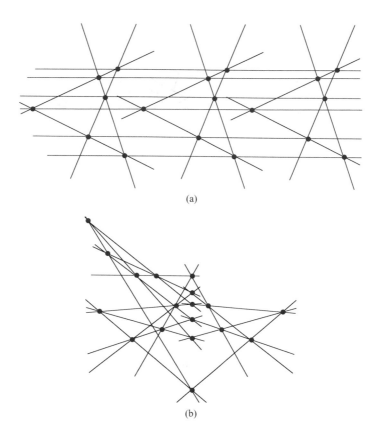

(a)

(b)

Figure 5.8.1. A pair of dual configurations (18_3). The configuration in (a) consists of three $(6_2, 4_3)$ subconfigurations (also known as "complete quadrilaterals") connected by six "parallel" lines. The dual configuration in (b) consists of three $(4_3, 6_2)$ subconfigurations (known as "complete quadrangles") and six points at which the three subconfigurations are joined.

that interchanges the green and blue labels. Since the configuration is very symmetric, it is reasonable to inquire about polarity. The correspondence between the labels actually establishes a more particular kind of selfduality. Although the polar of the configuration does not coincide with the configuration itself, it coincides with a copy reflected in a suitable mirror. In such a situation we shall say that the configuration is **oppositely selfpolar**. The necessity of using a mirror is obvious from the opposite orientations of the green and blue labels. The impossibility of finding a different selfduality that would not require a mirror follows from the observation that the outer ring of points gets mapped into the inner family of lines—but these are not aligned in the required way. Similar arguments will be applicable in several cases to be mentioned later.

Figure 5.8.2. The (24_4) astral configuration $12\#(5,4;1,4)$ is oppositely selfpolar; its polar needs to be reflected in order to coincide with the original.

The same configuration (24_4) exhibits **orbit transitivity** (or "ring transitivity"). This means that there is an automorphism that interchanges the (geometric) symmetry orbits. For a proof of this assertion it is enough to consider Figure 5.8.3, in which the labels indicate an isomorphism with Figure 5.8.2.

We shall see similar behavior with other configurations. However, the (24_4) configuration exhibits a more rare kind of symmetry: It is not just point-transitive and line-transitive under automorphisms, but it is **flag-transitive**. (By this is meant that all **flags**, each consisting of an incident pair point-line, are equivalent under automorphisms of the configuration.) This is made visible by considering another labeling of the configuration, shown in Figure 5.8.4. With this labeling (and the geometric symmetries of the configuration) it is easy to show that all flags form a single orbit under automorphisms. Obviously, this implies point-transitivity and line-transitivity.

In Section 3.6 we mentioned that there are six different astral configurations (36_4), and we presented them in Figure 3.6.3. The labels on three of these show how they are isomorphic. Each can be shown to be polar to the one near it; hence these are all isomorphic as well. In contrast to

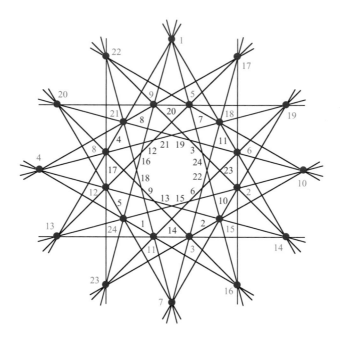

Figure 5.8.3. A different labeling of the (24_4) configuration from Figure 5.8.2. The identity map of the labels establishes that there is an automorphism of the configuration that maps the inner orbit of points onto the outer orbit of points, and similarly for lines.

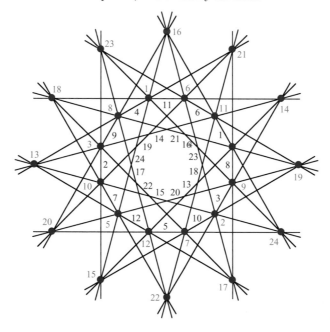

Figure 5.8.4. A labeling of the (24_4) configuration that can be used to establish the flag-transitivity of the configuration.

the configuration (24_4), none of these six configurations is selfpolar. Moreover, there are two orbits of points and two orbits of lines in each of these configurations; hence there is no transitivity of anything.

As mentioned in Section 3.6, the astral configurations (48_4) belong to four cohorts: $24\#\{\{11,1\},\{10,8\}\}$, $24\#\{\{9,3\},\{8,6\}\}$, $24\#\{\{8,2\},\{7,5\}\}$, $24\#\{\{10,2\},\{8,8\}\}$. This gives rise to seven distinct configurations, with symbols $24\#(8,7;2,5)$, $24\#(8,5;2,7)$, $24\#(11,10;1,8)$, $24\#(11,8;1,10)$, $24\#(9,8;3,6)$, $24\#(9,6;3,8)$, and $24\#(10,8;2,8)$. By the general results from Section 3.5, the last one is disconnected and consists of two copies of $12\#(5,4;1,4)$; hence we shall not be concerned with it here. Each of the other six is selfdual, but this does not translate into any geometric selfpolarity. In fact, the polars of $24\#(8,7;2,5)$, $24\#(11,10;1,8)$, $24\#(9,8;3,6)$ are $24\#(8,5;2,7)$, $24\#(11,8;1,10)$, and $24\#(9,6;3,8)$, respectively. Moreover, $24\#(8,7;2,5)$, $24\#(8,5;2,7)$, $24\#(11,10;1,8)$, $24\#(11,8;1,10)$ are all isomorphic, and $24\#(9,8;3,6)$ and $24\#(9,6;3,8)$ are isomorphic. These combinatorial symmetries are far from obvious. In Figures 5.8.5 and 5.8.6 we show the first two of these configurations, with labels that indicate the selfduality of each as well as the polarity between them. By switching the label colors, it also indicates isomorphism. In Figure 5.8.7 we show the configuration $24\#(11,10;1,8)$, with only the points labeled to show its isomorphism with the other two. (See Exercise 5.) In Figures 5.8.8 and 5.8.9 we show the configurations $24\#(9,8;3,6)$ and $24\#(9,6;3,8)$, labeled to show their selfpolarity and the polarity and isomorphism between them.

There is no additional information available at present concerning the combinatorial properties of other 2-astral 4-configurations. The large number of such configurations—already for (60_4) there are 15 configurations—make "experimental" progress unlikely without some new theoretical insights.

On the other hand, there is some knowledge of the situation concerning 3-astral 4-configurations. To begin with, by the general results of Sections 3.5 and 3.7, all such configurations are selfdual. Already in [**122**] it was noted that the smallest 3-astral configuration (21_4) (shown in Figure 3.7.1, as well as in Figure 5.8.10) is selfpolar, orbit-transitive, and flag-transitive. We show these relations in the two parts of Figure 5.8.10.

This highly symmetric behavior does not extend to the next larger 3-astral configuration. In Figure 5.8.11 we show the (24_4) configuration $8\#(3,2;1,3;2,1)$, with labels that show it is selfpolar. However, a count of symmetric trilaterals shows that there are three distinct orbits of points, hence also of lines.

The 3-astral configurations (27_4) show an interesting variety. There are six such configurations. Three of the trivial ones are isomorphic, as shown

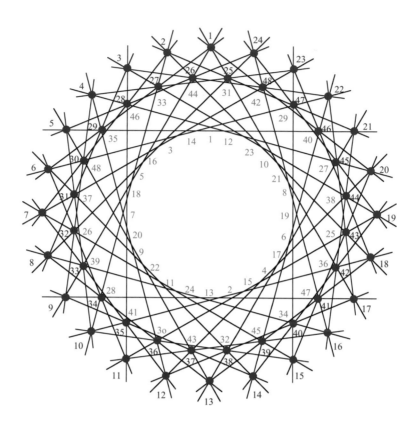

Figure 5.8.5. The configuration $24\#(8,7;2,5)$ labeled to show its selfduality.

in Figure 5.8.12; one of these is selfpolar, while the other two are selfpolar*. We use the term **selfpolar*** to indicate that the polar configuration needs to be reflected in the origin (that is, turned $180°$) in order to coincide with the original configurations. In each there are three orbits of points and three orbits of lines. There is one other trivial configuration, $9\#(4,2;1,4;2,1)$, which is also selfpolar* but is not isomorphic to any of the other (27_4) 3-astral configurations. It is shown in Figure 5.8.13, with labeling that enables a check of the orbit-transitivity of the configuration. The last two configurations of this family, $9\#(4,3;2,3;1,3)$ and $9\#(4,3;1,3;2,3)$, are shown in Figure 5.8.14. They are the first nontrivial 3-astral 4-configurations we encountered so far. In the terminology we used in Section 3.7, they belong to the family (2) with m divisible by 3. The two configurations are polars of each other and are point- and line-transitive.

There are seven 3-astral configurations (30_4). Four are trivial, and three are systematic. Among the trivial ones, $10\#(3,2;1,3;2,1)$ and $10\#(4,3; 1,4;3,1)$ are selfpolar, and the other two are oppositely selfpolar. In the notation of Section 3.7 the nontrivial configurations are in the cohort of the

Figure 5.8.6. The configuration $24\#(8,5;2,7)$ labeled to show its self-duality and its polarity and isomorphism with $24\#(8,7;2,5)$, the last by switching colors on labels.

family (1) with $q = 5$, $p = 1$, and $r = 2$. The configuration $10\#(4,3;1,2;1,3)$ is selfpolar, and the other two in this cohort are polars of each other. The details of their isomorphisms and transitivities have not been investigated, nor is any nonobvious information available concerning the ten 3-astral configurations (33_4) or any (n_4) with $n > 11$.

D. Marusic and T. Pisanski [**154**] investigated configurations in which the group of automorphisms does not act transitively on flags, but automorphisms together with dualities do act transitively on flags. They call such configurations **weakly flag-transitive** and prove that there are no weakly flag-transitive k-configurations for $k = 2$ and any odd k. To complement this, they construct several weakly flag-transitive 4-configurations; they show the smallest known such configuration, a (27_4) configuration. In retrospect, it is easy to check that this is a 3-astral configuration from the

Figure 5.8.7. Configuration $24\#(11, 10; 1, 8)$ with points labeled to show isomorphism with $24\#(8, 7; 2, 5)$ and hence with $24\#(8, 5; 2, 7)$.

trivial family, namely $9\#(2, 1; 4, 2; 1, 4)$. Similarly, their smallest known 3-astral configuration with an even number of vertices is the 3-astral configuration (42_4), namely $14\#(3, 1; 5, 3; 1, 5)$. Their final example is a trilateral-free 4-configuration (68_4), the sporadic $17\#(3, 1; 6, 2; 5, 4; 7, 8)$.

There is also a complete absence of information concerning the isomorphism and duality properties of k-astral configurations with $k \geq 4$. There is a whole family of unexplored problems, waiting for new initiatives and insights.

$$* * * * * *$$

Returning now to the more complicated situations possible for more general configurations, we first note that if a configuration C is selfdual under a selfduality δ, then obviously the inverse map δ^{-1} is also a selfduality. In most situations it is also true that the automorphism $\delta \circ \delta$ is the identity map ι on C. However, this does not happen in all cases; a larger number of repetitions of the map δ needs to be applied to reach the identity automorphism ι. We call the smallest such number the **rank** $r = r(\delta)$ of δ. In Figure

Figure 5.8.8. The configuration $24\#(9,8;3,6)$ with labels that show the selfduality.

5.8.15 we show a (5_2) configuration, with two selfdualities; one has rank 2, and the other has rank 10. The minimum of the ranks of all selfdualities of a selfdual configuration C is called the rank $r(C)$. Obviously, the rank of the configuration in Figure 5.8.15 is 2. In contrast, the rank of the selfduality τ of the octalateral (8_2) in Figure 5.8.16 is 16; however, although there are several other selfdualities of this configuration, they all have rank 16. Hence this is the rank of the configuration.

Several facts concerning rank can be established easily. First, the rank of any configuration C is a power of 2; that is, $r(C) = 2^k$ for some $k \geq 1$. It is also obvious that the case of the configuration in Figure 5.8.16 illustrates the general fact that the rank of any (2^k)-lateral is 2^{k+1}. Finally, if $r(C) = 2$, then one may use the same labels for the points and the lines of C. Conversely it is obvious that a selfduality that preserves the labels is of rank 2.

Regarding configurations (n_3), it is known (and easily verified) that for $n \leq 10$ all are selfdual of rank 2. According to Betten et al. [**17**], among the 31 configurations (11_3) there are 25 selfdual configurations of rank 2 and

Figure 5.8.9. The configuration $24\#(9,6;3,8)$ with labels that show its selfduality and its polarity with $24\#(9,8;3,6)$, as well as its isomorphism with that configuration (on interchanging the colors of the labels).

three pairs of mutually dual configurations—however, there is no identification of the dual pairs. In the same paper it is stated that among the 229 configurations (12_3) there are 95 that are selfdual of rank 2, the other forming dual pairs. Among the 2,036 configurations (13_3) there are reported to be 366 selfdual configurations; all but one of which have rank 2. There is no indication in [**17**] of what is the rank of that configuration, but it is possible to determine that the rank is 4. In [**17**] the authors also report that among the 21,399 configurations (14_3) there is one selfdual of rank greater than 2, and among the 245,342 configurations (15_3) there are three of rank greater than 2. For neither of these is the rank indicated, but their configuration tables are given; this makes it possible to find out their rank, if enough effort is put into it. One may conjecture that they all have rank 4.

In another direction of investigation of configurations, in [**3**] Ashley et al. constructed, for ever k, selfdual configurations (n_3) with rank 2^k. Moreover, one can find such configurations with $n \le 2^{k+5}$. The construction is quite complicated, and we shall not reproduce it here.

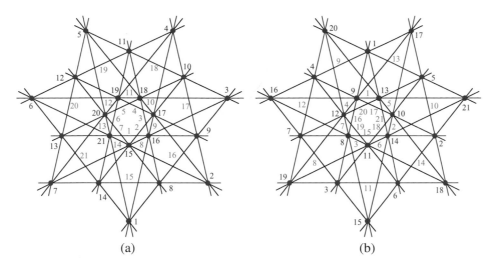

(a)   (b)

Figure 5.8.10. The 3-astral (21_4) configuration $7\#(3,2;1,3;2,1)$. (a) The labels indicate a selfpolarity. (b) Together with part (a), the labels indicate orbit-transitivity as well as flag-transitivity. (A more elegant labeling to exhibit these relations is used in [**122**].)

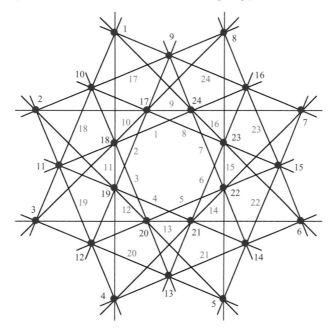

Figure 5.8.11. The 3-astral configuration (24_4) with symbol $8\#(3,2; 1,3;2,1)$ is selfpolar but has three orbits of points and three orbits of lines.

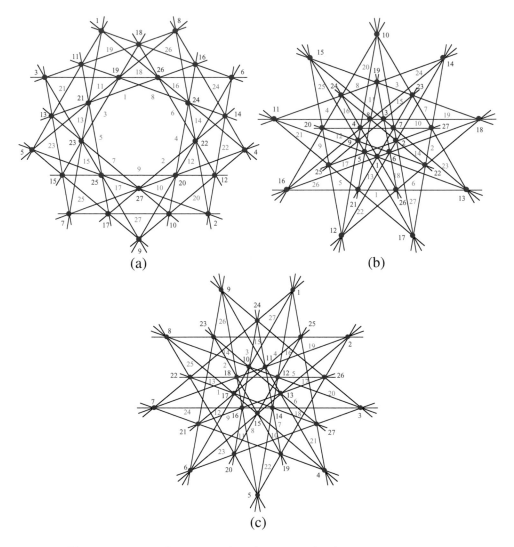

Figure 5.8.12. Three 3-astral configurations (27_4) that are isomorphic, with appropriate labels. (a) $9\#(3,2;1,3;2,1)$ and (b) $9\#(4,3;1,4;3,1)$ are selfpolar, while (c) $9\#(4,3;2,4;3,2)$ is selfpolar*.

There seems to be no information available concerning the rank of 4-configurations. However, we venture:

Conjecture 5.8.1. *Every selfdual geometric configuration of rank 2 has a realization such that the polar (in a suitable circle) is congruent to the original configuration.*

Conjecture 5.8.2. *For every $q \geq 1$ and every $k \geq 1$ there exist selfdual geometric q-configurations of rank 2^k.*

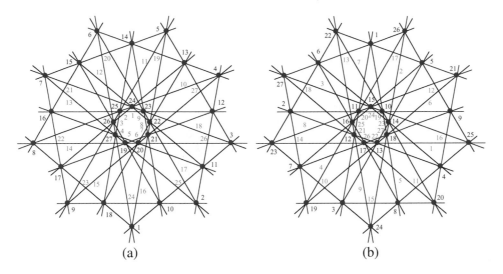

(a) (b)

Figure 5.8.13. (a) The 3-astral (27_4) configuration $9\#(4,2;1,4;2,1)$ is selfpolar*, as shown by the labels. (b) Comparison with (a) shows that the three point orbits are equivalent; hence there is point-transitivity and line-transitivity.

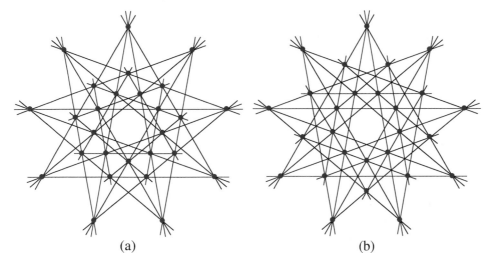

(a) (b)

Figure 5.8.14. The last two 3-astral configuration (27_4): (a) $9\#(4,3;2,3;1,3)$ and (b) $9\#(4,3;1,3;2,3)$. They are polar to each other, but there are no transitivity properties beyond that.

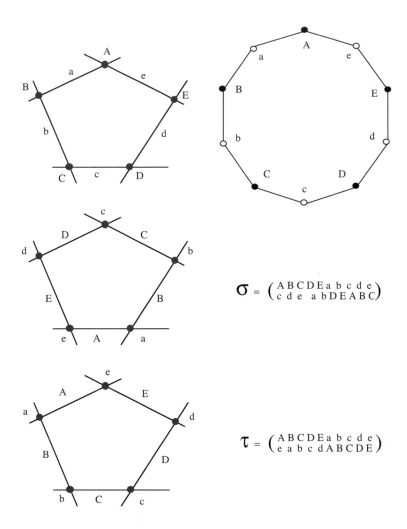

$$\sigma = \begin{pmatrix} A\,B\,C\,D\,E\,a\ b\ c\ d\ e \\ c\ d\ e\ \ \ a\ b\,D\,E\,A\,B\,C \end{pmatrix}$$

$$\tau = \begin{pmatrix} A\,B\,C\,D\,E\,a\ b\ c\ d\ e \\ e\ a\ b\ c\ d\,A\,B\,C\,D\,E \end{pmatrix}$$

Figure 5.8.15. A configuration (5_2) (that is, a pentalateral) and two of its selfdualities. Either from the permutation representation or by using the Levi graph shown, it is easy to verify that the rank of the selfduality σ is 2 and the rank of τ is 10.

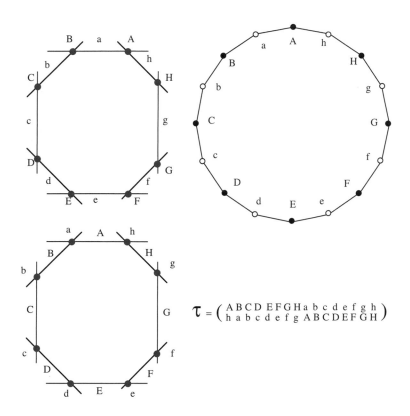

Figure 5.8.16. An octalateral (8_2) and one of its selfdualities. This selfduality, as well as all other selfdualities of the octalateral, have rank 16; this is therefore the rank of the octalateral as well.

Exercises and Problems 5.8.

1. Show that the astral topological configuration $11\#(5,4;1,4)$ shown in Figure 5.8.3 is selfpolar*. Investigate its transitivity properties (that is, orbit-, point-, line-, flag-transitivity).

2. Prove the assertions made above concerning the astral configurations (36_4).

3. The connectedness of a configuration has nothing to do with such properties as selfduality. Prove that the configuration $24\#(10,8;2,8)$ is selfpolar and flag transitive.

4. Provide arguments that prove that the configuration $24\#(8,7;2,5)$ is not selfpolar in any sense.

5. Draw the configuration $24\#(11,8;1,10)$, and show its isomorphism with $24\#(11,10;1,8)$.

6. Verify the claim that the 3-astral configuration (24_4) with symbol $8\#(3,2;1,3;2,1)$ has three orbits of points.

7. Find a selfduality of rank 2 of the topological configuration $(10_3)_4$.

5.9. Open problems

1. Is there a relationship between the maximal size of an independent family and the maximal sizes of independent families consisting of points only, or of lines only? Notice that in the examples in Figures 5.1.8 and 5.1.9 the first is strictly greater than the latter two.

2. If C is a 4-configuration and H a Hamiltonian multilateral in C, then each line is determined by two vertices of H and passes through two other vertices of C. Is it possible to obtain a new Hamiltonian multilateral H^* of C by using such pairs of points of C as vertices of a multilateral? If so, can this happen in a k-astral 4-configuration, or even in an astral 4-configuration?

3. It is conceivable that every connected configuration (n_k) can be presented as a family of mutually inscribed/circumscribed multilaterals; this naturally includes Hamiltonian multilaterals. Either a counterexample or an affirmation (at least for $k = 3$ or 4) would be very interesting.

4. Is there any relationship between the maximal size of an independent family in a configuration C and the dimension of C?

5. Are all movable k-astral 4-configurations 2-dimensional?

Postscript

Predicting the future is always risky, and the author does not presume to do so concerning configurations of points and lines. But two guesses appear quite safe to make.

In the first place, the interest in configurations is likely to persist. This is due to the fact that the topic is elementary—in the sense that its results and problems are accessible without years of graduate studies—as well as visually attractive and nontrivial. Not only do the problems mentioned in the present text deserve attention, but it is possible that completely new problems, approaches, methods, and directions may emerge from the current interest in the topic.

The second seemingly safe guess is that there will be interest in higher-dimensional analogues of the material described in this book. A few tentative steps in this direction occurred already in the "prehistoric period" (for example, in the works of Cayley), as well as in the "classical period" of the late nineteenth century. Quite clearly, this direction is necessarily less visual and more algebraic and topological than the theory of configurations of points and lines. However, at least some of the spirit of the latter can probably be translated into higher dimensions—with approaches and concerns different from the present emphases of algebraic geometry and topology.

Appendix

The Euclidean, projective, and extended Euclidean planes

For most readers of this book it is probably not necessary to explain what is meant by these concepts—it may even be presumptuous on the author's part to try to do so. However, his experience with the students in several courses on configurations has been such that he felt obliged to clarify what the three planes that we are concerned with are. The following pages are a version of the notes the author gave to his students. It is important to acquire the facility to switch from one model of one of these planes to another and to be aware of their relationships.

It is natural to start with the *Euclidean plane* E^2, since our acquaintance with it goes back to kindergarten. In more formal terms, we can take the Euclidean plane as defined by the well-known axioms of Hilbert. Of the concepts that enter Hilbert's system of axioms, we essentially need only "point", "line", and "incident". But besides these ways of thinking about the Euclidean plane, there are several other ways that present different but equivalent models, which are sometimes useful. In each such model, certain objects represent points, and certain other objects represent lines; naturally, circles and other planar figures can also be represented in any such model, but we shall not have occasion to do this.

To construct any such alternative model of the Euclidean plane E^2, we shall, therefore, have to state what its "points" are and what its "lines" are, as well as what "points" and "lines" are incident.

The best known such model of the Euclidean plane is the one given by a Cartesian coordinate system. In it, a "point" is an ordered pair of real numbers, $P = (x, y)$, and a "line" is a set of linear polynomials consisting of all non-zero multiples of a polynomial $L = L(u,v) = au + bv + c$ with $a^2 + b^2 > 0$. The "point" P and the "line" L are "incident" if and only if $L(x,y) = 0$. To most of us, most of the time, this is the Euclidean plane.

A more symmetric way of looking at essentially the same thing is the *homogeneous coordinates* model $\mathscr{H}$. In it, a point is the set of (ordered) triplets consisting of all non-zero multiples of an ordered triplet $P = (x,y,z)$ in which $z \neq 0$. A "line" is the set of (ordered) triplets consisting of all non-zero multiples of an ordered triplet $L = (a,b,c)$ in which $a^2 + b^2 > 0$ and P is "incident" with L if and only if $\langle P, L \rangle = ax + by + cz = 0$. A geometric interpretation of this model is as follows (see Figure A.1). We consider the Cartesian plane E^2 as being the plane $z = 1$ in an (x,y,z)-coordinate system in 3-dimensional space E^3. The point $P = (x,y,1)$ of E^2 is represented in the model $\mathscr{H}$ by the **line** through the origin $O = (0,0,0)$ and P; it consists, besides the origin O, of all the non-zero multiples of P. The line $L = L(u,v) = au + by + c$ is represented in $\mathscr{H}$ by the **plane** determined by O and L. Notice that the correspondence between the points and lines of E^2 and the lines and planes through O is a bijection **except** that the horizontal plane through O and the horizontal lines through O must be excluded.

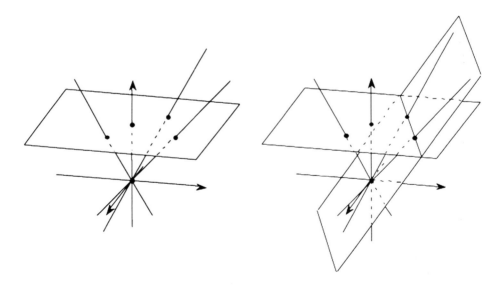

Figure A.1. An illustration of the model $\mathscr{H}$ of the Euclidean plane.

Another model of E^2 can be derived from $\mathscr{H}$. This is the **sphere** model $\mathscr{S}$, illustrated in Figure A.2, in which each "point" is a **pair** of **antipodal** (that is, diametrically opposite) points of a unit sphere S centered at O and each "line" is a great circle of S. (For easier visualization, the z-axis is pointed downwards in Figure A.2.) These "points" and "lines" are obtained

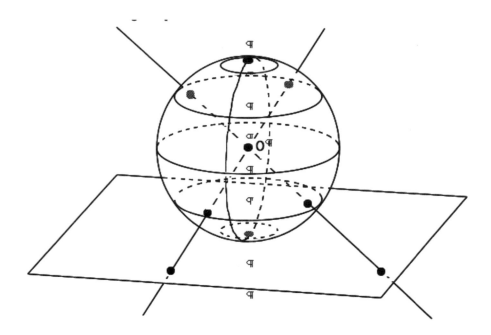

Figure A.2. An illustration of the **sphere** model $\mathscr{S}$ of the Euclidean plane.

by intersecting S with the "points" and "lines" of $\mathscr{H}$. This would be a very nice model, but it is complicated by the fact that the "equatorial" great circle is not included among the "lines" and the pairs of antipodal points of the equator are not included among the "points".

The **hemisphere** model $\mathscr{HS}$ of E^2 is obtained from S by taking a hemisphere H of S, tangent to the plane $z = 1$ at the "north pole"; the rim of H (which is the equator of S) is not included in H. To each point P of E^2 there corresponds in $\mathscr{HS}$ a single point. To each line of E^2 there corresponds in $\mathscr{HS}$ a semicircle (of a great circle of S) with endpoints on the rim of H. This model is illustrated in Figure A.3 where, as in Figure A.2, the z-axis is pointed downwards.

Now it is easy to give convenient descriptions of the **projective plane** P^2, simply by eliminating the bothersome restrictions in the above models of the Euclidean plane. Specifically, to obtain the $\mathscr{HP}$ model of the projective plane, add to $\mathscr{H}$ the horizontal plane and horizontal lines; equivalently, allow for "points" as well as for "lines" all triplets except $(0, 0, 0)$. For the spherical model $\mathscr{SP}$ of the projective plane, admit **all** antipodal pairs as "points" and **all** great circles as "lines". The hemisphere model $\mathscr{HSP}$ consists of the closed hemisphere H^* of S obtained by adding to H its rim, **but** identifying antipodal pairs of points on the rim to represent single "points" of $\mathscr{HSP}$; the rim itself represents a "line" of $\mathscr{HSP}$.

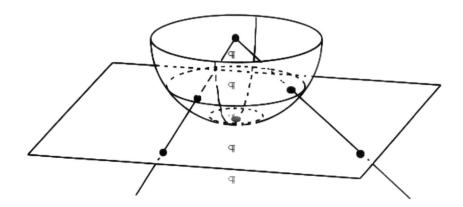

Figure A.3. An illustration of the hemisphere model $\mathscr{HS}$ of E^2.

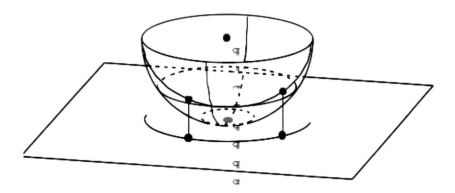

Figure A.4. An illustration of the $\mathscr{CP}$ model.

While these models of the (real) projective plane P^2 are useful and clearly illustrate the equivalent roles played by all its points (and by all its lines), they are cumbersome in that they are situated in 3-dimensional space. Hence other models that are less pleasing in some ways but which can be placed in a Euclidean plane are often more useful. One such model, which we shall denote by $\mathscr{CP}$, is illustrated in Figure A.4. We start from $\mathscr{HSP}$ and project H^* straight down onto the plane tangent to it at its south pole. This projection is a bijection between H^* and a closed circular disk C^* of radius 1; hence (see Figure A.5) the "points" of $\mathscr{CP}$ are the points of the interior of C^* together with antipodal pairs on the boundary of C^*, and the "lines" of $\mathscr{CP}$ are

   (i) **semiellipses** whose major axes are diameters of C^*,

   (ii) **diameters** of C^*,

   (iii) the **boundary circle** of C^*.

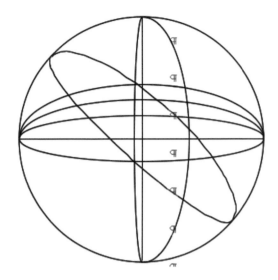

Figure A.5. Another illustration of the $\mathscr{CP}$ model.

It is worth noting that by considering only the interior C of C^*, a model $\mathscr{C}$ of the Euclidean plane can be obtained in the same way from $\mathscr{HS}$.

A very important property of the projective plane P^2 is that any two distinct lines in it have a (unique) point in common. It is important to consider how this happens in the various models. Although we have (for obvious "technical" reasons) not been using the $\mathscr{CP}$ model in the text, it is useful to see what happens in this case.

Figure A.5 can be used to understand how "lines" that are "parallel" in $\mathscr{C}$ acquire intersection "points" in $\mathscr{CP}$. With a little practice it becomes easy to see that the configuration of "points" and "lines" in $\mathscr{C}$ (or $\mathscr{CP}$) shown in Figure A.6(a) is a realization of a $(16_3, 12_4)$ configuration. Another view of this configuration is in Figure A.6(b), which can be understood in the Euclidean plane.

Consider now Figure A.7(a). Since it includes points on the rim of C^*, this must be understood as a configuration in $\mathscr{CP}$ (and not in $\mathscr{C}$), and pairs of antipodal points on the boundary have to be identified. Then this is a (12_3) configuration (which does not include the line represented by the rim of C^*).

Despite all that has been said so far, even $\mathscr{CP}$ is not really convenient. (Try redrawing in it—or in $\mathscr{C}$—the configurations in Figure 2.2.3!) But consider that if we have a certain figure F in the Euclidean plane, we can go through the procedure of representing it in the model $\mathscr{C}$ or $\mathscr{CP}$ but using a very large sphere instead of the unit sphere. Then the projections we used would entail very little distortion, and the parts of the ellipses that are used

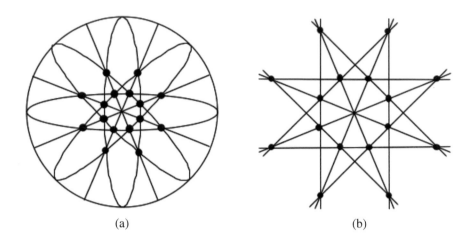

Figure A.6. Two views of a $(16_3, 12_4)$ configuration.

to represent the segments of the figure would be practically indistinguishable from straight line segments. This leads to the idea of still another model of P^2, which we call the **extended Euclidean plane** model and denote by E^{2+}: Let the points of $\mathscr{CP}$ that are also points of $\mathscr{C}$ be represented by points of E^2; the points on the rim of C^*, at which pencils of "lines" that are "parallel" in $\mathscr{C}$ meet in $\mathscr{CP}$, are represented by the **direction** of the corresponding parallel lines in E^2. (Here "direction" is meant to identify opposite orientations!) This leads to adding to each Euclidean line one "ideal point" (or "point at infinity" on that line) through which pass all other lines that are (in the Euclidean plane) parallel to that line. Moreover, the rim of C^*, which is formed by the totality of such points and is a "line" in $\mathscr{CP}$, can then be made to correspond to the totality of the points-at-infinity in E^{2+} and be the "ideal line" (or "line-at-infinity") of E^{2+}. The points that are not ideal points are called "finite points", and all lines except the ideal one are also called finite. Graphically, an ideal point of E^{2+} is often represented by a detached dot near a line for which it is the "point-at-infinity", or by a pair of such dots (as in Figure A.7(b), or by a dot or a pair of dots with arrows to indicate for which pencil of parallel lines the dot represents the ideal point. This is, in fact, the representation of the projective plane which we have almost invariably used. After some experience, it becomes very convenient. Moreover, it enables one to use concepts and results from Euclidean geometry and apply them to projective geometry. In particular, this enabled us to use Euclidean symmetries—extended as appropriate—in E^{2+}.

As an example, in Section 4.4 we have seen constructions of geometric $[3, 5]$-configurations, which were used in the proof of Theorem 4.4.1.

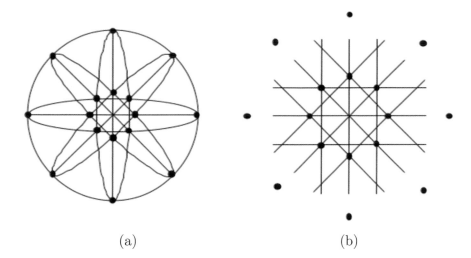

(a) (b)

Figure A.7. Two illustrations of the same (12_3) configuration.

Exercises.

1. Find a linear transformation of E^3 which maps the points $A = (1, 0, 0)$, $B = (0, 1, 0)$, $C = (0, 0, 1)$, $D = (1, 1, 1)$ of the $\mathscr{HP}$ model of the projective plane onto the points $A^* = (0, 0, 1)$, $B^* = (1, 0, 1)$, $C^* = (1, 1, 1)$, $D^* = (0, 1, 1)$, respectively.

2. Using the model of your choice, prove the following basic result of projective geometry: Given any four points A, B, C, D in the projective plane P^2, no three of which are collinear, and four points A^*, B^*, C^*, D^*, no three of which are collinear, there is a homeomorphism ϕ of the projective plane which maps lines onto lines and is such that $\phi(X) = X^*$ for all $X \in \{A, B, C, D\}$. (Such homeomorphisms are called **projective transformations**.)

3. Show that given a circle C in the extended Euclidean model E^{2+} of P^2 and points A, B in the interior of C, there is a projective transformation that maps C onto itself and A onto B.

4. In the spherical model $\mathscr{SP}$ of the projective plane, inscribe a regular dodecahedron D in the sphere. Consider the vertices of D as the points of a configuration **D** and the lines of $\mathscr{SP}$ determined by the edges of D as the lines of **D**. Give a configuration table for **D** and a realization in the plane by a linear configuration.

References

Each entry in the list of references ends in a list of pages in which this reference is mentioned. In order to avoid clutter, only the first occurrence of a reference in a section is listed; the presence of additional mentions of the same reference within the same section is signaled by the + sign.

1. B. Alspach and C. Q. Zhang, Hamilton cycles in cubic Cayley graphs on dihedral groups. Ars Combinat. 28(1989), 101–108. [319]

2. R. Artzy, Self-dual configurations and their Levi graphs. Proc. Amer. Math. Soc. 7(1956), 299–303. [150]

3. J. Ashley, B. Grünbaum, G. C. Shephard, and W. Stromquist, Self-duality groups and ranks of self-dualities. *Applied Geometry and Discrete Mathematics: The Victor Klee Festschrift*, P. Gritzmann and B. Sturmfels, eds., DIMACS Series in Discrete Mathematics and Theoretical Computer Science, Vol. 4, pp. 11–50. Amer. Math. Soc., 1991. [368]

4. A. T. Balaban, Trivalent graphs of girth nine and eleven and relations among cages. Rev. Roumaine Math. Pure Appl. 18(1973), 1033–1043. [336]

5. P. Barbarin, Georges Brunel. Enseignement Math. 3(1901), 237–239. [156]

6. L. W. Berman, A characterization of astral (n_4) configurations. Discrete Comput. Geom. 26 (2001), no. 4, 603–612. [203+, 281]

7. L. W. Berman, Astral Configurations. Ph.D. thesis, Univ. of Washington, Seattle, 2002. [203]

8. L. W. Berman, Even astral configurations. Electron. J. Combin. 11 (2004), Research Paper 37, 23 pp. (electronic). [240, 253]

9. L. W. Berman, Some results on odd astral configurations. Electron. J. Combin. 13(2006), Research Paper 27. [235]

10. L. W. Berman, Movable (n_4) configurations. Electron. J. Combin. 13(2006), Research Paper 104. [36, 352]

11. L. W. Berman, Symmetric simplicial pseudoline arrangements. Electr. J. Combinatorics 15(2008), #R13. [286]

12. L. W. Berman, Astral (n_4) configurations of pseudolines. Preprint, April 2007. [281+]

13. L. W. Berman, A new class of movable (n_4) configurations. Preprint, July 2007. [352+]

14. L. W. Berman and J. Bokowski, Astral (n_5) configurations. Europ. J. Combinatorics (to appear). [126+, 235]

15. L. W. Berman, J. Bokowski, B. Grünbaum, and T. Pisanski, Geometric "floral" configurations. Canad. Math. Bull. (to appear). [262+]

16. A. Betten and D. Betten, Tactical decompositions and some configurations v_4. J. Geom. 66(1999), 27–41. [82, 157+]

17. A. Betten, G. Brinkmann, and T. Pisanski, Counting symmetric configurations v_3. Discrete Appl. Math. 99(2000), 331–338. [81+, 91, 333, 343, 367+]

18. D. Betten and U. Schumacher, The ten configurations 10_3. Rostock Math. Kolloq. 46(1993), 3–10. [10]

19. M. Boben, Uporaba teorije grafov pri kombinatorcnih in geometricnih konfiguracijah. (In Slovenian) [The use of graph theory in the study of combinatorial and geometric configurations]. Ph.D. thesis, University of Ljubljana, November 2003. [91]

20. M. Boben, Irreducible (v_3) configurations and graphs. Discrete Math. 307(2007), 331-344. [91]

21. M. Boben, B. Grünbaum, and T. Pisanski, What did Steinitz prove in his Thesis? (in preparation). [103]

22. M. Boben, B. Grünbaum, T. Pisanski, and A. Zitnik, Small triangle-free configurations of points and lines. Discrete and Computational Geometry 35(2006), 405–427. [82, 91, 121, 329+]

23. M. Boben, B. Grünbaum, and T. Pisanski, Multilateral-free configurations (in preparation). [342]

24. M. Boben and T. Pisanski, Polycyclic configurations. Europ. J. Combin. 24(2003), 431–457. [40, 137+, 190+, 221+]

25. J. Bokowski, Computational Oriented Matroids. Cambridge Univ. Press, 2006. [32, 339]

26. J. Bokowski, B. Grünbaum, and L. Schewe, Topological configurations (n_4) exist for all $n \geq 17$. Europ. J. Combinatorics (in press; electronic version available from Elsevier). [162+]

27. J. Bokowski and L. Schewe, There are no realizable 15_4- and 16_4-configurations. Rev. Roumaine Math. Pures Appl. 50(2005), no. 5-6, 483–493. [161]

28. J. Bokowski and L. Schewe, On the finite set of missing geometric (n_4) configurations (in preparation). [162+, 173]

29. J. Bokowski and B. Sturmfels, Computational Synthetic Geometry. Lecture notes in Mathematics #1355, Springer, New York, 1989. [10+, 20+, 65+, 336]

30. G. Bol, Beantwoording van Prijsvraag no. 17, 1931. Nieuw Archief voor Wiskunde (2)18, 14–66 (1933). [210]

31. P. B. Borwein and W. O. J. Moser, A survey of Sylvester's problem and its generalizations. Aequat. Math. 40(1990), 111–135. [7]

32. P. Brass, W. Moser, and J. Pach, Research Problems in Discrete Geometry. Springer, New York, 2005. [3+, 64]

33. R. A. Brualdi, Introductory Combinatorics. 4th ed., Prentice Hall, Englewood Cliffs, NJ, 2004. [94, 158, 254]

34. G. Brunel, Polygones autoinscrits. Proc. Verb. Sánces Soc. Sci. Phys. Nat. Bordeau, 1895/96, pp. 35–39. [68, 156, 309, 324]

35. G. Brunel, Polygones à multiple. Proc. Verb. Séances Soc. Sci. Phys. Nat. Bordeau, 1897/98, pp. 43–46. [11, 156+, 161]

36. W. Burnside, On the Hessian configuration and its connection with the group of 360 plane collineations. Proc. London Math. Soc. (2) 4(1907), 54–71. [156+]

37. S. A. Burr, B. Grünbaum, and N. J. A. Sloane, The Orchard Problem, Geom. Dedicata 2(1974), 397–424. [8, 246+, 253]

38. B. Bydzovsky, Über eine ebene Konfiguration $(12_4, 16_3)$. Vestník Královské Česk. Spolecnosti Nauk. Trída Matemat.-Prírodoved., 1939, 8 pp. (1940). [249]

39. H. G. Carstens, T. Dinski, and E. Steffen, Reduction of symmetric configurations n_3. Discrete Appl. Math. 99(2000), 401–411. Erratum (by E. Steffen, T. Pisanski, M. Boben, and N. Ravnik), ibid. 154(2006), 1645–1646. [90]

40. W. B. Carver, Proof of the impossibility of the construction of one of the Kantor $(3,3)_{10}$ configurations. Johns Hopkins Univ. Circ. 22(1902), No. 160, pp. 3–4. [10, 76]

41. A. Cayley, Sur quelques théorèmes de la géométrie de position. J. Reine und Angew. Math. 31(1846), 213–226 = Collected Mathematical Papers vol. 1, pp. 317–328. [239+, 250, 253, 323]

42. W. K. Clifford, Synthetic proof of Miquel's theorem. Oxford, Cambridge and Dublin Messenger of Math. 5(1871), 124–141. [287]

43. A. M. Cohen and J. Tits, On generalized hexagons and a near octagon whose lines have three points. Europ. J. Combinat. 6(1985), 13–27. [337]

44. J. Colannino, Circular and modular Golomb rulers (2003). `http://cgm.cs.mcgill.ca/~athens/cs507/Projects/2003/JustinColannino/#References`. [234]

45. H. S. M. Coxeter, Configurations and maps. Reports of a Math. Colloq. (2) 8(1949), 18–38. [12, 27]

46. H. S. M. Coxeter, Self-dual configurations and regular graphs. Bull. Amer. Math. Soc. 56(1950), 413–455. (= Twelve Geometric Essays, Southern Illinois Univ. Press, Carbondale, IL, 1968 = The Beauty of Geometry, Dover, Mineola, NY, 1999. pp. 106–149. [12, 17+, 328+]

47. H. S. M. Coxeter, Introduction to Geometry. Wiley, New York, 1961. Second ed., 1981. [8, 287]

48. H. S. M. Coxeter, Desargues configurations and their collineation groups. Math. Proc. Cambridge Philos. Soc. 78(1975), 227–246. [12]

49. H. S. M. Coxeter, The Pappus configuration and the self-inscribed octagon. I, II, III. Nederl. Akad. Wetensch. Proc. Ser. A 80 = Indag. Math. 39(1977), pp. 256–269, 270–284, 285–300. [12]

50. H. S. M. Coxeter, My graph. Proc. London Math. Soc. (3) 46(1983). 117–136. [12, 156]

51. H. S. M. Coxeter, Twelve Geometric Essays, Southern Illinois Univ. Press, Carbondale, IL, 1968. Second ed., The Beauty of Geometry, Dover, Mineola, NY, 1999. [12]

52. H. S. M. Coxeter and S. L. Greitzer, Geometry revisited. Random House, New York, 1967. [8]

53. T. T. Croft, K. J. Falconer, and R. K. Guy, Unsolved Problems in Geometry. Springer, New York, 1991. [7]

54. R. Daublebsky von Sterneck, Die Configurationen 11_3. Monatshefte Math. Phys. 5(1894), 325–330 + 1 plate. [81]

55. R. Daublebsky von Sterneck, Die Configurationen 12_3. Monatshefte Math. Phys. 6(1895), 223–255 + 2 plates. [40+, 81+]

56. R. Daublebsky von Sterneck, Über die zu den Configurationen 12_3 zugehörigen Gruppen von Substitutionen. Monatshefte Math. Phys. 14(1903), 254–260. [82]

57. M. Daven and C. A. Rodger, (k,g)-cages are 3-connected. Discrete Math. 199(1999), 207–215. [339]

58. J. de Vries, Über gewisse ebene Configurationen. Acta Math. 12(1888), 63–81. [15, 249]

59. J. W. Di Paola and H. Gropp, Hyperbolic graphs from hyperbolic planes. Congressus Numerant. 68(1989), 23–44. [28, 53+]

60. J. W. Di Paola and H. Gropp, Symmetric configurations without blocking sets. Mitt. Math. Seminar Giessen 201(1991), 49–54. [321]

61. I. V. Dolgachev, Abstract configurations in algebraic geometry. Proc. Fano Conference, Torino, 2002, A. Collino, A. Conte, and M. Marchisio, eds., Torino, 2004, pp. 423–462. [10, 82]

62. H. L. Dorwart, The Geometry of Incidence. Autotelic Instructional Materials Publishers, New Haven, CT, 1966. [12, 324]

63. H. L. Dorwart and B. Grünbaum, Are these figures oxymora? Mathematics Magazine 65(1992), 158–169. [82, 92, 309]

64. P. Duhem, Notice sur la vie et les travaux de Georges Brunel (1856–1900). Mémoirs Soc. Sci. Phys. et Nat. Bordeaux (6) 2(1903), I–LXXXIX. [156+]

65. M. N. Ellingham and J. D. Horton, Non-hamiltonian 3-connected cubic bipartite graphs. J. Combinat. Theory Ser. B 34(1983), 330–333. [310+]

66. L. S. Evans, Some configurations of triangle centers. Forum Geometricorum 3(2003) 49–56. [19]

67. H. Eves, A survey of Geometry. Allyn & Bacon, Boston. Vol. 1, 1963; Vol.2, 1965. [287]

68. G. Fano, Sui postulati fondamentali della geometria proiettiva in uno spazio lineare a un numero qualunque di dimensioni. Giornale di Matematiche 30(1892), 106–132. [61]

69. G. Feigh, Review of [H1]. Jahrbuch Fortschritte Math. 58(1932), 597. [12]

70. H. L. Fu, K. C. Huang, and C. A. Rodger, Connectivity of cages. J. Graph Theory 24(1997), 187–191. [339]

71. J. P. Georges, Non-hamiltonian bicubic graphs. J. Combinat. Theory B 46(1989), 121–124. [310+]

72. D. G. Glynn, On the anti-Pappian 10_3 and its construction. Geom. Dedicata 77(1999), 71–75. [10, 76]

73. D. G. Glynn, On the representation of configurations in projective spaces. J. of Statistical Planning and Inference 86(2000), 443–456. [105+]

74. S. W. Golomb, Algebraic constructions for Costas arrays. J. Combinat. Theory (A) 37(1984), 13–21. [234]

75. S. W. Golomb, Construction of signals with favourable correlation symmetries. Surveys in Combinatorics 1991. London Math. Soc. Lecture Notes Series 166, A. D. Keedwell, ed., Cambridge Univ. Press, 1991, pp. 1–39. [234]

76. J. E. Goodman, Proof of a conjecture of Burr, Grünbaum and Sloane. Discrete Math. 32(1980), 27–35. [164]

77. J. T. Graves, On the functional symmetry exhibited in the notation of certain geometrical porisms, when they are stated merely with reference to the arrangement of points. The London and Edinburgh Philos. Magazine and J. of Science, Ser 3, Vol. 15(1839), 129–136. [322+]

78. H. Gropp, "Il methodo di Martinetti" (1887) or Configurations and Steiner systems $S(2,4,25)$. Ars Combinatoria 24B(1987), 179–188. [83, 90]

79. H. Gropp, On the existence and nonexistence of configurations n_k. J. Combinatorics, Information and System Science 15(1990), 34–48. [90, 157, 234]

80. H. Gropp, Configurations and the Tutte conjecture. Ars Combinat. 29A(1990), 171–177. [309+]

81. H. Gropp, On the history of configurations. Internat. Sympos. On Structures in Math. Theories, A. Diez, J. Echeverria, and A. Ibarra, eds., Bilbao, 1990, pp. 263–268. [13, 81+]

82. H. Gropp, Blocking sets in configurations n_3. Mitt. Math. Seminar Giessen 201(1991), 59–72. [17, 321]

83. H. Gropp, The history of Steiner systems $S(2,3,13)$. Mitt. Math. Ges. Hamburg 12(1991), 849–861. [254]

84. H. Gropp, Configurations and Steiner systems $S(2,4,25)$ II. Trojan configurations n_3. Combinatorics '88, Vol. 1 (Ravello, 1988). Res. Lecture Notes Math., Mediterranean, Rende, 1991, pp. 425–435. [82, 91]

85. H. Gropp, The construction of all configurations $(12_4, 16_3)$. Fourth Czechoslov. Symp. on Combinatorics, Graphs and Complexity, J. Nesetril and M. Fiedler, eds., Elsevier, 1992, pp. 85–91. [244+]

86. H. Gropp, Enumeration of regular graphs 100 years ago. Discrete Math. 101(1992), 73–85. [13]

87. H. Gropp, The history of configurations $(12_4, 16_3)$. Österr. Symp. Math. Gesch. (1992), 6 pp. [244]

88. H. Gropp, Non-symmetric configurations with deficiencies 1 and 2. In "Combinatorics '90", A. Barlotti et al., eds., Elsevier, 1992, pp. 227–239. [235, 254+]

89. H. Gropp, On Golomb birulers and their applications. Math. Slovaca 42(1992), 517–529. [234]

90. H. Gropp, Configurations and graphs, Discrete Math. 111(1993), 269–276. [317, 330]

91. H. Gropp, Nonsymmetric configurations with natural index. Discrete Math. 124(1994), 87–98. [255+]

92. H. Gropp, The drawing of configurations. In "Graph Drawing", F. J. Brandenburg, ed., Lecture Notes in Computer Science, No. 1027, Springer, 1995, pp. 267–276. [17]

93. H. Gropp, Configurations. CRC Handbook of Combinatorial Designs, C. J. Colbourn and J. H. Dinitz, eds., CRC Press, Boca Raton, 1996, pp. 253–255. [235]

94. H. Gropp, Configurations and graphs–II, Discrete Math. 164(1997), 155–163. [71]

95. H. Gropp, Blocking set free configurations and their relations to graphs and hypergraphs. Discrete Math. 165/166(1997), 359–370. [321]

96. H. Gropp, Configurations and their realizations. Discrete Math. 174(1997), 137–151. [82, 90+]

97. H. Gropp, On combinatorial papers of König and Steinitz. Acta Applicandae Math. 52(1998), 271–276. [13, 17, 93]

98. H. Gropp, On configurations and the book of Sainte-Laguë. Discrete Math. 191(1998), 91–99. [13]

99. H. Gropp, Die Configurationen von Theodor Reye in Straßburg nach 1876. Mathematik im Wandel, Math. Gesch. Unterr. 3, M. Toepell, ed., Franzbecker, Hildesheim-Berlin, 2001, pp. 287–301. [13]

100. H. Gropp, "Réseaux réguliers" or regular graphs—Georges Brunel as a French pioneer in graph theory. 6th International Conference on Graph Theory. Discrete Math. 276 (2004), no. 1-3, 219–227. [13, 156]

101. H. Gropp, Configurations between geometry and combinatorics. Discrete Appl. Math. 138(2004), 79–88. [13, 90]

102. H. Gropp, Existence and enumeration of configurations. Beyreuther Math. Schriften 74(2005), 123–129. [235]

103. H. Gropp, Nonisomorphic configurations n_k, Electronic Notes in Discrete Math. 27(2006), 43–44. [28, 234]

104. J. L. Gross, Voltage graphs. Discrete Math. 9(1974), 239–246. [40]

105. J. L Gross and T. W. Tucker, Topological Graph Theory. Wiley, New York, 1987. [40]

106. J. Gross and J. Yellen, Graph Theory and its Applications. CRC Press, Boca Raton, 1998. [40]

107. B. Grünbaum, Convex Polytopes. Wiley, New York, 1967. Second ed., Springer, New York, 2003. [8, 20]

108. B. Grünbaum, The importance of being straight. "Time Series and Stochastic Processes; Convexity and Combinatorics." Proc. Twelfth Bienn. Seminar Canad. Math. Congr., R. Pyke, ed., Canad. Math. Congress, Montreal, 1970, pp. 243–254. [23]

109. B. Grünbaum, Notes on configurations. Lectures presented in the "Combinatorics and Geometry" seminar, Univ. of Washington, Seattle, 1986. [13]

110. B. Grünbaum, Astral (n_k) configurations. Geombinatorics 3(1993), 32–37. [35, 119, 203]

111. B. Grünbaum, Astral (n_4) configurations. Geombinatorics 9(2000), 127–134. [35, 190, 208]

112. B. Grünbaum, Which (n_4) configurations exist? Geombinatorics 9(2000), 164–169. [62+]

113. B. Grünbaum, Connected (n_4) configurations exist for almost all n. Geombinatorics 10(2000), 24–29. [62]

114. B. Grünbaum, Connected (n_4) configurations exist for almost all n—an update. Geombinatorics 12(2002), 15–23. [62+]

115. B. Grünbaum, Small configurations with many incidences. Geombinatorics 14(2005), 200–207. [258]

116. B. Grünbaum, A 3-connected configuration (n_3) with no Hamiltonian circuit. Bull. Institute of Combinatorics and Applications 46(2006), 12–26. [310]

117. B. Grünbaum, Configurations of points and lines. The Coxeter Legacy. Reflections and Projections. C. Davis and W. W. Ellers, eds., Amer. Math. Soc., Providence, RI, 2006, pp. 179–225. [35, 82, 91, 108, 156, 235]

118. B. Grünbaum, Connected (n_4) configurations exist for almost all n—second update. Geombinatorics 16(2006), 254–261. [62+, 172]

119. B. Grünbaum, A catalogue of simplicial arrangements in the real projective plane. Ars Mathematica Contemporanea 2(2009), 1–25. (Preliminary version available at `http://hdl.handle.net/1773/2269`.) [173]

120. B. Grünbaum, Musings on an example of Danzer's. Europ. J. Combinatorics 29(2008), 1910–1918. [157]

121. B. Grünbaum, [4, 3]-configurations with many symmetries. Geombinatorics 18(2008), 5–12. [250]

122. B. Grünbaum and J. F. Rigby, The real configuration (21_4). J. London Math. Soc. (2) 41(1990), 336–346. [13, 53+, 157, 161+, 190, 207, 212, 235, 281, 363+]

123. P. Hall, On representatives of subsets. J. London Math. Soc. 10(1935), 26–30. [93]

124. O. Hesse, Über Curven dritter Ordnung und die Kegelschnitte, welche diese Curven in drei verschiedenen Puncten berühren. J. Reine Angew. Math. 36(1848), 143–176. [243+]

125. D. Hilbert, The Foundations of Geometry. Authorized translation by E. J. Townsend. Open Court, Chicago, 1902. [16]

126. D. Hilbert and S. Cohn-Vossen, Anschauliche Geometrie. Springer, Berlin, 1932. English translation: Geometry and the Imagination, Chelsea, New York, 1952. Second ed., Springer, Berlin, 1996. [12+, 17, 70+]

127. M. Hladnik, D. Marusic, and T. Pisanski, Cyclic Haar graphs. Discrete Math. 244(2002), 137–152. [319+]

128. D. Ismailescu, Restricted point configurations with many collinear k-tuplets. Discrete Comput. Geom. 28(2002), 571–575. [7]

129. M. J. Kalaher, Review of [P4]. Math. Reviews MR2146456 (2006e:51004). [24]

130. S. Kantor, Ueber eine Gattung von Configurationen in der Ebene und im Raume. Wien. Ber. LXXX (1879), 227. [2]

131. S. Kantor, Ueber die configurationen (3, 3) mit den Indices 8, 9 und ihren Zusammenhang mit den Curven dritter Ordnung. Wien. Ber. LXXXIV(1881), 915–932. [9+, 61+, 70+]

132. S. Kantor, Die Configurationen $(3, 3)_{10}$. Wien. Ber. LXXXIV(1881), 1291–1314 + plate. [9+, 73+, 309]

133. F. Kárteszi, Su una analogia sorprendente. Ann. Univ. Sci. Budapest. Sect. Math. 29(1986), 257–259. [212]

134. L. M. Kelly and W. O. J. Moser, On the number of ordinary lines determined by n points. Canad. J. Math. 10(1958), 210–219. [64]

135. L. M. Kelly and R. Rottenberg, Simple points in pseudoline arrangements. Pacif. J. Math. 40(1972), 617–622. [64]

136. A. K. Kelmans, Cubic bipartite cyclic 4-connected graphs without Hamiltonian circuits. [In Russian.] Uspekhi Mat. Nauk 43, no. 3 (1988), 181–182. English translation: Russian Math. Surveys 43, no. 3 (1988), 205–206. [317]

137. A. K. Kelmans, Constructions of cubic bipartite 3-connected graphs without Hamiltonian cycles. Amer. Math. Soc. Translations (2) 158(1994), 127–140. [309+]

138. R. Killgrove, R. Sternfeld, and R. Tamez, Quadrangle completions and the anti-Desargues configuration. Congr. Numer. 127(1997), 57–66. [76, 150]

139. F. Klein, Ueber die Transformationen siebenter Ordnung der elliptischen Funktionen. Math. Ann. 14(1879), 428–471. [156+, 161]

140. W. Kocay and R. Szypowski, The application of determining sets to projective configurations. Ars Combinatoria 53(1999), 193–207. [105+]

141. D. König, Über Graphen und ihre Anwendung auf Determinantentheorie und Mengenlehre. Math. Ann. 77(1916), 453–465. [93]

142. E. K. Lampe, Review of 152. Jahrbuch Fortschr. Math. 19(1887), 587–589. [89]

143. R. Laufer, Die nichkonstruirbare Konfiguration (10_3). Math. Nachrichten 11(1954), 303–304. [10, 27, 76]

144. F. Lazebnik, V. A. Ustimenko, and A. J. Woldar, New upper bounds on the order of cages. Electronic J. Combinatorics Vol. 4(2) (1997), # R13. [328]

145. F. Levi, Geometrische Konfigurationen. Hirzel, Leipzig, 1929. [11, 26, 63+, 70+, 92, 280]

146. F. W. Levi, Finite Geometrical Systems. University of Calcutta, Calcutta, 1942. [28]

147. G. de Longchamps, Note de géometrie. Nouvelle Corresp. Mathémat. 3(1877), 306–312 and 340–347. [287]

148. M. S. Longuet-Higgins, Inversive properties of the plane n-line, and a symmetric figure of 2×5 points on a quadric. J. London Math. Soc. (2) 12(1976), 206–212. [287]

149. M. S. Longuet-Higgins and C. F. Parry, Inversive properties of the plane n-line, II: An infinite six-fold chain of circle theorems. J. London Math. Soc. (2) 19(1979), 541–560. [287]

150. A. Lupinski, K. Petelczyc, and K. Prazmowski, Tresses of polygons. Demonstratio Math. 40(2007), 419–439. [324]

151. V. Martinetti, Sopra alcune configurazioni piane. Annali di Matematica Pura ed Applicata (2) 14(1886), 161–192. [328+, 343]

152. V. Martinetti, Sulle configurazioni piane μ_3. Annali di Matematica Pura ed Applicata (2) 15(1887), 1–26. [66, 70+, 81, 89, 158, 309]

153. V. Martinetti, Sulle configurazioni n_3 piane, atrigone. Giornale di Matematiche di Battaglini 54(1916), 174–182. [335]

154. D. Marusic and T. Pisanski, Weakly flag-transitive configurations and half-arc transitive graphs. Europ. J. Combinatorics 20(1999), 559–570. [365]

155. R. A. Mathon, K. T. Phelps, and A. Rosa, Small Steiner triple systems and their properties. Ars Combinatoria 15(1983), pp. 3–110. [254]

156. N. S. Mendelsohn, R. Padmanabhan, and B. Wolk, Planar projective configurations. I. Note di Matem. 7(1987), 91–112. [249]

157. N. S. Mendelsohn, R. Padmanabhan, and B. Wolk, Designs embeddable in a plane cubic curve. (Part 2 of Planar projective configurations). Note di Matem. 7(1987), 113–148. [249]

158. N. S. Mendelsohn, R. Padmanabhan, and B. Wolk, Straight edge constructions on planar cubic curves. C. R. Math. Rep. Acad. Sci. Canada 10(1988), 77–82. [28, 249]

159. E. Merlin, Sur les configurations planes n_4. Bull. Cl. Sci. Acad. Roy. Belg. 1913, 647–660. [31, 157+, 161]

160. J. Metelka, On certain $(12_4, 16_3)$ configurations in the plane. [In Czech.] Vestník Královské Česk. Společnosti Nauk. Trída Matemat.-Prírodoved., 1944, 8 pp. (1946). [249]

161. J. Metelka, Über ebene Konfigurationen $(12_4, 16_3)$. [In Czech, with German and Russian summaries.] Časopis pro Pestovani Matematiky 80(1955), 133 - 145. [249]

162. V. Metelka, Über gewisse ebene Konfigurationen $(12_4, 16_3)$ welche mindestens einen D-Punkt enthalten. [In Czech, with German and Russian summaries.] Časopis pro Pestovani Matematiky 80(1955), 146–151. [249]

163. V. Metelka, Über ebene Konfigurationen $(12_4, 16_3)$ welche mindestens einen D-Punkt enthalten. [In Czech, with German and Russian summaries.] Časopis pro Pestovani Matematiky 82(1957), 385–439. [249]

164. V. Metelka, Über ebene Konfigurationen $(12_4, 16_3)$, die mit einer irreduziblen Kurven dritter Ordnung inzidieren. Časopis pro Pestovani Matematiky 91(1966), 261–307. [249]

165. V. Metelka, Über gewisse ebene Konfigurationen $(12_4, 16_3)$, die auf den irreduziblen Kurven dritter Ordnung endliche Gruppoide bilden und über die Konfigurationen C_{12}. Časopis pro Pestovani Matematiky 95(1970), 23–53. [249]

166. V. Metelka, Über gewisse ebene Konfigurationen $(12_4, 16_3)$ die B-, C- und E-Punkte enthalten und über singulare Konfigurationen. [In Czech, with German summary.] Časopis pro Pestovani Matematiky 102(1980), 219–255. [249]

167. V. Metelka, On certain planar configurations $(12_4, 16_3)$ containing B, C and E points, and on singular configurations. [In Czech, with German summary.] Časopis pro Pestovani Matematiky 105(1980), 219–255. [249]

168. V. Metelka, On two special configurations $(12_4, 16_3)$. [In Czech, with German and Russian summaries.] Časopis pro Pestovani Matematiky 110(1985), 351–355. [245+]

169. D. Michelucci and P. Schreck, Incidence constraints: A combinatorial approach. Internat. J. Comput. Geom. & Appl. 16(2006), 443–460. [20, 169]

170. N. Miller, Euclid and His Twentieth Century Rivals. Diagrams in the Logic of Euclidean Geometry. Center for the Study of Language and Information, Stanford, CA, 2007. [117]

171. A. Miquel, Mémoire de géométrie. J. Math. Pures Appl. 9(1844), 20–27. [281]

172. A. F. Möbius, Kann von zwei dreiseitigen Pyramiden eine jede in Bezug auf die andere um—und eingeschrieben zugleich heisen? J. Reine Angew. Math. 3(1828), 273–278 = Gesammelte Werke 1(1885), 439–446. [61+, 109]

173. G. Myerson, Rational products of sines of rational angles. Aequationes Math. 45(1993), 70–82. [210+, 214]

174. M. H. Noronha, Euclidean and Non-Euclidean Geometries. Prentice Hall, Upper Saddle River, NJ, 2002. [16, 254]

175. J. Novák, Maximal systems of triples of 12 elements. [Czech, with German summary.] 1970 Mathematics (Geometry and Graph Theory), pp. 105–110. Univ. Karlova, Prag, 1970. [254]

176. J. J. O'Connor and E. F. Robertson, Ernst Steinitz. MacTutor History of Mathematics. `http://www-history.mcs.st-andrews.ac.uk/Biographies/Steinitz.html`. [11]

177. W. O'Keefe and P. K. Wong, A smallest graph of girth 10 and valency 3. Journal of Combinatorial Theory (B) 29 (1980), 91–105. [336]

178. W. Page and H. L. Dorwart, Numerical patterns and geometrical configurations. Math. Magazine 57(1984), 82–92. [93]

179. K. Petelczyc, Series of inscribed n-gons and rank 3 configurations. Beiträge zur Algebra und Geom. 46(2005), 283–300. [24, 324]

180. T. Pisanski, Strong and weak realizations of configurations. Lecture Notes from the Klee-Grünbaum Festival of Geometry, Ein Gev, Israel, April 9–16, 2000. [104]

181. T. Pisanski, Dimension of unsplittable incidence structures. Abstract for the Discrete and Computational Geometry session of the Summer 2005 meeting of the Canad. Math. Soc., Waterloo 2005. [305]

182. T. Pisanski, Yet another look at the Gray graph. New Zealand J. of Math. 36(2007), 85–92. [2, 328]

183. T. Pisanski, M. Boben, D. Marusic, A. Orbanic, and A. Graovac, The 10-cages and derived configurations. Discrete Math. 275(2004), 265–276. [336+]

184. B. Polster, A Geometrical Picture Book. Springer, New York, 1998. [330+]

185. B. Poonen and M. Rubinstein, The number of intersection points made by the diagonals of a regular polygon. SIAM J. Discrete Math. 11(1998), 135–156. [210]

186. M. Prazmowska, Multiple perspectives and generalizations of the Desargues configurations. Demonstratio Math. 39(2006), 887–906. [299, 324]

187. T. Reye, Geometrie der Lage. I. Second ed., 1876. [8+]

188. T. Reye, Das Problem der Configurationen. Acta Math. 1(1882), 93–96. [3+, 16, 61]

189. J. F. Rigby, Multiple intersections of diagonals of regular polygons, and related topics. Geom. Dedicata 9(1980), 207–238. [210]

190. J. F. Rigby, Half-turns and Clifford configurations in the inversive plane. J. London Math. Soc. (2) 15(1997), 521–533. [287]

191. J. F. Rigby, Two $12_4, 16_3$ configurations. Mitt. Math. Seminar Giessen 165(1984), 135–154. [190]

192. F. S. Roberts, Applied Combinatorics. Prentice Hall, Englewood Cliffs, NJ, 1984. [94, 254]

193. C. Rodenberg, Review of [K4]. Jahrbuch Fortschr. Math. 13(1881), 460. [10, 249]

194. A. Schönflies, Ueber einige ebene Configurationen und die zugehörigen Gruppen von Substitutionen. Nachr. Ges. Wiss Göttingen 1887, 410–417. [324, 343]

195. A. Schönflies, Ueber die regelmässigen Configurationen n_3. Math. Ann. 31(1888), 43–69. [61+, 67, 309, 324, 343]

196. A. Schönflies, Bemerkung zur Theorie der regelmässigen Configurationen n_3. Math. Ann. 42(1883), 595–597. [324]

197. A. Schönflies, Ueber regelmäßige Configurationen n_3 auf den Curven dritter Ordnung. Nachr. Ges. Wiss Göttingen 1889, 334–344. [324]

198. A. Schönflies, Ueber Configurationen, welche sich aus gegebenen Raumelementen durch blosses Schneiden und Vebinden ableiten Lassen. Jahresber. Deutsch. Math.-Vereiningung 1(1892), 62–63. [324+]

199. H. Schröter, Ueber lineare Konstruktionen zur Herstellung der Konfigurationen n_3. Nachr. Ges. Wiss Göttingen 1888, 193–236. [61+, 71, 91, 109, 289]

200. H. Schröter, Die Theorie der ebenen Curven dritter Ordnung. Teubner, Leipzig, 1888. [65]

201. H. Schröter, Über die Bildungsweise und geometrische Construction der Configurationen 10_3. Nachr. Ges. Wiss Göttingen 1889, 239–253. [10, 27, 72+, 81, 109, 280, 309]

202. A. E. Schroth, How to draw a hexagon. Discrete Math. 199(1999), 161–171. [337+]

203. H. Schubert, Review of [131] and [132]. Jahrbuch Fortschr. Math. 13(1881), 460. [10]

204. H. Schubert, Review of [240]. Jahrbuch Fortschr. Math. 21(1888), 535. [153]

205. H. A. Schwarz, Beispiel einer stetigen Funktion reellen Argumentes, für welche der Grenzwert des Differentialquotienten in jedem Teile des Intervalles unendlich oft gleich Null ist. Berl. Ber. 1910, 592–593. [72]

206. J. B. Shearer, Golomb ruler table. 1996, `http://www.research.ibm.com/people/s/shearer/grtab.html`. [234]

207. T. Q. Sibley, The Geometric Viewpoint. Addison Wesley Longman, Reading, MA, 1998. [16]

208. L. A. Sidorov, Configuration. In "Encyclopaedia of Mathematics", SpringerLink, http://eom.springer.de/C/c024670.htm#c024670_00f2. [64]

209. S. Stahl, Geometry: From Euclid to Knots. Pearson Education, Inc., Upper Saddle River, NJ, 2003. [16]

210. E. Steinitz, Über die Construction der Configurationen n_3. Ph.D. Thesis, Breslau, 1894. [18, 92+, 102+, 298]

211. E. Steinitz, Über die Unmöglichkeit, gewisse Configurationen n_3 in einem geschlossenen Zuge zu durchlaufen. Monathefte Math. Phys. 8(1897), 293–296. [309]

212. E. Steinitz, Konfigurationen der projektiven Geometrie. Encyklopädie der Math. Wissenschaften, Vol. 3 (Geometrie), Part IIIAB5a, pp. 481–516, 1910. [9, 72, 90+, 109, 324, 343]

213. E. Steinitz, Über Konfigurationen. Archiv Math. Phys., 3rd Ser., 16(1910), 289–313. [324, 343]

214. E. Steinitz and E. Merlin, Configurations. French translation of [212], incomplete. Encyclopédie des Sciences Mathématiques, edition française. Tome III, Vol. 2 (1913), pp. 144–160. [90+]

215. R. Sternfeld, D. Koster, D. Kiel, and R. Killgrove, Self-dual confined configurations with ten points. Ars Combinat. 67(2003), 37–63. [10, 76]

216. B. Sturmfels and N. White, Rational realizations of 11_3- and 12_3-configurations. In "Symbolic Computations in Geometry", by H. Crapo, T. F. Havel, B. Sturmfels, W. Whiteley, and N. L. White, IMA Preprint Series #389, Univ. of Minnesota, 1988, pp. 92–123. [81+]

217. B. Sturmfels and N. White, All 11_3- and 12_3-configurations are rational. Aequat. Math. 39(1990), 254–260. [13, 81+]

218. E. Togliatti, Review of [235]. Zentralblatt Math. 43(1952), p. 358. [76]

219. J. van de Craats, On Simonis' 10_3 configuration. Nieuw Archief voor Wiskunde 4(1983), 193–207. [76, 120, 141, 290]

220. H. van Maldeghem, Slim and bislim geometries. In *Topics in Diagram Geometry*, A. Pasini, ed., Quaderni di Matematiche 12, Aracne, Roma, 2003, pp. 227–254. [15+, 324]

221. M. P. van Straten, The topology of the configurations of Desargues and Pappus. Reports of a Math. Colloquium (2) 8(1949), 3–17. [27]

222. E. Visconti, Sulle configurazioni piane atrigone. Giornale di Matematiche di Battaglini 54(1916), 27–41. [334+]

223. E. W. Weisstein, Configuration. http://mathworld.wolfram.com/Configuration.html. [24]

224. D. Wells, The Penguin Dictionary of Curious and Interesting Geometry. Penguin, London, 1991. [329+]

225. Wikipedia, Projective configuration. http://en.wikipedia.org/wiki/Projective_configuration. [24]

226. Wikipedia, Möbius-Kantor graph. (As of 2-7-2008) http://en.wikipedia.org/wiki/Möbius-Kantor_configuration. [322]

227. Wikipedia, Ernst Steinitz. http://en.wikipedia.org/wiki/Ernst_Steinitz. [103]

228. P. K. Wong, On the smallest graphs of girth 10 and valency 3. Discrete Math. 43(1983), 119–124. [336]

229. P.K. Wong, Cages—a survey. Journal of Graph Theory 6 (1982), 1–22. [330, 336]

230. I. M. Yaglom, Complex Numbers in Geometry. Academic Press, New York, 1968. [287]

231. M. Zacharias, Untersuchungen über ebene Konfiguration $(12_4, 16_3)$. Deutsche Math. 6(1941), 147–170. [249]

232. M. Zacharias, Eine neue ebene Konfigurationen $(12_4, 16_3)$. Math. Nachrichten 1(1948), 332–336. [249]

233. M. Zacharias, Neue Wege zur Hesseschen Konfiguration $(12_4, 16_3)$. Math. Nachrichten 2(1949), 163–170. [249]

234. M. Zacharias, Streifzüge im Reich der Konfigurationen: Eine Reyesche Konfiguration (15_3), Stern- und Kettenkonfigurationen. Math. Nachrichten 5(1951), 329–345. [119]

235. M. Zacharias, Die ebenen Konfigurationen (10_3). Math. Nachrichten 6(1951), 129–144. [76]

236. M. Zacharias, Konstruktionen der ebenen Konfigurationen $(12_4, 16_3)$. Math. Nachrichten 8(1952), 1–6. [249]

237. M. Zacharias, Bemerkung zu meiner Arbeit: "Die ebenen Konfigurationen (10_3)". Math. Nachrichten 12(1954), p. 256. [76]

238. H. Zeitler, Über einen Satz von Karteszi. Elemente der Math. 42(1987), 15–18. [212]

239. P. Ziegenbein, Konfigurationen in der Kreisgeometrie. J. Reine Angew. Math. 183(1941), 9–24. [287]

240. K. Zindler, Zur Theorie der Netze und Configurationen. Wien. Ber. 98(1889), 499–519. [153]

Index

Titles in This Series

For a complete list of titles in this series, visit the
AMS Bookstore at **www.ams.org/bookstore/**.